Requirements Engineering: Laying a Firm Foundation

James A. Crowder • Curtis W. Hoff

Requirements Engineering: Laying a Firm Foundation

Springer

James A. Crowder
CAES Advanced Technology and Engineering
Colorado Springs, CO, USA

Curtis W. Hoff
Larkspur, CO, USA

Additional materials can be downloaded at https://link.springer.com/book/9783030910761

ISBN 978-3-030-91076-1 ISBN 978-3-030-91077-8 (eBook)
https://doi.org/10.1007/978-3-030-91077-8

This Springer imprint is published by the registered company Springer Nature Switzerland AG
The registered company address is: Gewerbestrasse 11, 6330 Cham, Switzerland

Preface

Since mankind first began to create buildings or construction of any kind there has been at least an implicit need for requirements. Without at least orally provided requirements, construction of megaliths like the pyramids or Stonehenge could not have been constructed. Below is a carving illustrating a Roman building project from 113 A.D. This depicts Roman soldiers building a fortress. Someone gave them information on what needed to be built (i.e., requirements).

One of the first explicit examples of written down requirements is in the Hebrew Bible in the book of Genesis. Whether you believe the Noah flood story or not, Genesis records very clear requirements given for building the ark:

1. Make it out of gopher wood.
2. Put compartments into the ark.
3. Cover all surfaces with pitch (i.e., tar).
4. Dimensions should be:
   a. 300 cubits long (about 450').
   b. 50 cubits wide (about 75').
   c. 30 cubits high (about 45').
5. The ark should have a window.
   a. Top of window should be 1 cubit from the top (about 18").
6. The ark should have a door.
7. There should be three decks to the ark.

The point here is that requirements are an important part of any project, large or small. Even for small projects you do yourself, you have a notion of what you want to do and what it will be used for. Even if you do not write them down, you have a notion of requirements. Throughout the history of humankind, requirements engineering and requirements management have been an integral part of humankind's overall evolution and progress throughout the centuries. Even cooking has requirements, only we call them recipes. The earliest known recipe is from 3300

BC, pictured below, and was, of course, for beer[1]. Again, the point is that without requirements very little is accomplished, or at least accomplished well.

Throughout history, projects have succeeded or failed based on the ability to create, communicate, and manage requirements needed to create something. This book is the second in the series on Multidisciplinary Systems Engineering. The first book, published in 2016, introduces the notion that systems engineering must be a multidisciplinary endeavor. Requirements engineering is no different. To handle the fast paced, rapidly changing technology environments of today, requirements engineering must be brought into the modern era of multidisciplinary engineering. As we deal with climate change and a host of other impending issues facing mankind, eliminating waste and project failure are more important than ever. It is our hope that this book is part of the process of revamping, improving, and innovating requirements engineering for the future.

Colorado Springs, CO, USA James A. Crowder
Larkspur, CO, USA Curtis W. Hoff

[1] The World's Oldest Known Beer Recipe Comes from Mesopotamia (realmofhistory.com)

Contents

1 Introduction to Multidisciplinary Requirement Engineering (MDRE)

Abbreviations

AC	Alternating current
aka	Also known as
DC	Direct current (also 0 Hz)
DFM	Design for Manufacturing
DREX+	Digital receiver-exciter
EIA	Engineering Industries Alliance
F	Fahrenheit
IEEE	Institute for Electrical and Electronic Engineers
IOT&E	Initial Operational Test and Evaluation
MBSE	Model-based systems engineering
MDRE	Multidisciplinary requirements engineering
MDSE	Multidisciplinary systems engineering
MIL-STD	Military Standard
NASA	National Aeronautics and Space Administration
OSHA	Occupational Safety and Health Administration
RAM	Reliability, availability, and maintainability
SI&T	Systems Integration and Test
SoS	System of Systems
TCO	Total cost of ownership
vac	Voltage AC

1.1 Requirements Engineering: The Backbone of Multidisciplinary Systems Engineering

Systems engineering has continued to evolve but has had issues keeping up with changes in system/software design methodologies. However, with the movement toward multidisciplinary systems engineering, systems design and requirements engineering has entered a new era of common practice among designers [2]. The complexity of systems and information today requires expertise in many disciplines and domains, leaving engineering, just like during the industrial revolution, without the tools and level of understanding required to engineer across disciplines, much less obtain knowledge outside of their silos of expertise. Therefore, new systems engineering concepts, mechanisms, and implementations are required to facilitate both the development and competency of the systems engineering discipline and the complex systems themselves, to ensure proper operation, and, in some cases, autonomous operation, self-healing, and critical self-management of knowledge and real-time operational self-awareness [3]. Presented in this second of a series of books into multidisciplinary systems engineering are new requirements engineering concepts, processes, methodologies, notional architectures, and tools to support the engineering discipline in understanding and evolving engineering of systems across the full spectrum of disciplines required today. The materials include the rationale for multidisciplinary requirements engineering (MDRE) as the standard for systems engineering and requirements development; ensuring new System of Systems (SoS) designs are successful and avoid the failures introduced by complexities of the information overloads facing our customer and management teams. The mission/business analysis community typically attempts to keep analysis distinct from technical design. It is not inherently different or an incorrect term for development of a solution to a system design. In the end, the activities which we would call design are nothing different from the activities required to create the "To-be" requirements [4]. Throughout the book, we will dive into the details of each type of requirements, their use, composition, decomposition, and overall need, based on a complete life cycle view of systems. Many systems are designed with basic capability requirements and ignore inclusion of things like supportability, maintainability, and sustainability until the system is designed and built. This may expedite the original development and delivery of the system, but greatly increases the overall cost, or total cost of ownership (TCO) to the eventual customer, as design modifications are required to include these capabilities. One of the purposes of this book is to drive home the need for complete requirements engineering and management throughout a system's life cycle. The book is

J. A. Crowder, C. W. Hoff, *Requirements Engineering: Laying a Firm Foundation*,
https://doi.org/10.1007/978-3-030-91077-8_1

organized to provide a logical progression through the multidisciplinary requirements engineering process. It is composed of three main sections and 18 chapters. A short description of each section and each chapter is provided below. Throughout the book, we will utilize a set of case studies for a new digital receiver-exciter system called the DREX+ as an example of requirements management and its use across two sample programs, a possible Naval application of long-range radar and a possible application for a radar used for medium range border sentry surveillance.

1.2 Chapter Organization

1.2.1 Section I: The Basics of Requirements Engineering

Requirements engineering is the heart and soul of systems engineering and engineering design. The entire purpose of a program is to meet a given set of requirements, either internally or externally generated, and deliver a system that meets those requirements. Requirements engineering is based on the need for a system to meet certain objectives [5]. From the initial set of requirements, requirements are refined, expanded, and allocated down to lower levels of design through an iterative process explained in detail throughout the book. Requirements are typically decomposed and allocated down to a level where a single software/hardware/systems engineer is responsible for a given set of low-level tasks. This section lays out the basic decomposition, allocation, and management techniques for a system's life cycle requirements management.

1.2.1.1 Chapter 2 The Need for Requirements Engineering

The world around us is filled with projects that have failed due to improper or lack of requirements management. From shuttle disasters to projects 10–100 times over budget, the failure to manage requirements correctly is responsible for the failure of 70% of projects that ultimately fail.[1] Of all the phases the systems, software, and product support development life cycle, requirements management is arguably the most critical. Here we will explore and describe the issues associated with good and bad requirements management, regardless of the development style (e.g., waterfall vs. agile) [6].

1.2.1.2 Chapter 3 MDRE Systems Supportability Analysis

Most developed systems must provide a needed capability, meet user needs, including operational effectiveness, operational suitability, and most must be affordable. While operational effectiveness addresses the degree of mission accomplishment in the intended environment, operational suitability addresses the degree to which a system can be satisfactorily placed in use, given reliability, availability, maintainability (RAM), supportability, and ownership cost, among other factors. These requirements are tested and quantified prior to fielding by the initial operational test and evaluation (IOT&E) process and assessed against defined criteria [7]. This chapter will address the issues of mission capability requirements and how they play into the overall life cycle cost of a fielded system.

1.2.1.3 Chapter 4 MDRE: Designing for Manufacturability

Systems that are intended to be produced in numbers greater than one or two must be designed with the manufacturing process in mind. Requirements for these systems must include requirements that allow multiple versions of identical systems to be produced. This may include requirements for how circuit boards are manufactured, how parts are procured to ensure consistency between copies, etc. This involves the discipline of Design for Manufacturability (DFM). Design for Manufacturability (DFM) involves designing parts or products so that they meet the critical needs of an application while simultaneously being designed for optimal, efficient, and cost-effective manufacturing. Creating an optimal design requires an understanding of the many variables that can go into its manufacture. Design engineers have a solid knowledge of all manufacturing methods, but most still see the value of a DFM exchange with manufacturers. It can be worth its weight in gold if it creates an opportunity to reduce the timeframe from concept to part—which will increase speed to market. This chapter will explore the notion of DFM and will utilize the DREX+ case studies to illustrate the process.

1.2.1.4 Chapter 5 MDRE: The Requirements Engineering Process

Requirement engineering is the process of defining, documenting, and maintaining the set of overall requirements that will be utilized for system design, development, test, deployment, and operation throughout the systems life cycle. Requirement engineering provides methods and tools to understand the desires (objectives) of the customer, which includes feasibility assessment. In this chapter we will discuss and present the overall processes required for successful requirements engineering to ensure the system meets the customers' requirements and functions as designed. Here we will discuss the following requirements processes [8]:

1. Feasibility analysis.
2. System architecture development.
3. Requirements elicitation.
4. Requirements analysis.
5. Requirements derivation.
6. Requirements management.
7. Requirements metrics.

[1] Flawed Requirements Trigger 70% of Project Failures | Info-Tech Research Group

1.2.1.5 Chapter 6 Systems Engineering and Requirements

To effectively deal with the complexity of modern systems, formalized methods and tools are utilized to allow the creation, documentation, communication, and management of system requirements. Requirements process standards have been developed within the multidisciplinary systems engineering discipline with established baselines for requirements management and ultimately for system design. One example is the IEEE 29148–2011—Systems and Software Engineering Lifecycle Processes—Requirements Engineering Standards. Others include the IEEE 12207—Systems and Software Engineering—Software Lifecycle Processes [9]. These are international standards for establishing a common requirements framework for both systems and software development. The EIA 632 Processes for Systems/Requirement Engineering was established by the Engineering Industries Alliance (EIA) to focus on standards for conceptualization and implementation of complex systems through comprehensive and common requirements management practices. Chapter 6 is devoted to understanding such standards and how they play out to allow thorough, efficient, and proper requirements management regardless of the size of the system. Chapter 6 discussion will focus on [10]:

1. Formal vs. information requirements.
2. Contractual (customer provided) vs. derived requirements.
3. Statements of work/statements of objectives.
4. Interface control documents.
5. Guidance and compliance policies.
6. Concepts of Operation.
7. Compliance and standards documents (e.g., MIL-STD).

1.2.1.6 Chapter 7 Developer Derived/Decomposed Requirements

Decomposition of the customer's requirements is the first step toward defining the architecture after the systems analysis has been completed [11]. The requirements decomposition process is not set-in stone and is influenced by the type of system being designed, the domain in which the system must operate, the system constraints, and other factors that may be specific to any individual system. The requirements decomposition process radically affects the optimization of the architecture, system design, and system implementation, since different decompositions drive different internal interfaces, different information and data flows, and a different set of Enterprise Services to execute the functionality laid out in the systems analysis and subsequent decomposition [12]. Each of the different decomposition possibilities drives a trio of different time, quality, and cost profile for the development of the system. These must be balanced against the customer technical, performance, and quality requirements, customer-driven schedule, and available cost profile. Chapter 7 will delve into the requirements derivation/allocation process in detail, providing methods, tools, and practices for requirements that are:

1. Derived into specific specifications.
2. Derived into development guidelines.

Each of these has very different effects on the overall design of the system.

1.2.1.7 Chapter 8 New Development Vs. Existing System, Extension, or Upgrade

The derivation/allocation and overall management of requirements varies depending on whether a brand-new system is being developed (aka a revolutionary system), or there are extensions and/or upgrades to an existing system (aka an evolutionary system). Chapter 8 will discuss the process of analyzing an enhancement to an existing system or process. In the traditional mission/business analysis approach, the "As-Is" state is analyzed as a starting point to capture processes, data, and interactions that must be kept for any new solution. The requirements/functionality of the legacy system is identified, along with the processes, data, system interfaces, and applicable displays. The design process comes into play when determining how to integrate them with the new features being added. These new features become requirements/functionality that must be added to design the overall new/enhanced system. These types of analysis and processes are not necessarily required for all new development. Current requirements may not be of concern for new systems; however there may be legacy constraints and considerations that still come into play as national or international laws and agreements may affect any system design and deployment.

1.2.1.8 Chapter 9 System Tiers

Ensuring proper requirement traceability is essential to integrating and testing any system, throughout the system development and maintenance life cycle. Traceability is concerned with documenting the parent/child relationships between requirements at each level of the architecture. It also documents the relationships between the requirements and other architectural and design artifacts. The purposes of traceability are threefold [13–15]:

1. Manage engineering changes across the system development.
2. Understand the decomposition of the system into each architecture tier.
3. Manage the overall quality of the developed system.

Requirement traceability allows understanding of the objectives, goals, aims, expectations, and needs of the system under development at every tier in the architecture. It allows understanding of the relationship between the architecture tiers. Chapter 9 will explore how allocation/derivation and traceability across the architecture is essential to the requirements management process.

1.2.2 Section II: The Mechanics of Requirements Engineering

The overall life cycle of requirements engineering is a multifaceted, multidisciplinary process. It provides the methods, tools, and mechanisms to understand not only the customer's needs but also the customer's desires (wants vs. must haves) and provides the analytical tools to formalize and analyze the needs and wants, and the documentation and formalization to be able to negotiate a solution with the customer that is reasonable. The requirements engineering process also provides clarity in how the system is specified, validated/verified, and managed not only during design, development, and deployment, but also throughout the sustainment, support, and maintenance of the system [16]. The process of requirements engineering includes such things as:

1. Requirements feasibility—which includes technical, operational, and economic feasibility.
2. Requirement management method—this may be different for the systems vs. software requirements.
3. Requirements documentation and tracing.

This section includes the following chapters:

1.2.2.1 Chapter 10 Multidisciplinary Systems Engineering and Requirements Management

Many books and many engineers will tell you that systems engineering is much bigger than requirements management. And while they are completely correct, if requirements management is not done correctly, the rest of systems engineering will be compromised—and the development and deployment of the system will fail. Systems do NOT happen successfully by accident. The purpose of Chap. 10 is lay out the requirements management thread that runs completely through the multidisciplinary systems engineering process [17]. Driving home the purpose of requirements: "***Elaboration of the Target System to Meet Customer Expectations***."

1.2.2.2 Chapter 11 Categories of Requirements

Requirements fall into a host of categories, most of which are ignored by most systems engineers and system architects. Many put requirements into three main categories [18]:

1. Functional requirements.
2. Nonfunctional requirements.
3. Performance requirements.

And while these certainly are important categories of requirements, they are not the only ones, nor possibly the most important ones. Chapter 11 will explore 20 different categories of requirements that must be gathered, derived, allocated, and verification criteria created for all of them. This includes requirements categories like [19]:

1. Maintainability.
2. Sustainability.
3. Reliability.
4. Availability.
5. Manufacturability.
6. Behavioral (this comes into play in cognitive systems).
7. Human-centered engineering.
8. Compliance.

These and others will be presented and discussed.

1.2.2.3 Chapter 12 Requirements Composition

Many systems fail because the requirements were not composed correctly. This may come range from incorrect requirements to vague requirements, to compound requirements. Part of Requirements Management is to make sure the requirements at every level in the systems are [20]:

- **Defined Precisely**: Technical Requirements (all requirements really) should be clearly worded, sufficiently detailed to eliminate ambiguity, necessary (related to the mission/business needs), and testable/verifiable in relation to the tier of the system.
- **Analyzed for Feasibility**: Part of Requirements Planning is to determine the feasibility of the requirements at each level of the System of Systems, ensuring that there are no conflicts in the requirements that will make execution of the system problematic. This includes an impact analysis of the major requirements and their impacts on the overall architecture.

The purpose of Chap. 12 is to discuss and lay out in simple terms the steps for creating a proper requirement, along with an example of what happens when requirements are not done properly. Another impact of proper requirement composition is to facilitate requirements verification and validation during the system integration and test and culminates in sell off the system to the customer.

1.2.2.4 Chapter 13 Requirements Traceability

One of the most important parts of the overall requirements management process is requirements traceability throughout the life cycle of a developed systems [21]. At each level of the system design, the multidisciplinary systems engineering organization and the customer must have full visibility into the traceability of all requirements. The development engineering team must work closely with the system integration and test (SI&T) team in the development and integration of requirements for integration, maintenance, and sustainability to ensure the operational personnel can operate and maintain the system once it is handed off by the development team. Chapter 13 will lay out tools, methods, and processes for proper requirements traceability [22].

1.2.3 Section III: Requirements Management: Change Is Inevitable

Once we have worked through what requirements are, how to make sure they are feasible, deployable, and maintainable, we need to understand that requirements can and will change over time. In any system that takes more than a week to develop, odds are needs, objectives, and or wants will change and someone will want to change one or more of your requirements [23]. Do not panic. If you are working a project using Agile development methods, change is a part of the process. Section III is devoted to understanding how to manage the requirements change process.

1.2.3.1 Chapter 14 Requirements Change Management

One of the most important aspects of choosing an architecture framework for a given system design is ensuring that the system architecture documentation keeps up with changes in the system design. This involves an effective change management system. Without it, the multidisciplinary systems engineering process can be in store for spectacular failures. In Chap. 14 we will explore the requirements change management process, including a dramatic (real-life) example of change management failure [24].

1.2.3.2 Chapter 15 Requirements Agility

As we have discussed, the Agile development methodology is all about allowing requirements and development volatility throughout the system/software development process. Chapter 15 discusses the agile methods and how requirements management can be made to fully support the agile/extreme development process, including how to ensure systems engineering keeps up with the agile software development changes [25].

1.2.3.3 Chapter 16 Requirements Antipatterns

When projects go bad or fail, bad requirements are often the reason and are responsible for project failure(s). Here we discuss the most common requirements antipatterns that lead to project failure. These represent conditions and/or design methodologies that lend themselves to requirements issues throughout the system life cycle. A few of them include [26]:

1. Legacy requirements levied on a new system.
2. Political infighting dominating requirements discussions.
3. Design by committee—which generally reflect the social interactions of the group—not good requirements definitions.

There are many more that will be discussed in Chap. 16.

1.2.3.4 Chapter 17 New Trends in Requirements Management: Model-Based Systems Engineering (MBSE)

Model-based systems engineering (MBSE) is a new formalized methodology used to support requirements management associated with the development of complex systems. Unlike classical systems engineering which is "document-centric" and depends on many, many views, and systems documents, many of which are never looked at after creation, MBSE is "system-design centric," utilizing modeling and simulation techniques and environments to ensure proper requirements are created, maintained, and managed, including requirements change management [27]. NASA adopted the practice in 2020 and it is gaining hold across the industry as a means of tracking complex system development, deployment, and sustainment. Chapter 17 will lay out the basics of MBSE and how it might be adopted by a wider audience in the future.

1.2.3.5 Chapter 18 Conclusions and Discussion: Putting it all Together

All the methods, procedures, processes, and techniques discussed in the book are intended to lead the multidisciplinary systems engineer to "Systems Thinking." Systems thinking is a way of looking at a system under development (or under consideration for development) as a System of System and not as a self-contained entity. Almost all systems operate in an environment where there are other systems around it (ok—Voyager out in interstellar space may be an exception). The systems thinker observes systems from a distance and explores the boundaries, system contexts, and system life cycle, noting not only its behavior, but also how it is likely to influence the behavior of other systems. Chapter 18 will wrap up the book and hopefully put it all into perspective for the multidisciplinary systems engineer.

1.2.3.6 Assignments

The appendices at the end of each chapter provide a requirement design problem with assignments that will be available for professors and students, based on each chapter's content. By the end of the book, the students should have a complete set of requirements and analysis for an alternative energy system capable of providing alternative energies to a 500-family neighborhood and biosphere for production of fruits and vegetables.

1.2.4 Design Problem

The philosophy of the assignment development for this textbook was to create a set of though-provoking questions, problems, and design assignments that help drive home the principles, practices, and methodologies of multidisciplinary requirements engineering (MDRE). We endeavored to provide the opportunity for the student to practice the art and science of requirements engineering, helping the student to understand that there is rarely one answer to a requirements engineering problem, and it is only through study and practice that successful requirements engineers are developed. MDRE is a lifelong journey, and anyone who tells you they have nothing left to learn does not truly understand the discipline or has never been involved in the development of a complex system. With the recent interest in alternative energies and talk of the "Green New Deal," along with the recent issues with ice storms in Texas taking out the power grid, the assignments for this textbook will follow the theme of effectively dealing with both local alternative energy power grids and the national infrastructure required to facilitate large-scale alternative energy production, sustainment, and management.

1.2.4.1 Design Problem Description: Alternative Energy Power Grid

The purpose of this project is to, over the course of the semester, develop a feasibility study, requirements analysis, requirements elicitation, derivation, and allocation, for a system that will facilitate an "local" alternative energy power grid capable of powering a neighborhood of 500 homes and a large biosphere capable of supporting growing year-round fruits and vegetables for the 500-home neighborhood. The student will have to deal with issues of choosing multiple alternative energies that can provide power throughout all seasonal difficulties, from intense summer heat to freezing winter temperatures and possible ice storms. Control, viability, sustainability, and maintainability will be major factors to understand the entire possibly 30-year life cycle of such a system. How to effectively utilize, store, and manage power distribution, along with when to upconvert "extra" energy to the national power grid, will be issues that drive many of the assignments. Simply relying on the current technologies and energy infrastructures will not move the country and world forward. Hopefully, this will provide insight to the students using this textbook to understand some of the issues that face everyone looking out over the next many decades.

1.2.4.2 DC Power Grid Requirements Discussion

In the initial days of power generation and distribution, DC power was far less practical than the AC distribution system that soon supplanted it. However, even after Tesla's AC system became the predominant system (and rightly so), Edison's DC power did not become obsolete for some time. Eventually power plants eliminated their delivery of DC power, but the overall concept of a DC power grid is not dead. Even our computer systems (large server systems) have uninterruptible power supplies which contain batteries that are charged off the grid and used to upconvert to AC to power devices when the AC power grid goes off-line. So, the notion of a DC power grid is not dead, but generally used as a backup system. The point is, the notion of a DC power grid has remained and now that renewable energies are viable, we should re-examine their use for widespread, local, power distribution.

One of the advantages of a locally generated DC station is energy savings as well as allowing a facility to run more easily off various DC sources, such as Solar Arrays. And for power-hungry installations, the notion of generating the power locally is growing in popularity. Part of the reason is that waste heat from the power generation generators (solar, geothermal, etc.) can be used to warm nearby buildings. By combining the production of heat and power, local power grids can squeeze much more useful energy out of the renewable energy they use, so this approach will certainly become more widespread as time goes on. The purpose of this design problem is to look ahead and provide an architecture and specification for a local DC power generation station that can serve a local community.

Currently many systems exist to control and handle alternative energies. There are systems to control and manage solar power, wind power, geothermal power, and others, but very few are designed to control, manage, store, and distribute energy from a power grid consisting of multiple alternative energy sources. As we saw the issues recently caused by the intense ice storms in Texas, one alternative energy source is not the answer to the growing energy needs/wants in our society. Each alternative energy source has limitations and issues associated with its use. By harnessing multiple alternative energies, and managing the storage of the energies produces, communities can be worry free from power generation. In order to facilitate utilizing multiple alternative energy sources, a control center must be included in the design that can manage the energy creation, distribution, storage, and return unused energy to the power grid.

1.2.4.3 High-Level Technical Description of the DC Power Grid System

The purpose of the multiple alternative energies DC power grid is to:

- Supply DC power to a customer community of 500 homes and a biosphere for growing food.
- Provide a storage device to store energy in case of a system failure.
- Determine the amount of surplus energy that is above and beyond what is required to power the grid and charge the backup system, and then deliver that surplus energy to the national power grid systems.

Throughout the semester you will be asked to analyze, design, and architect different components of the overall DC power grid system. This will include research into alternative designs, like lithium nanowire storage systems for backup power and/or supercapacitors (also called ultracapacitors). Fig. 1.1 illustrates these technol-

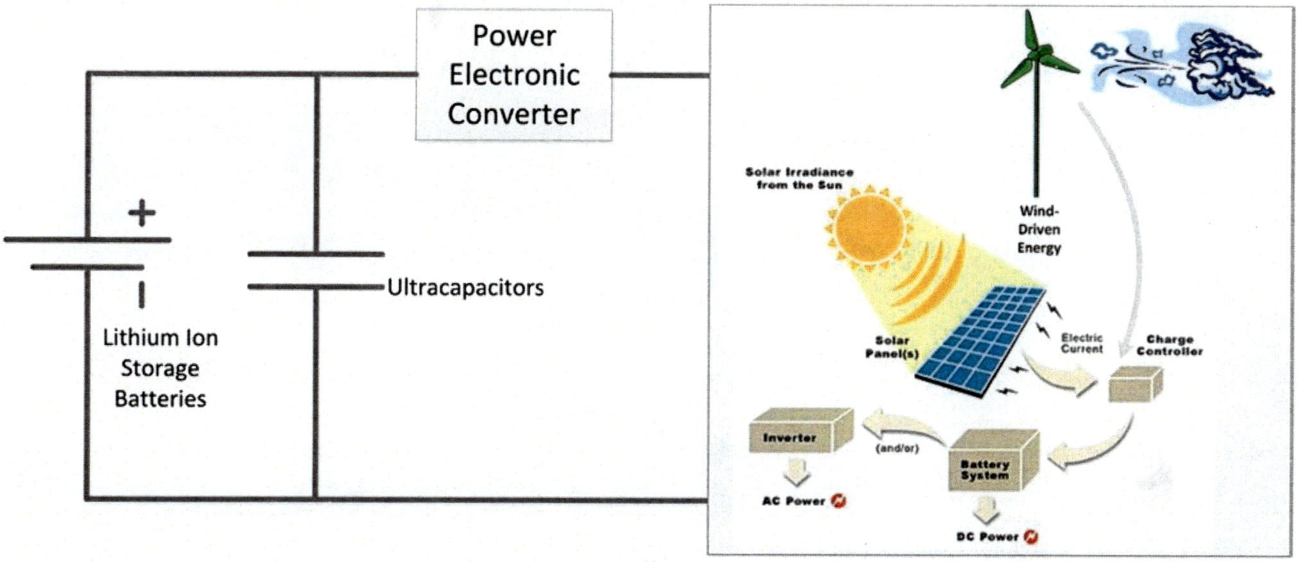

Fig. 1.1 Alternative power management system overview

ogies connected to the DC Power Grid Regulator Control Center.

1.2.4.4 High-Level Technical Requirements for the DC Grid System

- Design shall utilize at least 2 alternative energies, with a maximum of 4.
- Total Power Generation shall be 3.0 megawatts minimum.
- Each of the alternative energy sources shall be capable of generating at least 60% of minimum wattage.
- DC power grid shall generate DC power, with Grid voltage being 48, 24, or 12 volts.
- DC power grid shall be capable of servicing at least 500 single family homes and one 4-acre biosphere with hydroponics capable of growing vegetables year-round. (See Fig. 1.2 for a block diagram of a DC-powered home).
- All lights, appliances, and facilities in each home and the biosphere shall be configured for DC operations, if possible. Things like electric ovens may need DC/AC conversion to accommodate their 220v. requirements.
- Only allowed exceptions are the HVAC system and the Washer/Dryer, although DC is preferred.
- There shall be energy storage capability capable of operating the grid of 500 single family dwellings for 7 days if all alternative forms of energy are not producing.
  - Compliance Criteria: Energy storage devices shall be capable of producing 750,000 watts for the entire 7-day period. Nominal energy storage should be batteries or supercapacitors.
- The DC power grid system shall include controlling h/w and s/w to manage the multi-alternative energy strategies and production.
- All energy produced above the required home and energy storage usage shall be upconverted to 220vac for release to the national power grid.
- The DC power grid shall be capable of operating in a temperature range of −40 degrees F. to 125 degrees F. without degradation of power generation.
- The DC power grid must meet OSHA regulations.

1.2.4.5 Assignments

- Research requirements engineering and you will find multiple definitions. Pick 3 different definitions you find in your research and describe:
- How they are similar?
- How they are different?
- How might the requirements and subsequent system designs be different if each of the definitions were utilized?
- What role does requirements engineering play in the overall success of modern engineering systems?
- What role does human systems engineering play in requirements derivation/allocation, and what is the result of not deriving human-centered systems requirements in the overall usability of products/systems?
- How do you believe requirements engineering enhances the development of advanced Communications and Information Systems?

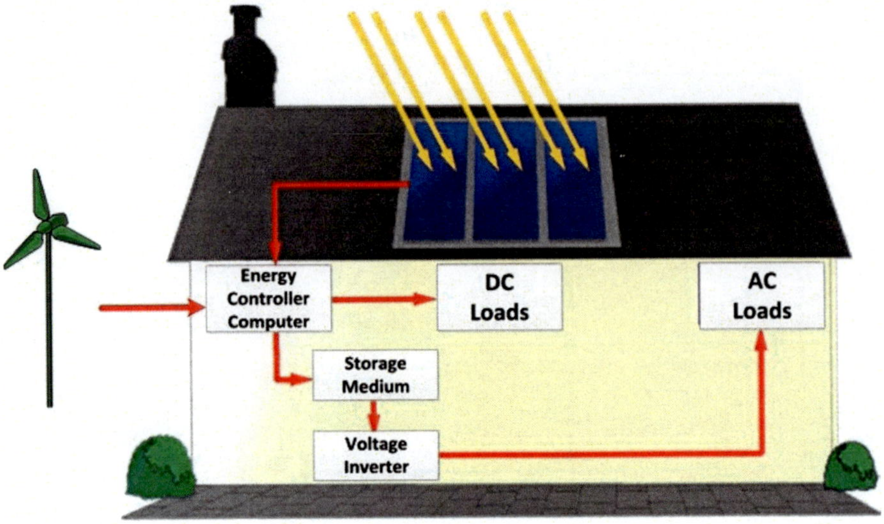

Fig. 1.2 Example of a whole house power grid

References

1. Crowder, J., Carbone, J., & Demijohn, R. (2016). *Multidisciplinary systems engineering: Architecting the design process*. Springer publishing. ISBN 978-3319223971.
2. Kotonya, G., & Sommerville, I. (1998). *Requirements engineering: Processes and techniques*. John Wiley and Sons.
3. Beck, A., Boeing, G., & Shannon, D. (2014). Systems and methods for analyzing requirements. US Patent 8650186.
4. Berenback, B., Paulish, D., Katzmeier, J. and Rudorfer, A. 2009. Software & systems requirements engineering: In practice. : McGraw-Hill Professional. ISBN 0-07-160547-9.
5. David, H. (2003). *Requirements analysis: From business views to architecture* (1st ed.). Prentice Hall. ISBN 0-13-028228-6.
6. Laplante, P. (2009). *Requirements Engineering for Software and Systems* (1st ed.). CRC Press. ISBN 1-4200-6467-3.
7. McConnell, S. (1996). *Rapid development: taming wild software schedules* (1st ed.). Microsoft Press. ISBN 1-55615-900-5.
8. Nuseibeh, B. and Easterbrook, S. (2000). Requirement engineering: a roadmap (PDF). ICSE'00. Proceedings of the conference on the future of Software engineering. pp. 35–46. CiteSeerX 10.1.1.131.3116. doi:https://doi.org/10.1145/336512.336523. ISBN 1-58113-253-0.
9. Stellman, A., & Greene, J. (2005). *Applied software project management*. O'Reilly Media. ISBN 0-596-00948-8.
10. Wiegers, K., & Beatty, J. (2013). *Software requirements* (3rd ed.). Microsoft Press. ISBN 978-0-7356-7966-5.
11. Alencar, F., Castro, J., Cysneiros, G., & Mylopoulos, J. (2000). From early requirements Modeled by the i* technique to later requirements modeled. In U. M. L. Precise (Ed.), *Anais do III Workshop em Engenharia de Requisitos* (pp. 92–109).
12. Akao, Y. (1995). *Quality function deployment: Integrating customer requirements into product design*. Productivity Press.
13. Bresciani, P., Giorgini, P., Giunchiglia, F., Mylopoulos, J., & Perini, A. (2003). *TROPOS: An agent-oriented software development methodology. Journal of autonomous agents and multi-agent systems*. Kluwer Academic Publishers.
14. Caire, G., Leal. F., Chainho, P., Evans, R., Gómez-Sanz, J., Pavón, J., Kearney, P., Stark, J. and Massonet. P. (2000). Project p907, deliverable 1: Initial methodology. Technical Information Final version, European Institute for Research and Strategic Studies in Telecommunications (EURESCOM). Available from http://www.eurescom.de/public/projectresults/P900-series/907d1.asp
15. Caire, G., Coulier, W., Garijo, F., Gómez-Sanz, J., Pavón, J., Kearney, P., & Massonet, P. (2004). In F. Bergenti, M.-P. Gleizes, & F. Zambonelli (Eds.), *MESSAGE: A methodology for development of agent-based applications, to appear at methodologies and software engineering for agent systems*. Kluwer Academic Publishing (New York).
16. Castro, J., Kolp, M., & Mylopoulos, J. (2002). Towards requirements-driven information systems engineering: The Tropos Project. In *Information systems*. Elsevier.
17. Ceruzzi, L. and Rossi, G. (2002). "On the evaluation of agent-oriented methods", Agent Oriented Methodology Workshop.
18. Cysneiros, L., & Leite, J. (2004). *Non-functional requirements: From elicitation to conceptual models*. IEEE Trans. On Soft. Eng.
19. Dam, K., & Winikoff, M. (2003). Comparing agent-oriented methodologies, 5th international workshop on agent-oriented. *Information Systems, AOIS'03*, 52–59.
20. Dam, K. (2003). Evaluating and comparing agent-oriented software engineering methodologies, Master thesis of Applied Science in Information Technology. School of Computer Science and Information Technology of RMIT University, Australia.
21. Evans, R., Kearney, P., Stark, J., Caire Garijo F. J., Gómez-Sanz, J., Pavón, J., Leal, F., Chainho, P. and Massonet, P. (2001). Project p907, deliverable 3: Methodology for agentoriented software engineering. Technical Information Final version, European Institute for Research and Strategic Studies in Telecommunications (EURESCOM), Available from http://www.eurescom.de/public/projectresults/P900series/907ti1.asp.
22. Feather, M., Fickas, S., Finkelstein, A., & van Lamsweerde, A. (1997). Requirements and specification exemplars. *Automated Software Engineering, 4*, 4.

23. Garzetti, M., Giorgini, P., Mylopoulos, J. and Sannicolò, F. (2002). Applying Tropos Methodology to a real case study: Complexity and Criticality analysis. In Proceeding of the Italian workshop on "Dagli OGGETTI agli AGENTI - Dall'informazione alla Conoscenza (WOA02)", Milano.
24. Giorgini, P., Perini, A., Mylopoulos, J., Giunchiglia, F and Bresciani, P. 2001. Agent-oriented software development: A case study. Proceedings of the Thirteenth International Conference on Software Engineering & Knowledge Engineering (SEKE01), Buenos Aires, Argentina.
25. Iglesias, C., & González, J. (1998). A survey of agent-oriented methodologies. In *Proceedings of the 5th International Workshop on Agent Theories, Architectures and Languages (ATAL'98), LNAI n1555* (p. 317330). Springer Verlag.
26. Kolp, M., Giorgini, P. and Mylopoulos, J. (2001). A goal-based organizational perspective on multi-agents architectures, In Proceedings of the Eighth International Workshop on Agent Theories, architectures, and languages (ATAL-2001), Seattle, USA.
27. Parnas, D. (1972). On the criteria to be used in decomposing systems into modules reprinted from. *Communications of the ACM, 15*(12), 10531058.

2 The Need for Requirements Engineering

Abbreviations

AC	Alternating current
CONOPS	Concept of Operations
DC	Direct current (also 0 Hz)
DIA	Denver International Airport
F	Fahrenheit
hp.	Horsepower
HVAC	Heating ventilation and air conditioning
LRU	Line replaceable unit
MBSE	Model-based systems engineering
MDRE	Multidisciplinary requirements engineering
mph	Miles per hour
NASA	National Aeronautics and Space Administration
OSHA	Occupational Safety and Health Administration
RDBMS	Relational database management system
ROM	Rough order of magnitude
ROM	Rough order of measure
rpm	Revolutions per minute
SEI	Software Engineering Institute
TCO	Total cost of ownership
V	Volts
vac	Voltage, AC

2.1 Requirements Engineering: A Multidisciplinary Process

Multidisciplinary systems engineering has continued to evolve but has had issues keeping up with changes in system/software design methodologies. However, with the movement toward multidisciplinary systems engineering, systems design and requirements engineering has entered a new era of common practice among designers [2]. The complexity of systems and information today requires expertise in many disciplines and domains, leaving engineering, just like during the industrial revolution, without the tools and level of understanding required to engineer across disciplines, much less obtain knowledge outside of their silos of expertise. Therefore, new systems engineering concepts, mechanisms, and implementations are required to facilitate both the development and competency of the systems engineering discipline and the complex systems themselves, to ensure proper operation, and, in some cases, autonomous operation, self-healing, and critical self-management of knowledge and real-time operational self-awareness [3].

The mission/business analysis community typically attempts to keep analysis distinct from technical design. It is not inherently different or an incorrect term for development of a solution to a system design. In the end, the activities which we would call design are nothing different from the activities required to create the "To-be" requirements. Throughout the book, we will dive into the details of each type of requirements, their use, composition, decomposition, and overall need, based on a complete life cycle view of systems. Many systems are designed with basic capability requirements and ignore inclusion of things like supportability, maintainability, and sustainability until the system is designed and built [4]. This may expedite the original development and delivery of the system, but greatly increases the overall cost, or total cost of ownership (TCO) to the eventual customer, as design modifications will be required to include these capabilities. One of the purposes of this book is to drive home the need for complete requirements engineering and management throughout a system's life cycle.

Question to ponder

What is the difference between development cost and total cost of ownership (TOC)? Why do you think TOC is important?

J. A. Crowder, C. W. Hoff, *Requirements Engineering: Laying a Firm Foundation*,
https://doi.org/10.1007/978-3-030-91077-8_2

2.2 Multidisciplinary Requirements Engineering

Requirements engineering is a discipline that is used throughout industry and appears in all industry standards, including military, aerospace, railway systems, automotive, nuclear, electrical systems, etc. Multidisciplinary requirements engineering (MDRE) is a recognition that requirements engineering is a multidisciplinary endeavor due to the widely varied nature of the requirement types needed to adequately describe and lay out a complete set of system requirements for today's complex systems [5]. MDRE comprises a multitude of disciplines, as illustrated in Fig. 2.1.

All industries have horror stories about projects that failed due to improper or a lack of requirements management. Of all the phases of systems and software development life cycle, requirements management is arguably the most crucial [6].

2.3 The MDRE Process

Like other life cycle develop processes, the MDRE development process requires coordinated activities involved in the development, management, and maintenance of the set of requirements required to specify a system [7]. There is not one requirement process that is applicable to every size system, but there are general MDRE guidelines/tasks that should be followed for MDRE to result in a successful system development, test, and sustainment.

2.3.1 The MDRE Feasibility Study

The MDRE process begins with a feasibility study that will identify the components, functions, etc. that are required to instantiate the system to be developed [8]. Initiating the systems feasibility study and subsequent feasibility report(s) will lay out:

1. Does the system contribute to the overall high-level mission objectives and requirements? Every system operates in an environment and interacts with something else. Even the Curiosity Rover on Mars must fit into NASA's overall mission and objectives. During this step, high-level goals and objectives (both functional and nonfunctional) are explored to determine the scope of the system to be developed in a system-of-systems context.
2. Alternate solutions that will all meet the basic requirements called for by the customer [9].

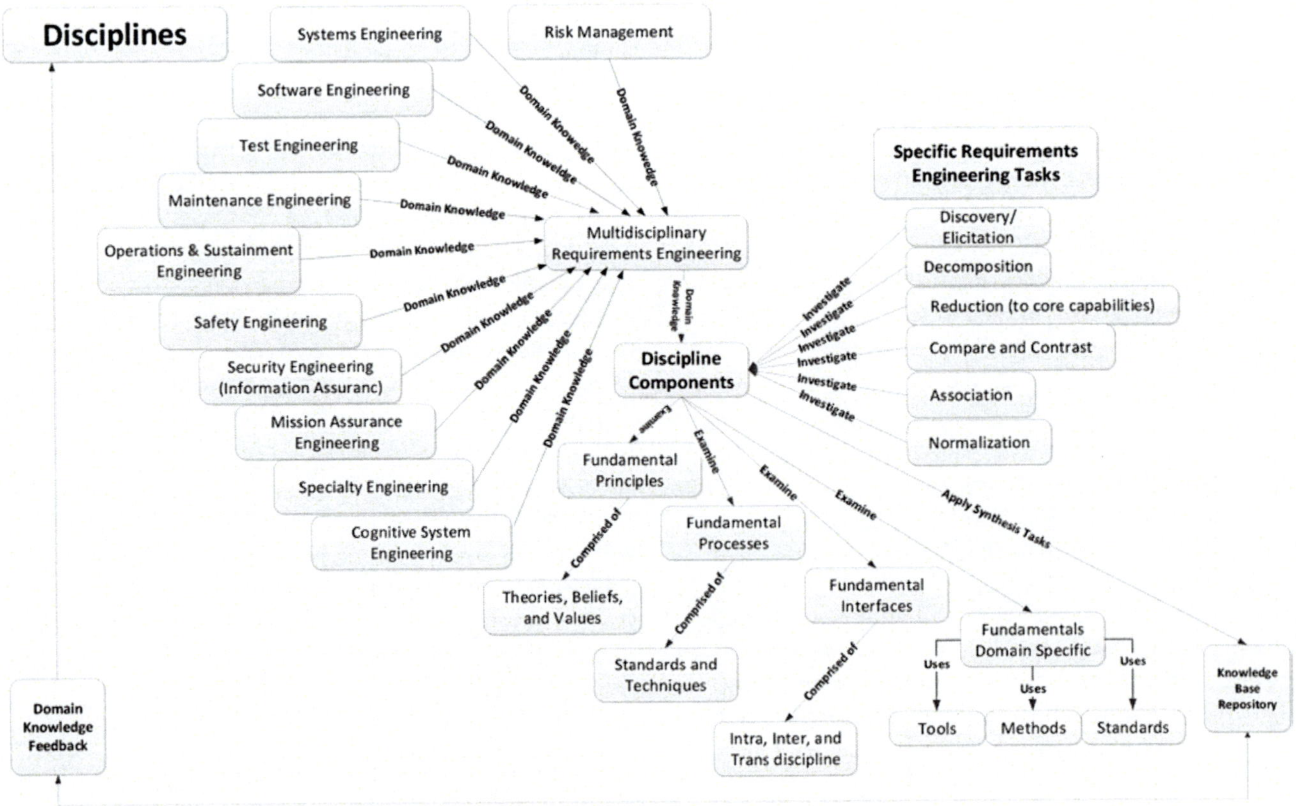

Fig. 2.1 Overview of multidisciplinary requirements engineering flow

3. High-level cost estimates: the feasibility study should include a "rough order of magnitude" (ROM) for each alternative solution to understand which solutions are economically feasible to the customer. During this step, many people will assume a dollar amount and make sure the estimates fit the envisioned cost profile. This often leads to incorrect or "idealized" costs that can never be realized during actual development. That is why many government agencies have gone away from "lowest bidder" and have adopted the "best value" assessment for contract award.
4. Explore potential complications with development/deployment/sustainment for the system.
5. Understand the customer's expectations and "feelings" about the system.
6. Estimate the likelihood for success for both the initial development, integration into the target system/environment, as well as sustainment/maintenance of the system [10].
7. Understand the resources that will be required for development, deployment, and sustainment of the systems. This includes an estimate of how the system will be maintained, what constitutes a "Line Replaceable Unit" (LRU) for the system, and the skill level(s) required to maintain the system [11]. Also included will be a first-order estimate of what maintenance can/will be performed during depot level maintenance, and which actions will require factory maintenance.
8. Create a prospective timeline (how long is reasonable to build and deploy the system).
9. Help write the overall business plan for the system.

Question to ponder

Why do you feel a feasibility study is important? What problems could occur if a feasibility study is not done before development?

The output of the analysis, the Feasibility Report, should, at least, contain [12]:

1. Definition of the system to be developed.
2. High-level view of alternative solutions, challenges, competitors, and consequences.
3. High-level set of requirements.
4. An executive summary.

The feasibility study feeds the initial requirements analysis required to create a requirements definition and requirements specification for the system to be developed. The MDRE requirements analysis process provides the initial set of requirements for the system.

Question to ponder

What questions should a Feasibility Study answer?

2.3.2 The Requirements Analysis Process

The purpose of the MDRE requirements analysis processes is to create an initial set of requirements and determine whether the requirements set represents requirements that are clear, complete, non-ambiguous, and non-contradictory [13]. The resultant set of requirements definitions and requirements specification should drive a set of system models that are used at a high level to validate that the requirements constitute a system that will indeed meet the customer's requirements. Fig. 2.2 illustrates the overall information flow for the requirements analysis process. The requirements analysis process is needed to determine the overall quality of the requirements and resolve requirements issues (e.g., contradictory requirements) before the design process ensues. System modeling is used to document (in different formats) the requirements to allow presentation of the system requirement set across organizational levels within both the development company (or, in some cases, companies) and the customer [14].

All requirements that are allocated and/or derived must trace back up to a given customer requirements document. Traceability is essential to ensure that the system meets the customer's original requirements and objectives [15]. This applies to functional, nonfunctional, interface, sustainment, etc. requirements as illustrated in Fig. 2.3. This will also indicate the types of inputs that are taken in by the customer in creation of the high-level requirements that are utilized within the MDRE process to arrive at a complete set of systems/software requirements for system development.

The MDRE requirements analysis process is a continuously feedback-driven system, since throughout the requirements management process additional analysis must be performed when requirements change, are found to be insufficient, or conditions change like environmental laws or conditions [16]. Overall, the MDRE process establishes the needs/services the customer requires from the systems under development and the constraints under which it must operate [17]. The main goal of MDRE is the creation, sustainment, and management of the system/software requirements. MDRE is concerned with discovering, analyzing, specifying, and documenting requirements for the systems. MDRE deserves more attention than almost any other aspect of system development, since without adequate requirements the system has little chance of success. Table 2.1 below lays out the basics of the modern MDRE feedback-driven continuous requirements process [18].

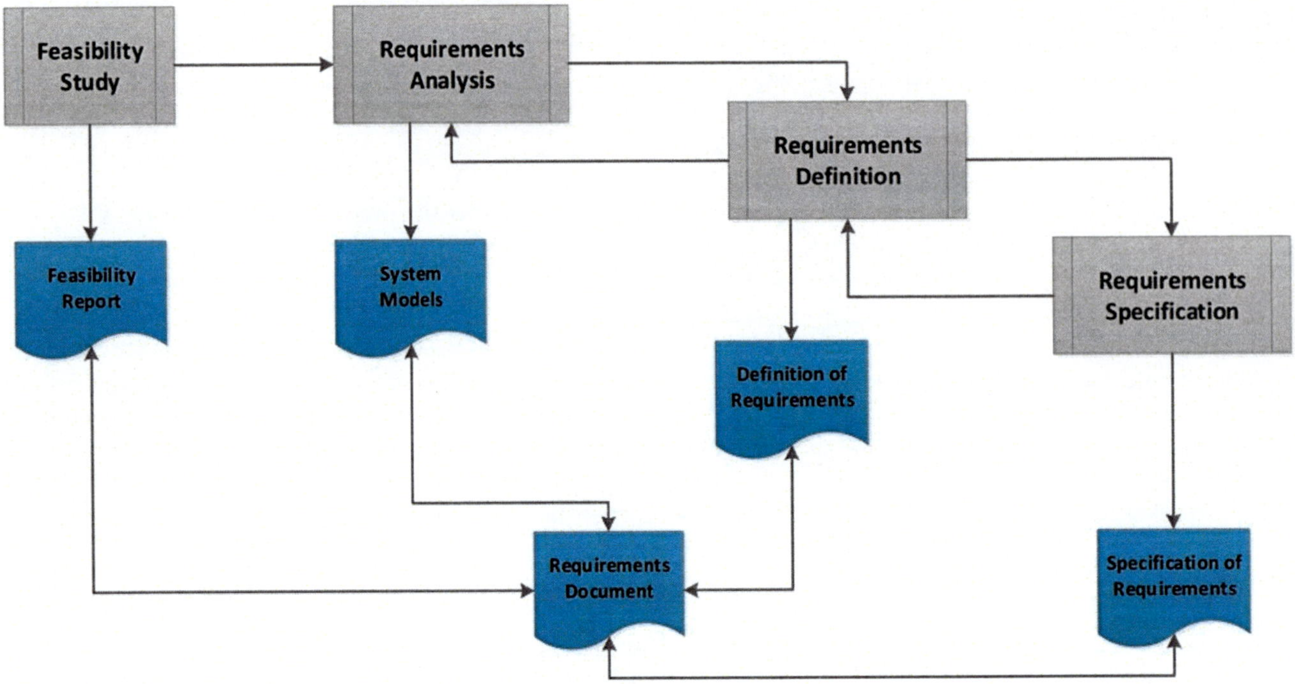

Fig. 2.2 The MDRE requirements analysis data flow

> **Question to ponder**
>
> What role do nonfunctional requirements (e.g., size) have on the technology decisions (functional requirements)?

The complete feedback-driven requirements analysis process includes development of system models, prototypes, and mock-ups to ensure the requirements will result in a system that meets the intent of the user/customer community [19]. Fig. 2.4 illustrates the overall requirements data flow process. The MBSE designation in Fig. 2.4 represents a new set of procedures in MDRE called model-based systems engineering and is discussed in detail in Chap. 17. The MDRE requirements data flow process takes advantage of the user scenarios (sometimes called use cases) as well as customer feedback and Concept-of-Operations (CONOPS) documentation to use as validation of the system models and prototypes. All this feedback is folded back into the MDRE Requirements Analysis before the system/software specification is released [20]. It is also monitored continually throughout the development, test, validation, and sustainment cycles (i.e., the entire system life cycle) to understand the impacts of not only changing requirements but also changes to environments, interactions with legacy and new systems, or any other changes or conditions that might affect the operational parameters of the system. The MDRE requirements data flow process begins with the Requirements Analysis task and drives the entire MDRE feedback-driven process clear though life cycle sustainment and maintenance of the system.

2.4 MDRE Technical Requirements

Technical requirements are one of the most, if not the most, important aspects of system engineering. The task to build, manage, and maintain them are not appreciated until satisfaction of stakeholder expectations fails, at which point, technical requirements become the most critical aspect of any discrepancy discussion [21]. Some may argue this point, depending on their role within the development project, but without the technical requirements understood, the non-technical requirements do not help, since there is not definition of what is being developed.

> **Question to ponder**
>
> Why are User Scenarios (also called Use Cases) so important to the overall requirements analysis process? What do you think would be the result of skipping User Scenario development?

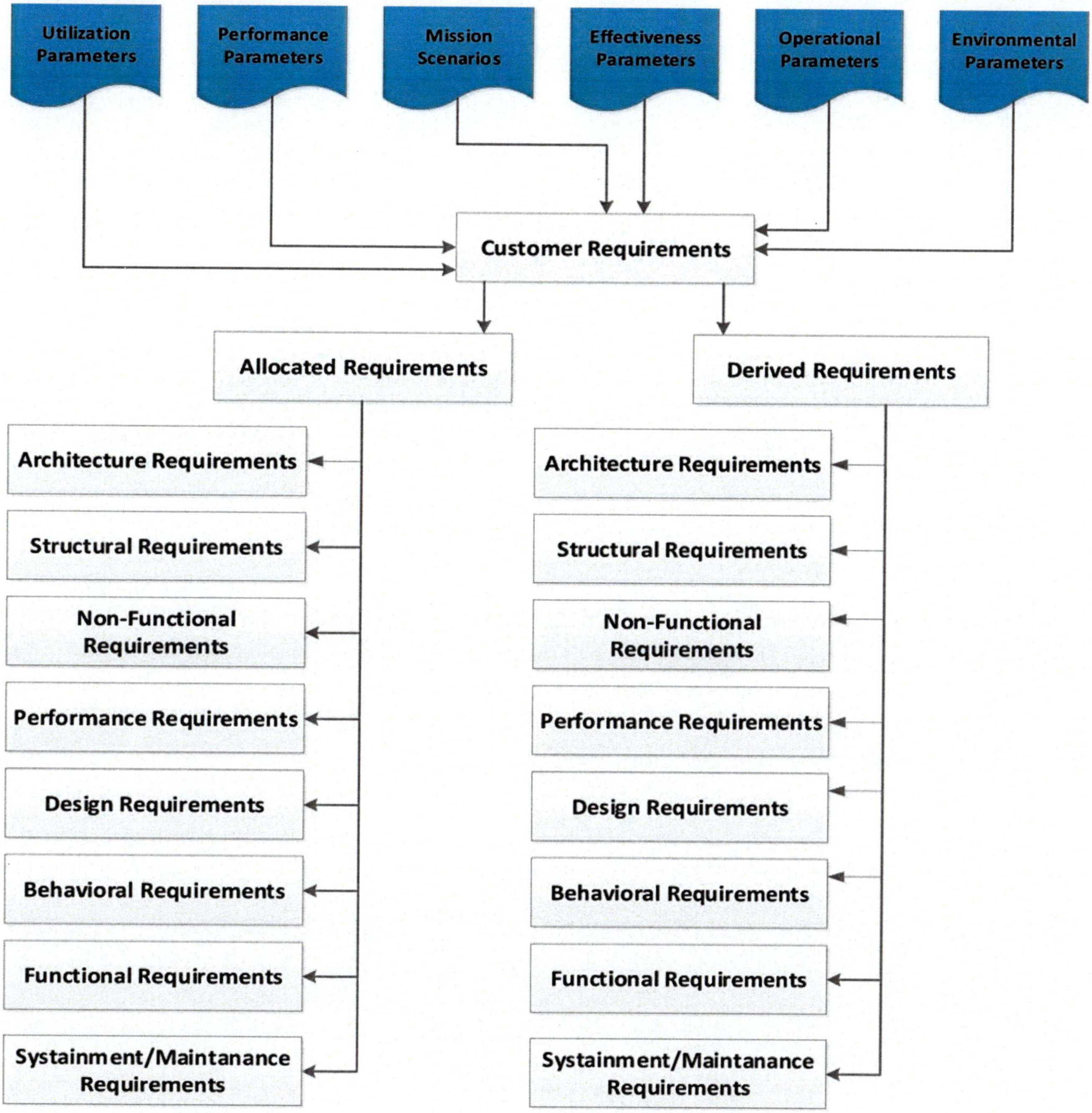

Fig. 2.3 MDRE requirements decomposition/derivation/allocation flow

While significant support data is included with requirements, such as use cases, architectural materials, design guides, and alike, the technical requirements are the first and foremost MDRE artifact that must be verified and validated. Externally, the technical requirements are those items that will be tested or demonstrated to verify that the system is acceptable to the customer. Internally, technical requirements form the "contract" between system engineering and the stakeholders, as well as system engineering and the developers of the system's capabilities [22].

Definition: Requirement—A statement that mandates that "something" must be provided, produced, accomplished, or changed. These are often referred to as "shall" statements.

Examples

- The system shall collect solar flare metrics continuously.
- The diesel engineer shall maintain 535 hp. at 2000 rpm at nominal cruising speed of 75mph.

Table 2.1 The basic MDRE process descriptions

MDRE Activity	MDRE Implementation
Requirements elicitation	• Requirements elicited through usage scenarios.
Requirements analysis	• Continuous process throughout development. • Customer prioritizes usage scenarios.
Requirements documentation	• Usage scenarios documented along with acceptance tests for each user scenario. • Continued updates throughout system life cycle.
Requirements validation	• Test driven development. • Acceptance test criteria development. • Continuous feedback.
Requirements management	• Short planning iteration in every phase of development/sustainment. • User scenarios used for tracking. • Continuous re-evaluation and re-factoring as requirements and environments change.

of the requirements [24]. Formal requirements are governed by processes and practices to ensure that the system being built, or procured, meets the end needs of the stakeholder/customer, and include formal change control processes. Informal requirements may or may not result in meeting the end needs of the stakeholders, and are measured using subjective methods of assessment, rather than distinct measurable criteria [14]. Informal requirements typically have informal change control associated with them, forcing the developers or vendors to satisfy a moving target. Informal control can be good when building simpler, low-cost, non-complex systems like a sandcastle on the beach, but do not work well in development activities when used to meet the end needs of a customer or management staff spending large sums of money on a large complex system [19, 25].

All developments or procurements should recognize when informal requirements can and cannot be used.

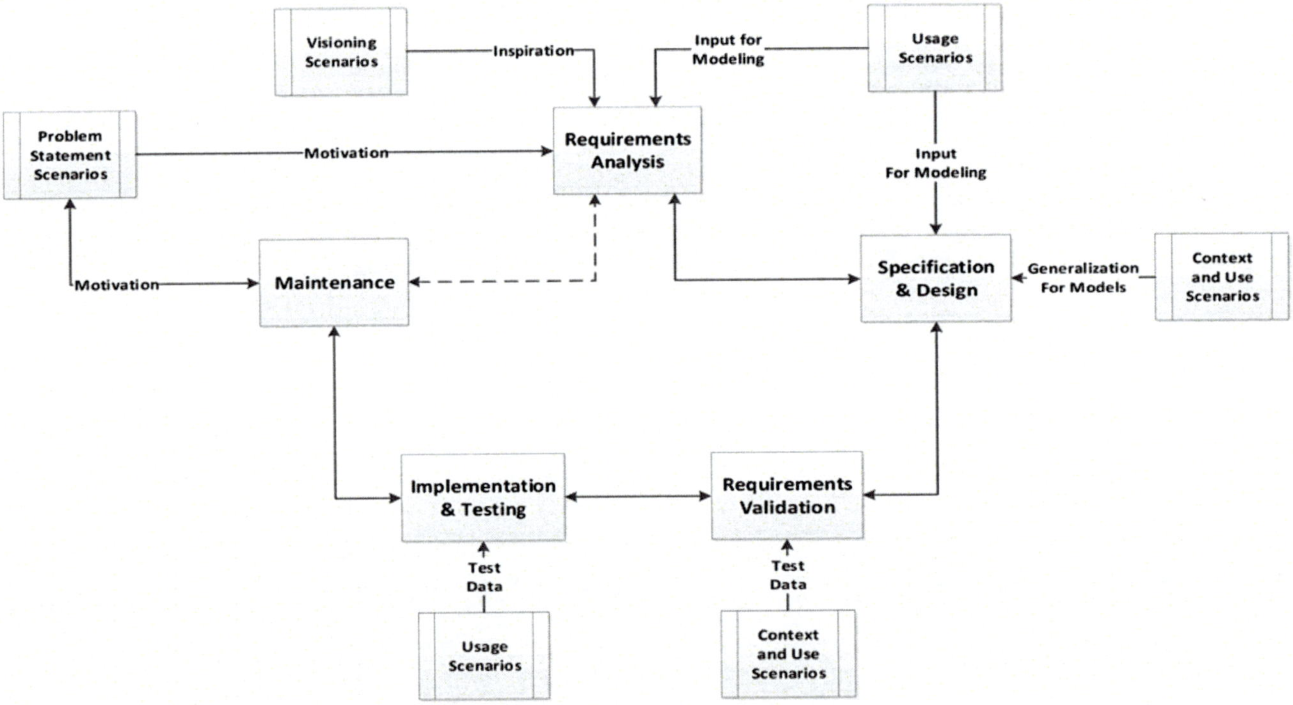

Fig. 2.4 The MDRE requirements

We will discuss technical requirements in detail later, but before discussion on the development of technical requirements, there are several general MDRE topical areas that must be addressed to provide some corollary insights into proper MDRE requirements generation [23].

2.5 Formal Vs. Informal Requirements

"*All systems have technical requirements.*" One can argue that this statement is not true; however, the distinction will be determined by comparison of formal verses informal control However, as a rule the MDRE staff must adeptly understand that outcomes of efforts employing informal requirements result in subjective interpretation and derivation and may be unreliable. Technical requirements must be formally controlled, where the organization uses a formal tracking mechanism (e.g., Relational Database Management System (RDBMS), managed word documents, spreadsheets, and/or like materials that have versioning capabilities) [26]. In general, the larger and more complex a system is, the fewer informal requirements will be allowed. Requirements management tools (databases, etc.) will be discussed in Chap. 14.

Another aspect to technical requirements is associated with development of new systems verses upgrading and/or replacing existing systems. Since time eternal, we have developed "new" systems, but rarely are the systems new, rather they are new applications of thought and/or use new technology to solve existing problems or are evolutionary improvements/capabilities of existing systems with newer/modern technologies [27]. The MDRE group must understand that a stakeholder, who wants a new system, must be prepared to accept a "new" development. This means that although the MDRE personnel works with the stakeholder(s) to determine customer needs, the MDRE group is free to define the requirements to meet the objectives of the stakeholders through the generation of derived requirements, rather than working to a predefined requirement set (primary and/or contractual requirements) that has been updated from prior developments and/or procurements [28]. Rarely is the system engineer allowed to define a new system purely from derived requirements outside of research and development, where new thought processes and applications of technology may be invented to solve the solution for a stakeholder [29].

Question to ponder

What are some of the differences between an evolutionary development (new capabilities to an existing system) project and a revolutionary development (new system) project?

The point being made here is that most systems are enhancements to existing systems and come with some predefined expectations (requirements) of what the stakeholders are looking for with the development and procurement activities. These existing requirements may have to be assessed for proper "Tiering" of the technical requirements set [30].

This may cause one to wonder what the effect of not utilizing a process like that depicted in Fig. 2.2 can have on the overall development of a complex system. One example of why having good requirements is crucial is the Denver International Airport (DIA) Baggage Handling System. According to the Software Engineering Institute (SEI) two top factors led to project failure; these are: (1) inadequate requirements specifications and (2) changes in requirements throughout the development cycle.

2.5.1 Case Study: The Denver International Airport Baggage Handling System

The Denver International Airport (DIA) is the largest airport in the United States (area), with the longest public runway in the United States. During the original construction of DIA, a new, state-of-the-art baggage system was to be built and was touted as the most advanced baggage handling system in the world. That system is now touted as one of the clearest examples of project failure in the last 10–15 years. The baggage system that was originally proposed was supposed to be designed to automate the entire baggage handling activities across all three airport concourses for both outgoing and incoming bags.

One major problem was the failure to properly understand and decompose the high-level requirements and understand the overall complexity of the system being proposed. In addition, the design team did not adequately assess the risks involved in the new system. As a result of improper requirements derivation/decomposition, there were major problems building the baggage handling system and these issues caused a 16-month delay in the opening of the airport, while the engineering team tried to salvage the system and find something that would work and was acceptable. The delay in construction and testing of the baggage handling system resulted in a $560 Million dollar (US dollars) overrun and was featured in many articles as a major failure of the new airport.

The final design solution that was implemented at DIA was far less capable than was originally envisioned, providing automated only the outbound baggage of one of DIA's three concourses. The rest of the baggage is handled with a manual system. Even the one concourse's automated baggage handling system had to be scrapped completely in 2005 due to improper functionality. Having not been designed properly for maintenance, the $one Million monthly maintenance costs proved too expensive and going to a completely manual system reduced the overall life cycle costs. In general, the major issues with the baggage system were:

- The complexity of the System of Systems architecture.
- Changes in system requirements.
- Underestimating cost and schedule.
- Improper risk analysis.
- Failure to design and factor in backup and recovery systems.

2.6 The MDRE Process: Enhancement of an Existing Legacy System

As discussed above, sometimes legacy/existing systems need to be modified or upgraded to provide either new capabilities or a technology refresh because the electronics of the system are outdated and no longer available and/or serviceable [31]. In this case current requirements for the system that will not change must be analyzed and coordinated with new requirements for new features, new equipment, etc. that will be put

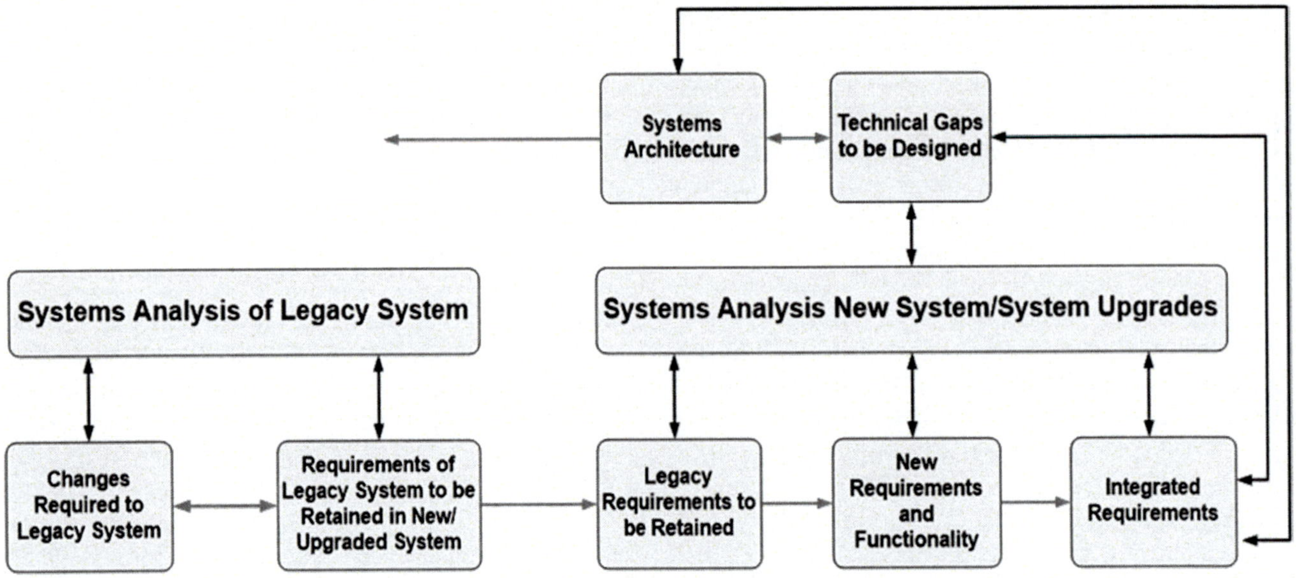

Fig. 2.5 Analytical process for reuse of legacy systems in a new or enhanced system

into the existing system. Fig. 2.5 illustrates the MDRE process for upgrading/updating a legacy system. In the traditional mission/business analysis approach, the "As-Is" state is analyzed as a starting point to capture processes, data, and interactions that must be kept for any new solution [32]. Those are shown as the "as-is" requirements in Fig. 2.5. The requirements/functionality of the legacy system is identified, along with the processes, data, system interfaces, and applicable displays. The design process comes into play when determining how to integrate them with the new features being added. These new features become requirements/functionality that must be added to design the overall new/enhanced system. This is not technical or physical design, but a logical design discussed earlier. Examples of new requirements/functionality include many traditional outputs including [33]: Once the architecture has been finalized and approved, the development process ensues, although this too is a feedback-driven process that may involve rethinking requirements as the development process progresses.

Process models showing the legacy processes to be reused as well as the required new processes [34].

Data models showing final new/enhanced system data requirements and mission/business rules relating to the relationships between entities.

Use Case models with interaction requirements and corresponding mission/business rules.

Prototypes and mock-ups indicating interface requirements, including presentation mock-ups [35].

Interfaces with other systems or parts of the system (e.g., subsystem interfaces).

Question to ponder

Do you think it is possible for new requirements for a system upgrade to conflict/contradict current requirements for the legacy system being upgraded? If so, how would you handle this?

2.6.1 Design Problem

The philosophy of the assignment development for this textbook was to create a set of though-provoking questions, problems, and design assignments that help drive home the principles, practices, and methodologies of multidisciplinary requirements engineering (MDRE). We endeavored to provide the opportunity for the student to practice the art and science of requirements engineering, helping the student to understand that there is rarely one answer to a requirements engineering problem, and it is only through study and practice that successful Requirements Engineers are developed. MDRE is a life-long journey, and anyone who tells you they have nothing left to learn does not truly understand the discipline or has never been involved in the development of a complex system. With the recent interest in alternative energies and talk of the "Green New Deal," along with the recent issues with ice storms in Texas taking out the power grid, the assignments for this textbook will follow the theme of effectively dealing with both local alternative energy power grids and the national infrastructure required to facilitate large-scale alternative energy production, sustainment, and management.

2.6.2 Design Problem Description: Alternative Energy Power Grid

The purpose of this project is to, over the course of the semester, develop a feasibility study, requirements analysis, requirements elicitation, derivation, and allocation, for a system that will facilitate an "local" alternative energy power grid capable of powering a neighborhood of 500 homes and a large biosphere capable of supporting growing year-round fruits and vegetables for the 500-home neighborhood. The student will have to deal with issues of choosing multiple alternative energies that can provide power throughout all seasonal difficulties, from intense summer heat to freezing winter temperatures and possible ice storms. Control, viability, sustainability, and maintainability will be major factors to understand the entire possibly 30-year life cycle of such a system. How to effectively utilize, store, and manage power distribution, along with when to up-convert "extra" energy to the national power grid, will be issues that drive many of the assignments. Simply relying on the current technologies and energy infrastructures will not move the country and world forward. Hopefully, this will provide insight to the students using this textbook to understand some of the issues that face everyone looking out over the next many decades.

2.6.3 DC Power Grid Requirements Discussion

In the initial days of power generation and distribution, DC power was far less practical than the AC distribution system that soon supplanted it. However, even after Tesla's AC system became the predominant system (and rightly so), Edison's DC power did not become obsolete for some time. Eventually power plants eliminated their delivery of DC power, but the overall concept of a DC power grid is not dead. Even our computer systems (large server systems) have uninterruptible power supplies which contain batteries that are charged off the grid and used to up-convert to AC to power devices when the AC power grid goes off-line. So, the notion of a DC power grid is not dead, but generally used as a backup system. The point is, the notion of a DC power grid has remained and now that renewable energies are viable, we should re-examine their use for widespread, local, power distribution.

One of the advantages of a locally generated DC station is energy savings as well as allowing a facility to run more easily off various DC sources, such as Solar Arrays. And for power-hungry installations, the notion of generating the power locally is growing in popularity. Part of the reason is that waste heat from the power generation generators (solar, geothermal, etc.) can be used to warm nearby buildings. By combining the production of heat and power, local power grids can squeeze much more useful energy out of the renewable energy they use, so this approach will certainly become more widespread as time goes on. The purpose of this design problem is to look ahead and provide an architecture and specification for a local DC power generation station that can serve a local community.

Currently many systems exist to control and handle alternative energies. There are systems to control and manage solar power, wind power, geothermal power, and others, but very few are designed to control, manage, store, and distribute energy from a power grid consisting of multiple alternative energy sources. As we saw the issues recently caused by the intense ice storms in Texas, one alternative energy source is not the answer to the growing energy needs/wants in our society. Each alternative energy source has limitations and issues associated with its use. By harnessing multiple alternative energies, and managing the storage of the energies produces, communities can be worry free from power generation. In order to facilitate utilizing multiple alternative energy sources, a control center must be included in the design that can manage the energy creation, distribution, storage, and return unused energy to the power grid.

2.6.4 High-Level Technical Description of the DC Power Grid System

The purpose of the multiple alternative energies DC power grid is to:

- Supply DC power to a customer community of 500 homes and a biosphere for growing food.
- Provide a storage device to store energy in case of a system failure.
- Determine the amount of surplus energy that is above and beyond what is required to power the grid and charge the backup system, and then deliver that surplus energy to the national power grid systems.

Throughout the semester you will be asked to analyze, design, and architect different components of the overall DC power grid system. This will include research into alternative designs, like Lithium Nanowire Storage systems for backup power and/or Super Capacitors (also called Ultracapacitors). Fig. 2.6 illustrates these technologies connected to the DC Power Grid Regulator Control Center.

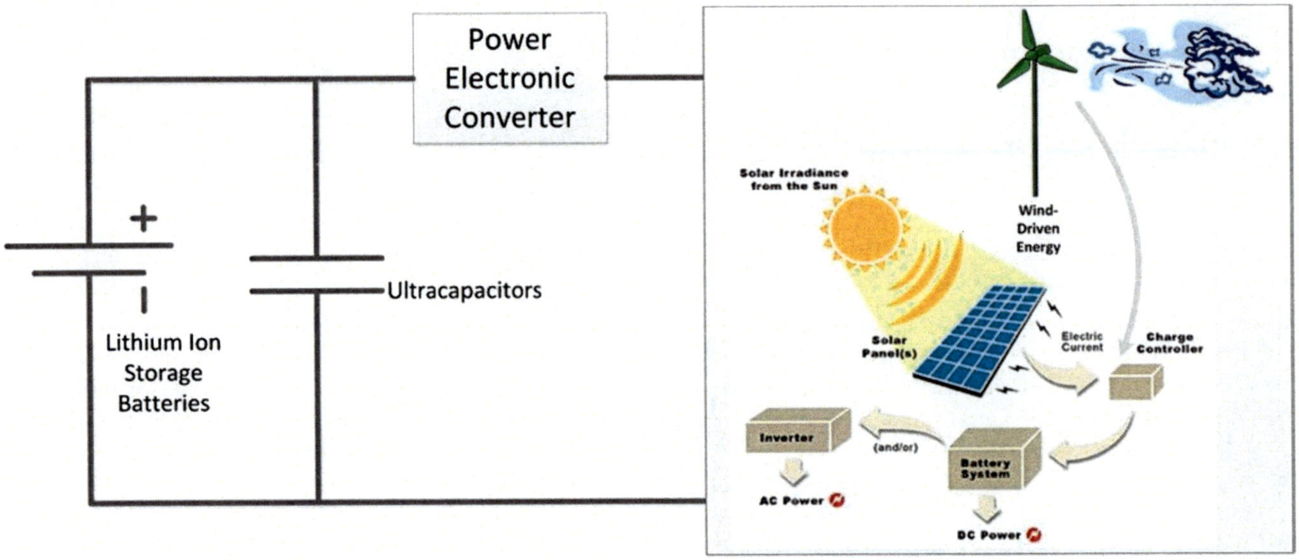

Fig. 2.6 Alternative power management system overview

2.6.5 High-Level Technical Requirements for the DC Grid System

- Design shall utilize at least two alternative energies, with a maximum of 4.
- Total Power Generation shall be 3.0 megawatts minimum.
- Each of the alternative energy sources shall be capable of generating at least 60% of minimum wattage.
- DC power grid shall generate DC power, with Grid voltage being 48, 24, or 12 volts.
- DC power grid shall be capable of servicing at least 500 single family homes and one 4-acre biosphere with hydroponics capable of growing vegetables year-round. (See Fig. 2.7 for a block diagram of a DC-powered home).
- All lights, appliances, and facilities in each home and the biosphere shall be configured for DC operations, if possible. Things like electric ovens may need DC/AC conversion to accommodate their 220v. requirements.
- Only allowed exceptions are the HVAC system and the Washer/Dryer, although DC is preferred.
- There shall be energy storage capability capable of operating the grid of 500 single family dwellings for 7 days if all alternative forms of energy are not producing.
  - Compliance Criteria: Energy storage devices shall be capable of producing 750,000 watts for the entire 7-day period. Nominal energy storage should be batteries or supercapacitors.
- The DC power grid system shall include controlling h/w and s/w to manage the multi-alternative energy strategies and production.
- All energy produced above the required home and energy storage usage shall be upconverted to 220vac for release to the national power grid.
- The DC power grid shall be capable of operating in a temperature range of −40°F to 125°F. without degradation of power generation.
- The DC power grid must meet OSHA regulations (Fig. 2.7).

2.6.6 Assignments

- Develop a feasibility study for the whole house DC system and answer the following questions:
- Is it practical to run everything in a normal household off DC?
- Which appliances, etc., need to run off AC?
- Which DC voltage makes sense for the system?
  - 12 V?
  - 24 V?
  - Other voltages?
- What storage medium makes practical sense?
- Batteries? What kind?
- Supercapacitors? What size?

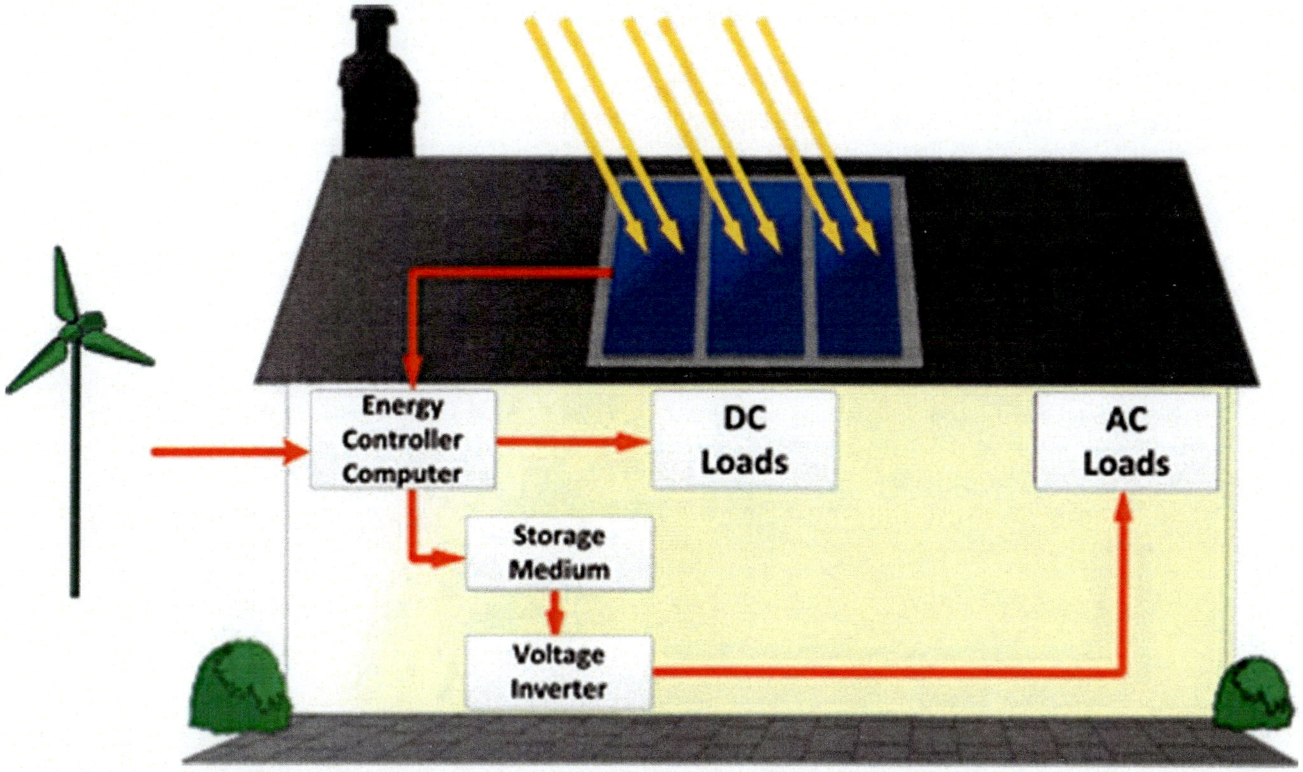

Fig. 2.7 Example of a whole house power grid

References

1. Pavóon, J. and Gómez-Sanz, J. (2003). Agent oriented software engineering with INGENIAS". In: Multi-Agent Systems and Applications III, Proc. of CEEMAS, LNCS 2691, Springer-Verlag, pp 394–403.
2. Bettencourt, D., & Penzenstadler, B. (2008). *Designing, documenting, and evaluating software architecture technical report TUM-I0818*. Technische Universität München.
3. Parnas, D. (1972). On the criteria to be used in decomposing systems into modules. *Reprinted from Communications of the ACM, 15*(12), 10531058.
4. Dwyer, M., Avrunin, G., and Corbett, J. (1999). Patterns in property specifications for finite-state verification. In Proceedings of the International Conference on Software Engineering, Los Angeles, CA.
5. Shehory, O. and. Sturm, A. (2001). Evaluation of modelling techniques for agent-based systems, 5th International Conference on Autonomous Agents (Agents 01), ACM Press, pp 624–631.
6. Sturm, A. and Shehory, O. (2003). A framework for evaluating agent-oriented methodologies, 5th International Workshop on Agent-Oriented Information Systems (AOIS'03), pp 60–67.
7. First International Workshop on Evaluation of Modeling Methods in Systems Analysis and Design http://www.ait.unl.edu/siau/conference/emmsad03-CFP.htm
8. Jain, H., Vitharana, P., & Zahedi, F. (2003). An assessment model for requirements identification in component-based software development. *ACM SIGMIS Database, 34*(4), 48–63.
9. Beck, A., Boeing, G., & Shannon, D. (2014). Systems and methods for analyzing requirements. US Patent 8650186.
10. Marjo, K., Matti, V., Jyrki, K., Sari, K., & Reijo, S. (2004). Implementing requirements engineering processes throughout organizations: Success factors and challenges. *Elsevier Information and Software Technology, 46*(14), 937–953.
11. Mathiassen L. (2008). Managing requirement engineering risks: An analysis and synthesis of literature, AIS Electronic Library, http://aisle.aisnet.org/sprouts_all/82.
12. Karlsson, L. (2007). Requirements engineering challenges in market-driven software development—An interview study with practitioners. *Elsevier Information and Software Technology, 49*(6), 588–604.
13. Mantere, T., & Jarmo, A. (2005). Evolutionary software engineering, a review. *Elsevier Applied Software Computing, 5*(3), 315–331.
14. Alexander, I., & Stevens, R. (2002). *Writing better requirements*. Addison Wesley.
15. Kaufman, L., Thebaut, S., Interrante, M. (2009). System modeling for scenario-based requirements engineering, SERC Technical Report, SERC-TR-33-F.
16. Loucopoulos, P., & Karakostas, V. (1995). *Systems requirements engineering*. McGraw-Hill.
17. Sutcliffe, A., & Maiden, N. (1998). The domain theory for requirements engineering. *IEEE Transactions on Software Engineering, 24*(3), 174–196. https://doi.org/10.1109/32.667878
18. Jackson, M. (2000). *Problem frames: Analyzing and structuring software development problems*. Springer Publishing. ISBN-13: 978-0201596274.
19. Goguen, J., Linde, C. (1993). Techniques for requirements elicitation. In: Proceedings of the IEEE International symposium on Requirements Engineering, San Diego, CA.
20. Martin, R. (2003). *Agile software development: Principles, patterns, and practices*. Prentice Hall.
21. Hickey, A. and Davis, A. (2003). Elicitation technique selection: How do experts do it?, Proceedings of the Eleventh IEEE International Requirements Engineering Conference, pp. 169–178, September 8–12, Monterey Bay, CA.

22. Maiden, N., Gizikis, A., & Robertson, S. (2004). Provoking creativity: imagine what your requirements could be like. *IEEE Software, 21*(5), 68–67.
23. Wood, D., Christel, M., and Stevens, S. (1994). A multimedia approach to requirements capture and modeling, Proceedings of the First International Conference on Requirements Engineering, pp. 53–56, April 18–22, Colorado Springs, CO.
24. Osborn, A. (1979). *Applied imagination*. Charles Scribner's Sons.
25. Burge, J., & Brown, D. (2008). Software engineering using RATionale. *The Journal of Systems and Software, 8*, 3.
26. Nuseibeh, B. and Easterbrook, S. (2000). Requirements engineering: A Roadmap, Proceedings of the Conference on The Future of Software Engineering, pp. 35–46, June 4–11, Limerick, Ireland.
27. Jackson, M. (2000). *Problem frames: analyzing and structuring software development problems*. Addison Wesley.
28. Akao, Y. (1995). *Quality function deployment: Integrating customer requirements into product design*. Productivity Press.
29. Sherif, K., & Vinze, A. (2002). Domain engineering for developing software repositories: A case study. *Elsevier, Decision Support Systems, 33*(1), 55–69.
30. Herela D. and Greenberg S. (1998). Using a groupware space for distributed requirements engineering. In: Proceedings of the 7th workshop on enabling technologies: Infrastructure for collaborative enterprises, June 17–19, Stanford, CA.
31. Hickey A. and Davis A. (2003). Elicitation technique selection: How do experts do it? In: Proceedings of the 11th IEEE International requirements engineering conference, Monterey Bay, CA.
32. Nuseibeh, B., & Easterbrook, S. (2000). *Requirements engineering: A roadmap*. ACM New York.
33. Shaw, M. (1990). Prospects for an engineering discipline of software. *IEEE Software, 7*(6), 15–24.
34. Stevens, R., Brook, P., Jackson, K., & Arnold, S. (1998). *Systems engineering: Coping with complexity*. Prentice Hall Europe.

MDRE System Supportability Analysis

3

Abbreviations

AC	Alternating current
ACQ	Acquisition
ADC	Analog-to-digital converter
AFRL	Air Force Research Labs
AI/ML	Artificial intelligence/machine learning
ASOE	Affordability System Operational Effectiveness
ATM	Automated teller machine
ATP	Authorization to Proceed
Az	Azimuth
BIT	Built-in Test
BSP	Board Supply Packages
CBM	Condition-based maintenance
CERN	European Organization for Nuclear Research
cm	Centimeter
COTS	Commercial off the shelf
COVID-19	Coronavirus Disease 2019
CPPI	Continuous process/product improvement
CSWaP	Cost, size, weight and power
CW	Continuous wave
DAC	Digital-to-analog converter
dB	Decibel
dBm	Decibel reference to a milliwatt
dBsm	Decibel per square meter
DC	Direct current (also 0 Hz)
DFM	Design for Maintenance
DFS	Design for Supportability
DMA	Damage Mode Analysis
DoD	Department of Defense
DREX+	Digital receiver/exciter
DSP	Digital signal processing
EKF	Extended Kalman filter
El	Elevation
ESD	Electrostatic discharge
EW	Electronic Warfare
F	Fahrenheit
FAA	Federal Aviation Administration
FAA	Federal Aviation Agency
FMEA	Failure Modes and Effect Analysis
FMECA	Failure Mode, Effects, and Critical Analysis
FOV	Field of view
FPGA	Field programmable gate array
ft.	Feet
Gbps	Giga bits per second
GHz	Giga-Hertz
GPP	General purpose processor
H/W	Hardware
HF	High frequency
HPC	High-performance computing
HVAC	Heating, ventilation, and air conditioning
I/O	Input/output
ID	Identification
IOT&E	Initial Operational Test and Evaluation
KPI	Key Performance Indicator
kW	Kilo-watt
LCC	Life cycle cost
LED	Light emitting diode
LHC	Large Hadron Collider
LRU	Line replaceable unit
m	Meter
MDA	Missile Defense Agency
MDRE	Multidisciplinary requirements engineering
MHz	Mega-Hertz
MIL-STD	Military Standard
MoE	Measures of Effectiveness
MOS	Metal oxide semiconductor
MTBF	Mean time between failure
MTBMA	Mean time between maintenance action
MTTR	Mean time to repair
MWC	Miner wearable component
nm or nmi	Nautical miles
O&M	Operational and Maintenance

J. A. Crowder, C. W. Hoff, *Requirements Engineering: Laying a Firm Foundation*,
https://doi.org/10.1007/978-3-030-91077-8_3

O&S	Operations and Support
OoI	Object of Interest
OSHA	Occupational Safety and Health Administration
PD	Pulse Doppler
pps	Pulse per second
R&D	Research and Development
RADAR	Radio detect and ranging
RAM	Reliability, availability, and maintainability
RARE	Radar Advanced Receiver Exciter
RCM	Reliability centered maintenance
RCS	Radar cross section
RF	Radio frequency
RPY	Roll, pitch and yaw
S/W	Software
SARS-COV-2	Severe acute respiratory syndrome coronavirus 2
SCA	Supply chain analysis
SCM	Supply chain management
SCOR	Supply chain operations reference
SDR	Software-defined radio
SIGINT	Signals intelligence
SNR	Signal-to-noise ratio
SSTX	Solid-state transmitter
TBD	To be determined
TCO	Total cost of ownership
TCP/IP	Transmission Control Protocol/Internet Protocol
TOC	Total ownership cost
ToF	Time of flight
TV	Television
UHF	Ultra-high frequency
USB	Universal Serial Bus
USTAR	Unmanned Search/Track Air Radar
UWB	Ultra-wideband
Vac	Volts AC
W	Watt
WHO	World Health Organization
WIP	Work in progress
WSARA	Weapon System Acquisition Reform Act

3.1 Introduction to Supportability Analysis

Complex sensor systems are utilized in a host of various military and commercial operations. Military, civil, and commercial systems must provide a needed capability, meet user needs as evidenced by operational effectiveness, suitability, and availability, while remaining affordable [1]. While operational effectiveness addresses the degree of mission accomplishment in the intended environment, operational suitability and availability addresses the degree to which a system can be satisfactorily placed in use, given reliability, availability, maintainability (RAM), supportability, and ownership costs, among other factors. These requirements are tested and quantified prior to fielding by the initial operational test and evaluation (IOT&E) process and assessed against defined criteria. As illustrated in Fig. 3.1, total ownership costs (TOC) incurred during the operations and support (O&S) phase may constitute 60–80% of total life cycle cost (LCC) [2]. Throughout this book on MDRE, we will utilize a radar transmit/receive system called DREX+ (Digital Receiver and EXciter) as an example of the requirements management required for any engineering project. We will utilize the unique characteristics of radar systems, so vital to both military and civil operations, to drive home the overall need for efficient and effective requirements allocation, derivation, and management. This drives home the need for MDRE within the overall development of both military and commercial systems [3]. Here we use the Navy to illustrate the need for supportability considerations within the overall MDRE process. One of the major sensors utilized by Naval assets is the radio detect and ranging (RADAR), capable of detecting Objects of Interest (OoI) and providing extremely accurate information on distance, direction, height, and speed of the objects. To keep defense and commercial (e.g., airports) operations safe, these radar systems must be maintained and sustained to provide necessary capabilities toward mission success. Any breakdown in the availability of parts, sub-assemblies, and other necessary equipment can put normal operations at risk. The manufacturability of components, circuit board assemblies, subsystems up to and including complete systems must be maintained and ready at any time to supply needed components to keep the systems that use them operational and safe [4]. This drives the need for requirements allocation and derivation that accommodate reliability, sustainability, supportability, and manufacturability aspects of systems, especially those that may in service across many decades (e.g., commercial, and military radar systems) [5].

Question to ponder

Why do you think supportability is so important to define requirements for up-front? What is the result of designing a system without supportability requirements?

The pandemic caused by the severe acute respiratory syndrome coronavirus 2 (SARS-CoV-2), known as the COVID-19 pandemic, has brought additional, significant,

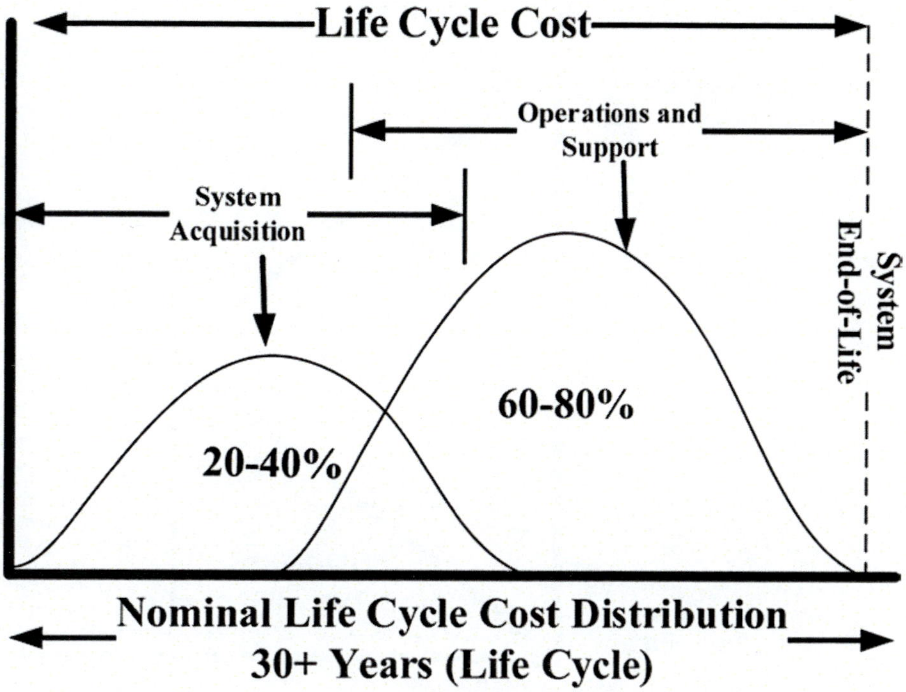

Fig. 3.1 Life cycle cost distribution across program lifetime

and long-lasting effects on the supply chain ecosystems on many industries, including the military and civil applications, and therefore affect requirements management across multiple systems. However, this does not negate the need to pursue aggressive modernization strategies across systems but does require caution in moving ahead to fast. For the foreseeable future requirements management will be crucial for deployable systems that extend the reach of current system capabilities.

How, then, do engineering systems like radar digital receiver and exciter (DREX+) systems address the problem of high TOC while still meeting the requirements for overall system supportability [6]. Here we focus on the causes of high TOC in both system design and system logistics footprint (quality and quantity). This will include the application of skills and processes in reliability, availability, and maintainability (RAM) to drive supportability [7]. This supportability analysis is part of an overall multidisciplinary systems engineering process that meets the requirements of the DoD 2009 Weapon Systems Acquisition Reform Act (WSARA) as well as civil requirements from civil groups like the Federal Aviation Agency (FAA). Supportability analyses measure the degree to which a system, in this case the DREX+, can be supported in terms of:

1. The inherent design characteristics pertaining to reliability and maintainability, such as mean time between maintenance actions (MTBMA), and mean time to repair (MTTR).
2. The efficacy of the various elements of the system product support, including spare parts, tools, and the training required to operate, maintain, and sustain the system.

Here, the DREX+ supportability will be analyzed as an example of how the MDRE process ensures the system is designed for supportability and the product support elements will be identified and made available to the user (e.g., Naval radar and potential border watch radar systems). We will look at the Affordability System Operational Effectiveness (ASOE) model, as it addresses both system design quality and the logistics footprint for manufacturability to assess the Total Cost of Ownership (TCO) for the DREX+ system. Supportability requirements analyses focus on two main components:

1. The system design for operational effectiveness, which assesses the impact of reliability and maintainability as design parameters within the overall system architecture to achieve operational effectiveness and mission suitability [8].
2. The Supply Chain Model (SCM) to assess the logistics activities that will allow the system to achieve effective sustainment across the systems life cycle [9].

Together, these analyses define the RAM/LCC trade space, which is illustrated in Fig. 3.2. Fig. 3.2 bounds the values of reliability and sustainment cycle times to minimize LCC for the system [10]. It requires continual balanc-

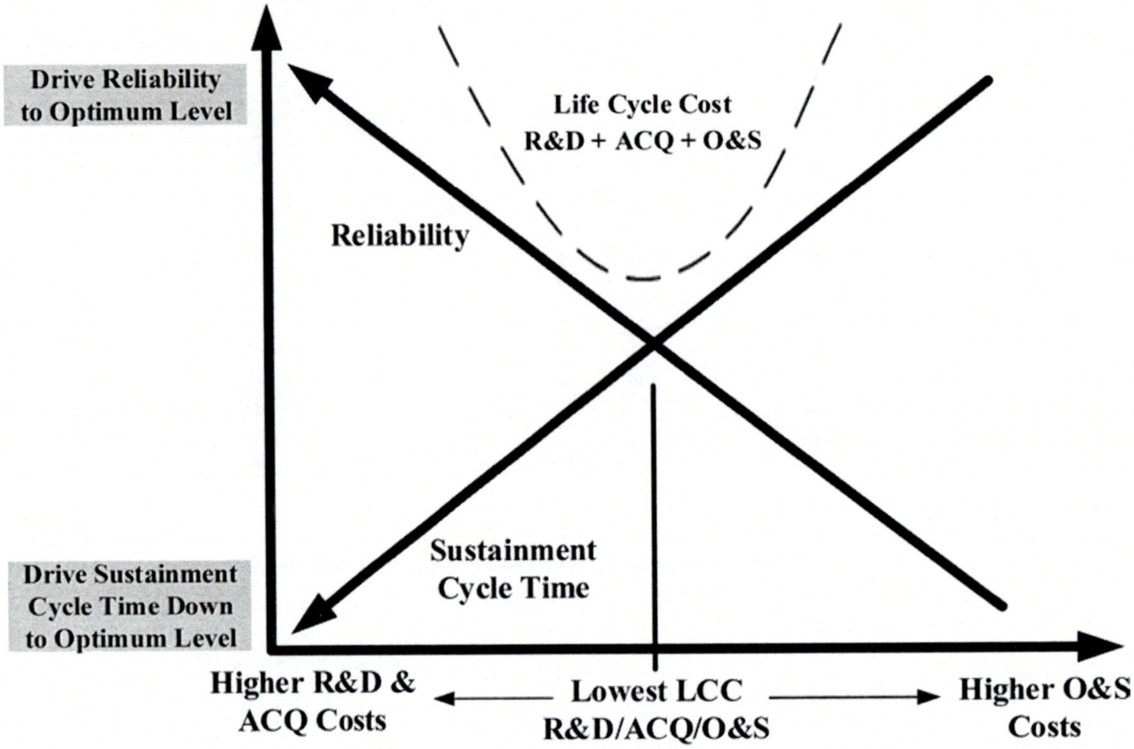

Fig. 3.2 RAM life cycle cost-trade space

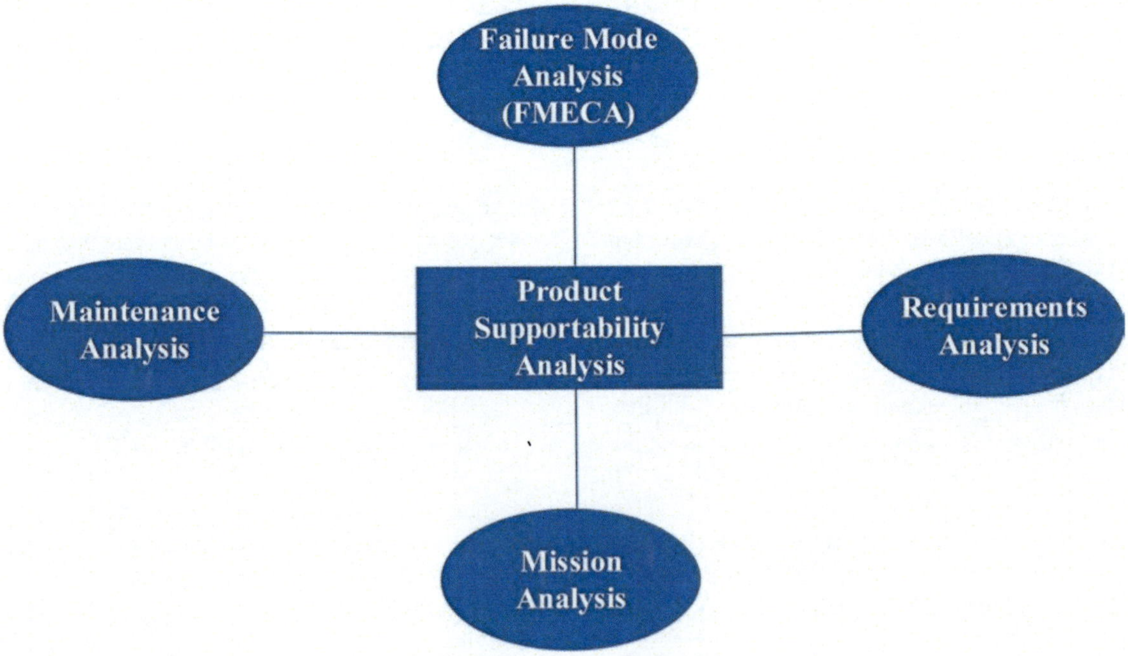

Fig. 3.3 Product supportability analysis components

ing throughout the life cycle of the system to ensure continued optimization of supportability. Supportability analyses comprise four main components, as illustrated in Fig. 3.3. These are:

1. **Mission/Operational Analysis**: Determining the applicability of the system across Naval/commercial operations and assets. Includes determining assumptions required for the new system compatibility, understanding

the limitation of the system and its applicability across platforms, identifies operational-level tasks and functions required of the system, provides a risk assessment of the use of the new system within current operational use [11].

2. **Requirements Analysis**: Understanding the requirements across assets and missions to assess what requirements are common and can be met with the new system. Includes data models showing the final new/enhanced system requirements and mission needs.
3. **Maintainability Analysis**: Analyzes the systems approach to influencing the maintenance activities through the design of the system, both hardware and software. These include determining MTBMA and MTTR for the system. The output is a set of guidelines required to apply Design for Maintenance for the new system, including the number of spares required, and what level of sparing is required—component, subsystem, or system levels. These guidelines will be in accordance with the definitions of what parts of the system are field repairable, depot repairable, or factory repairable.
4. **Failure Modes Analysis (FMECA)**: Understanding of how the system will fail and the severity of failure modes across the system is crucial to supportability analyses. Failure Mode, Effects, and Critical Analysis (FMECA) is an analytical process used to chart the probability of failure against the severity of the consequences [12].

Question to ponder

Think about the difference between supportability and maintainability. Why do you think both are important?

Each of these components of supportability will be discussed below in the context of the DREX+ system, using the two case studies, a Navy radar system and a potential border patrol radar system, both of which can utilize the DREX+ system. We will utilize the DREX+ in context with these two radar systems as case studies to provide concrete examples of requirements engineering within multidisciplinary requirements engineering (MDRE).

3.2 MDRE: DREX+ Mission Analysis

One of the first analyses that must be done within the MDRE process is analyzing the mission of the system to be developed to better understand its environment, components, and overall needs in context of possible solutions. Here we utilize the DREX+ to illustrate the concepts of mission analysis.

3.2.1 Use of Radar Systems for Military and Civil Surveillance Operations

Radar systems are used for early detection of surface or air Objects of Interest (OoIs) and to gather data, such as range, bearing, altitude, and speed of targets in land, sea, and air environments. They are also used for general purpose surveillance, navigation, and for controlling the ships own aircraft and small boats. Engineering systems designs like DREX+ are required to provide military and civilian operations the ability to push the technological envelope with its solutions and manufacturing partners [13].

As part of the overall discussion of multidisciplinary systems engineering requirements management, we provide the reader with a better understand how to manufacture systems like the DREX+ to meet the overall engineering requirements for rapid and reliable manufacturing.

3.2.2 Case Study #1: Naval Long-Range Radar Systems

The Navy is currently in the final development stage of a second-generation, universal DREX+ with RF and radar processing—scalable, software-defined radar (SDR) for HF-Ku bands with a first application to a technical refresh and upgrade for the US Navy's long-range, air-search radar. Components of the DREX+ were funded by Army, Navy, Air Force, and the Missile Defense Agency (MDA). This system is standard with eight transmit and receive channels but is scalable to any number of channels with the capability to implement direct RF sampling and super-heterodyne up/down-conversions. The DREX+ is broken into an RF and DSP conductive cooled enclosures. It leverages a high-performance computing (HPC) system called the Three-Dimensional Reconfigurable Advanced Rapid-Prototype Environment (3D RARE, or 3DR) technology. The combination has produced a software-defined radio/radar (SDR) that merges CEI's front-end radar developments with its back-end cognitive algorithm and data processing capabilities. The resulting SDR is an open-architecture system capable of being tailored to deliver radar protection across the RF spectrum. The DREX+ adds tremendous current and future performance capabilities while adding long-term sustainability that extends service life. This allows for unprecedented adaptability across various platforms such as aircraft, drones, and ships.

Studies were conducted on the DREX+ to understand how it could be adapted to the various Navy platforms targeted for this opportunity. This included a supportability analysis as well as an in-depth analysis on interfaces, signals, and environments to improve the DREX+ sustainability and adaptability. In conjunction with the supportability analysis, we studied the manufacturability and sustainability of the

RF Transmit/Receive Signal Conditioning Module

- **Input Frequencies DC to 14GHz**
- **Direct RF, Nyquist Zone 1, from DC to 6GHz**
- **Super-heterodyne Conversion from DC to 14 GHz**

Analog to Digital Converter (ADC) Module

- **#DR, Modular, High-Performance Lightning ADC Technology**
- **8-Channel Input**
- **Intel Arria 10 FPGA with OpenCL BSP**
- **Multiple ADC options available, including: 250MSPS, 3GSPS and 10GSPS**

Processor Model

- **3DR, Modular, High Performance Crestone Processor Module**
- **NXP (Freescale) T4240 Power PC (PPC) with 24 Virtual High-Performance Cores, Scaling to 1.8 GHz and AltiVec Floating Point Vector Math Accelerators**
- **Intel Stratix V FPGA, Optimized for Variable-Precision Digital Signal Processing (DSP) and Dynamically Reconfigurable Transceivers**
- **PCIe Gen 3, LVDS, SerDes and 4X 10GbE Interfaces**
- **NXP K61 Microcontroller Provides Total System Health Monitoring and Protection**

Enclosure

- **3-6U 19" Rack Mountable Server or Similar**
- **MIL-STD Ruggedization**
- **Approximately 30 lbs.**
- **Front Panel LED T.V. Screen for Status Control**

DREX+ Sample Enclosure

Fig. 3.4 The DREX+ high-level spec sheet

DREX+ system focused on increased quality, reliability, and reduced ownership cost [14]. Chapter 4 provides details of the manufacturability study. The overall specifications for DREX+ are shown in Fig. 3.4. The DREX+ provides eight receive channels with direct digital RF conversion/sampling capability up to 5 GHz in the first Nyquist zone and low 9 GHz at higher Nyquist zones. A selectable super-heterodyne circuit (not currently available for the Navy's use since it uses direct RF and not super-heterodyne) on the RF front-end module allows applications through 14 GHz band in the first Nyquist Zone; this also allows for lower sampling analog-to-digital converters (ADCs) and higher dynamic range. The eight-channel exciter can generate direct RF through 6 GHz and with the selectable super-heterodyne upconversion can generate waveforms up to 14 GHz band. This leverages a 12.6 GHz digital-to-analog converter (DAC) module. The RF DAC channels can feed the ADC inputs for testing, test targets, calibration, dithering, etc. The "+" indicates that the DREX has capabilities beyond the receiver and exciter: three bleeding edge Field Programmable Gate Arrays (FPGAs) and an advanced 24 core General Purpose Processor (GPP) with math acceleration provide unprecedented compute capability, including digital signal processing (DSP) and display system capability.

3.2.3 Case Study #2: Border Patrol Surveillance

Border patrol surveillance requires a medium range radar system. Here, DREX+ is coupled with a medium-range Unmanned Search/Track Air Radar (USTAR). A primary task is taking the C-band, planar, antenna and scaling it to X-band with increased power and larger aperture size for the tower-based border patrol version. While this DREX+ is designed to cover from HF to Ku band (14 GHz), limited only by the RF front ends, the first targeted system for the second-generation DREX+ is the radar for the Navy. The basic use is for a below deck design including a solid-state transmitter (SSTX). The USTAR DREX+ extended Kalman filter (EKF) based tracker is capable of handling roll, pitch and yaw (RPY) which will translate to handling beam pointing on targets when the surveillance radar is moving on a tower, ship, ground vehicle, or aircraft.

Since unmanned border patrol surveillance involves significant technology transfer from the USTAR and DREX+ systems, both solutions will be described in detail in the supportability and manufacturability reports [15], along with how the technology will be applied to the border patrol surveillance technical approach. Most of the Navy's technology transfer is classified and obviously cannot be discussed in this format. USTAR is an existing, proven airborne, search and track radar developed for the United States Air Force Research Laboratory (AFRL) (Fig. 3.5). During airborne testing, USTAR tracked 1 m^2 (0 dBm) RCS targets at 12 nmi and airliners at 30 nmi with only a 70 W peak transmit power and 20% duty cycle. USTAR is a lightweight, high-performance, and low-cost advanced radar system (Fig. 3.5). USTAR detects, tracks, and prioritizes multiple targets in a wide FOV within a strict Cost, Size, Weight, and Power (CSWaP) budget using the first-generation DREX+ technology. It is a mono-static, mono-pulse, PD, C-band phased array design. USTAR utilizes the first-generation DREX+, comprising backplane-less Radar Advance Receiver Exciter (RARE) technology: commercialized as "3DR." These modules are 6.25″ square and cover the back of the USTAR antenna array as shown in the L-shape in Fig. 3.5. Both generations of RARE/3DR were funded by the Army, MDA,

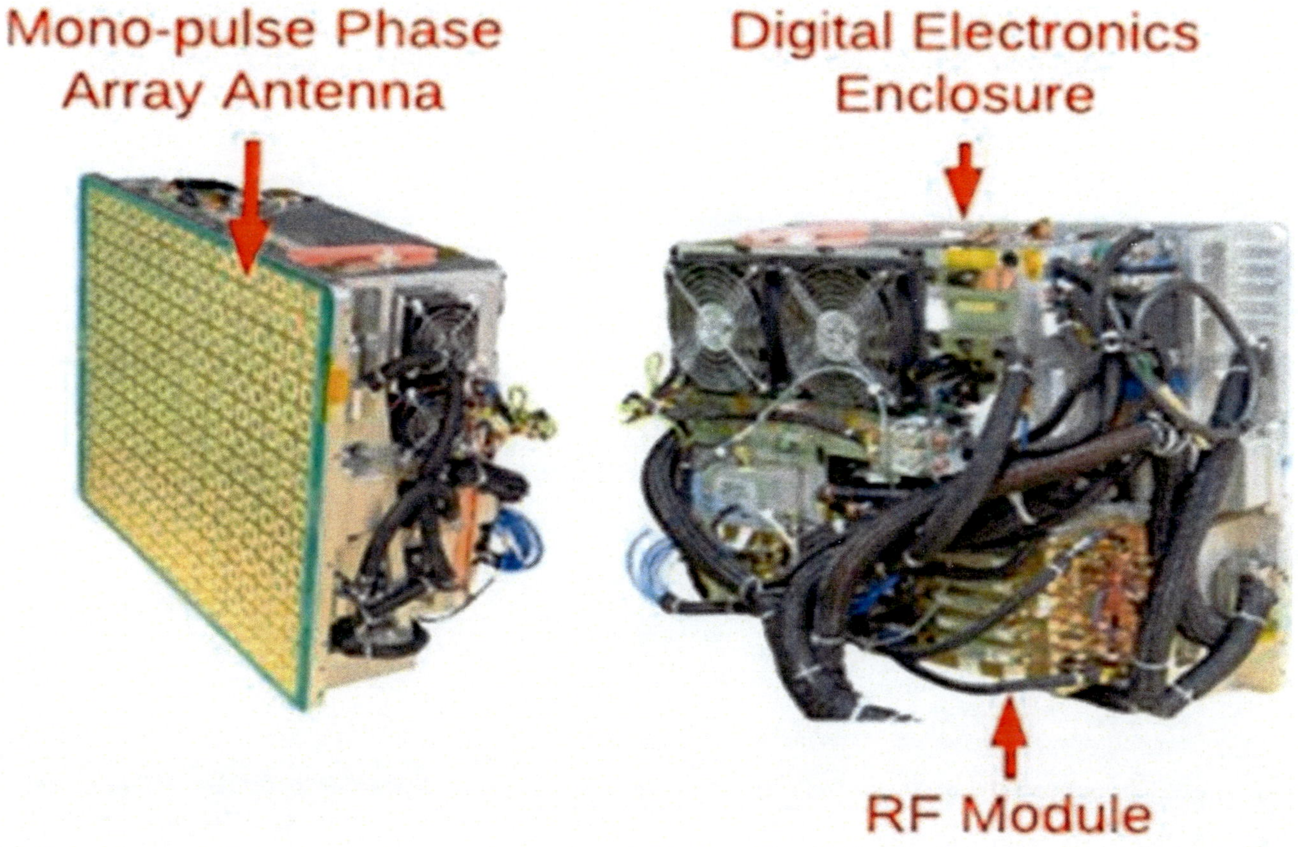

Fig. 3.5 The USTAR/DREX+ Radar for border patrol surveillance

Navy, and Air Force. Figure 3.5 provides a snapshot view of the border patrol surveillance version of the USTAR/DREX+ radar system.

3.2.4 MDRE: Data Mining and Discovery

Supportability, maintainability, and manufacturability cannot be successful unless data are captured, cataloged, metadata created for retrieval, and overall data management of operational assets (e.g., Air Traffic Control Radars) captured. Fig. 3.6 illustrates the proposed overall data mining/data discovery process. The description of each level within Fig. 3.6 is [16]:

- **Data Warehouse**: Data must be brought in, cataloged, and stored within the data warehouse.
- **Data Cleaning**: The data must be cleansed, i.e., headers stripped off (if applicable), ambiguous tags resolved, zeroed and missing data tagged and handled. The data are then transformed according to the data ontology model [17].
- **Data Selection and Transformation**: The data to be mined are selected and transformed via the ontology of the system performing the data mining.
- **Transformed Database**: Once the data has been selected and transformed, it comprises a "data mining database," based on the overall mining category, and transformed to the data model ontology for the system requesting the mining operation [18].
- **Data mining, Discovery Mining, and Visualization**: Data mining operations are performed, and any data relationships discovered during the mining operations are modeled, and this information is visualized for the operator requesting the mining operation.

Capturing maintenance, supportability, and manufacturability data is crucial to the overall success of Naval missions. Once data have entered the system and been processing through the data mining/discovery process, data are managed throughout the processing infrastructure. This process is illustrated in Fig. 3.7. The description of each level is [19]:

Level 0 (Data Refinement): This entails creating a belief system where the different and varying accuracy/belief semantics can be normalized into a semantic model that can then be used for data association, tracking, classification, etc.

Level 1 (Object Refinement): Consists of data association, target tracking, classification. Example: combat ID (i.e., declaration of friendly, hostile, or neutral) [20].

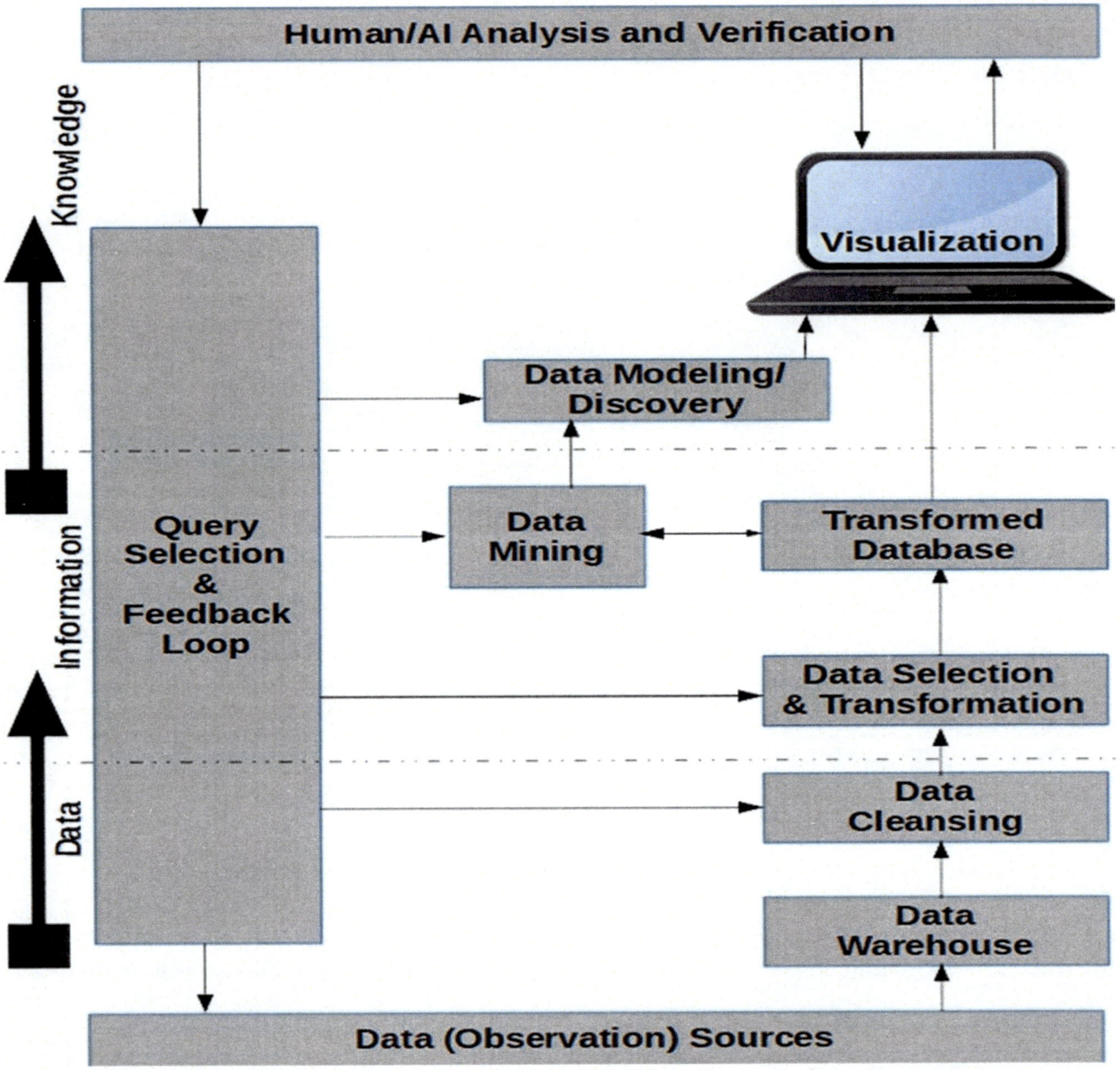

Fig. 3.6 The MDRE data mining/data discovery process

Level 2 (Situational Refinement): Includes track-to-track correlation and the inclusion of all relevant information into a common situational display.

Level 3 (Thread Assessment/Refinement): Consists of the assessment of activity of intent of entities. The assessment of intent is dependent on situational context [21].

Level 4 (Process Refinement/Resource Management): Enables process control and process enhancement. Resource allocation is required for effective movement of data through the battlespace and involves making performance trade-offs [22].

Level 5 (Knowledge, Decisions, and Actions): Allows threat information to be incorporated into the situational assessment/visualization through knowledge to create a choice of actions (Course of Action Support).

Question to ponder

Why is proper data management important in the overall systems design process? What can happen if data management requirements are missing during system design and implementation?

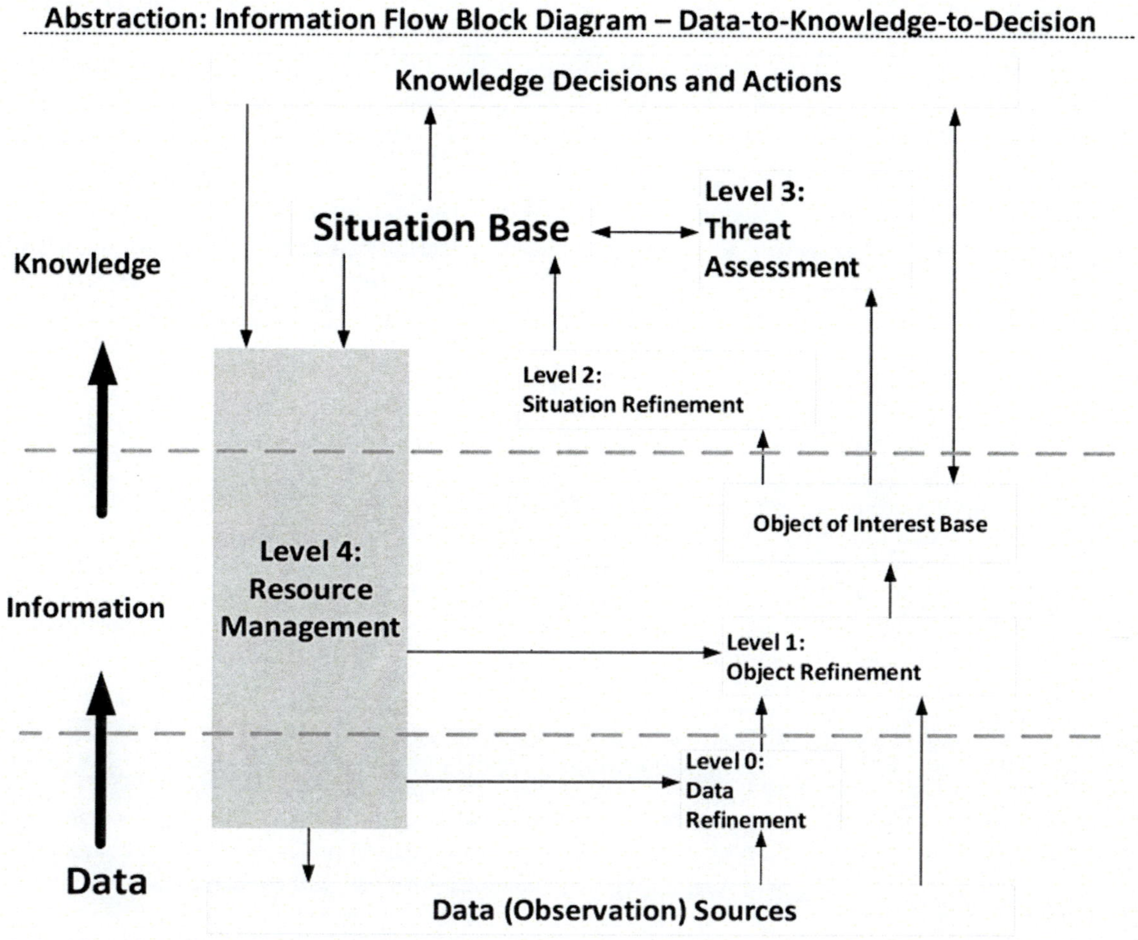

Fig. 3.7 The MDRE knowledge management process

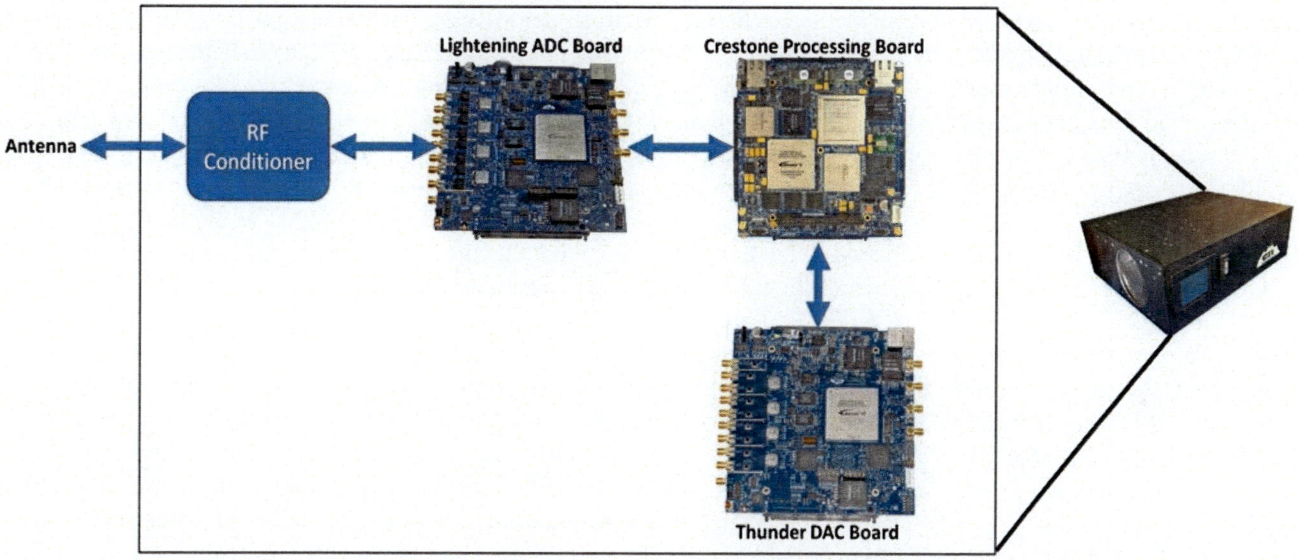

Fig. 3.8 The DREX+ components

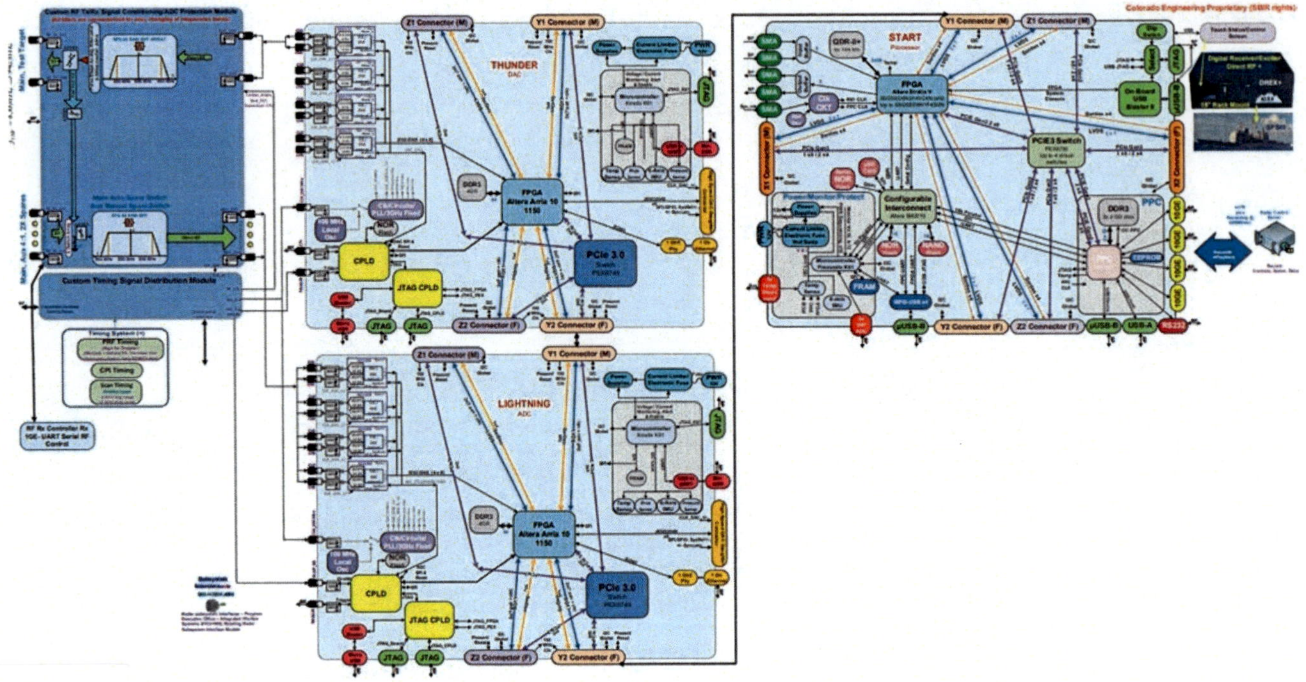

Fig. 3.9 The DREX+ detailed layout

3.2.5 MDRE Supportability, Maintainability, and Manufacturability Considerations

Figure 3.8 illustrates the components (boards) that are required for the DREX+ radar processing system. Three boards make up the heart of DREX+, the Lightening II Analog to Digital Converter (ADC) board, the Crestone processing board, and the Thunder Digital to Analog Convertor (DAC) board. Each of these components that make up the DREX+ must be analyzed separately as to their supportability, maintainability, and manufacturability to completely understand the overall supportability of the DREX+. Fig. 3.9 provides the schematic of the DREX+ using these three boards. Fig. 3.9 illustrates the detailed layout of the DREX+ radar processing system.

One of the critical issues to be understood to lay out the supportability Supportability, Maintainability, and Manufacturability and maintainability of the DREX+ are the failure modes expected. This will drive the analysis of parts/components that must be carried by Naval assets to properly maintain the DREX+ system. Table 3.1 illustrates examples of the failure modes expected for the DREX+ system.

3.2.6 DREX+ Mission/Operational Applicability

At present, the radar used for scanning large areas and for focused targeting are separate pieces of equipment, that are only integrated by the ship's combat system. Dual Band Radars will break from that approach by combining two different radar antennas, with the same back-end radar electronics and software driving both. The system will receive a single stream of data, and the radar itself will be able to mix and match its antennas as the situation requires. At the design tier, this approach allows fewer radar antennas, all flush-mounted with the superstructure for maximum stealth. At the tactical tier, integration at the radar level offers faster response time, faster adaptation to new situations, and better utilization of the ship's power, electronics, and bandwidth. At the life cycle maintenance tier, it allows one-step upgrades to the radar suite. The DREX+ radar processing system is capable of dual-band use, although for the long-range Navy application it is not required. What would be required is to move from the direct RF use to a super-heterodyne configuration to accommodate dual bands. For the Navy long-range radar system, the application offers 2D (range and heading only) very long-range scans in the L-band. It serves as the primary air search radar aboard a wide array of application. All these Naval assets could be configured to use DREX+ for their long-range radar systems, saving space, power, cooling, all with a massive increase in performance. We believe the DREX+ radar processing system has the potential to be the future of Naval radar processing systems.

Due to significantly leveraging/iterating existing radars systems/algorithm, hardware and software risks are low, allowing DREX+ to be targeting for a host of Naval applications. The border patrol surveillance thermal solution will

Table 3.1 Examples of the DREX+ failure modes

Function	Functional failure letter	Functional failure description	Failure rate (failures/million Hrs)	Local effects	End effects	Severity	Failure effect probability
EXC-236: The exciter/synchronizer shall source a clock signal to the receiver with a power NLT 2 dBm	F003	System clock <2 dBm	0.85	No system clock output to Signal Processor/Receiver	Loss of Radar Coverage	II	0.01 level C
EXC-237: The exciter/synchronizer shall source a clock signal to the receiver with an integrated phase noise NGT −75 dBc	F004	System clock integrated phase noise > −75 dBc	0.75	Noisy system clock output to Signal Processor/Receiver	Increased clutter returns	III	0.01 level C
EXC-230/231: Internal power generation	F002	Unable to generate internal power	2.19	No output to all subsystems. No Synchronizer SNMP Response to Radar Control	Loss of Radar coverage	II	0.02 level C
Exc-11: Excite transmitter 14 dBm ± 0.25 dB	F001	Low output	4.21	Exciter waveform output from Exciter/Synchronizer to Transmitter <13.75 dBm	Loss of Radar coverage	II	0.03 level C

need to support extended temperature environments as well as sealing requirements due to damp weather exposure that will be required. This is part of the supportability, maintainability analysis that will drive sustainability requirements into the system [23].

Question to Ponder

What are some of the contractual documents that are used to control and manage suppliers?

3.3 DREX+ Requirements Analysis

3.3.1 Requirements Decomposition

Decomposition of the Navy's requirements and applying them to the DREX+ was the first step toward defining the architecture of the overall system in which the DREX+ will be utilized [24]. The requirements decomposition process is not set-in stone and is influenced by the type of system being designed, the domain in which the system must operate, the system constraints, and other factors that may be specific to any individual system [25]. For the two case studies, the Navy long-range and border patrol surveillance radar systems, one is a legacy system whose overall architecture has been determined while for the border patrol surveillance program, the use of DREX+ is for a new system. The requirements decomposition process radically affects the optimization of the architecture, system design, and system implementation, since different decompositions drive different internal interfaces, different information and data flows, and a different set of Enterprise Services to execute the functionality laid out in the systems analysis and subsequent decomposition. Each of the different decomposition possibilities drives a trio of different time, quality, and cost profile for the development of the system. These must be balanced against the customer technical, performance, and quality requirements, customer-driven schedule, and available cost profile. Some of the factors that affect the balance of these three factors is the desire for system/subsystem/component/service reuse and supportability approach, different organizational structures required to execute (develop) the system, software development style (e.g., spiral, waterfall, agile), and technology base (DREX+). These are not the only considerations, but they are the major ones that will drive the development optimization of the DREX+ system. Fig. 3.10 illustrates this idea [26].

Question to Ponder

How does system quality drive system life cycle cost? Do you think it is possible for increased quality without an increase in cost?

To balance decomposition of the DREX+ requirements into optimized system functions at each level of decomposition (e.g., system, subsystem, or service group) the decomposition criteria was broken up into four main attribute groups: customer directives, functional directives, quality directives, and technical directives, which are illustrated in Fig. 3.11. The overall goal was to create a flexible set of requirements that could be utilized across a wide variety of Naval application without redesign. Understanding that

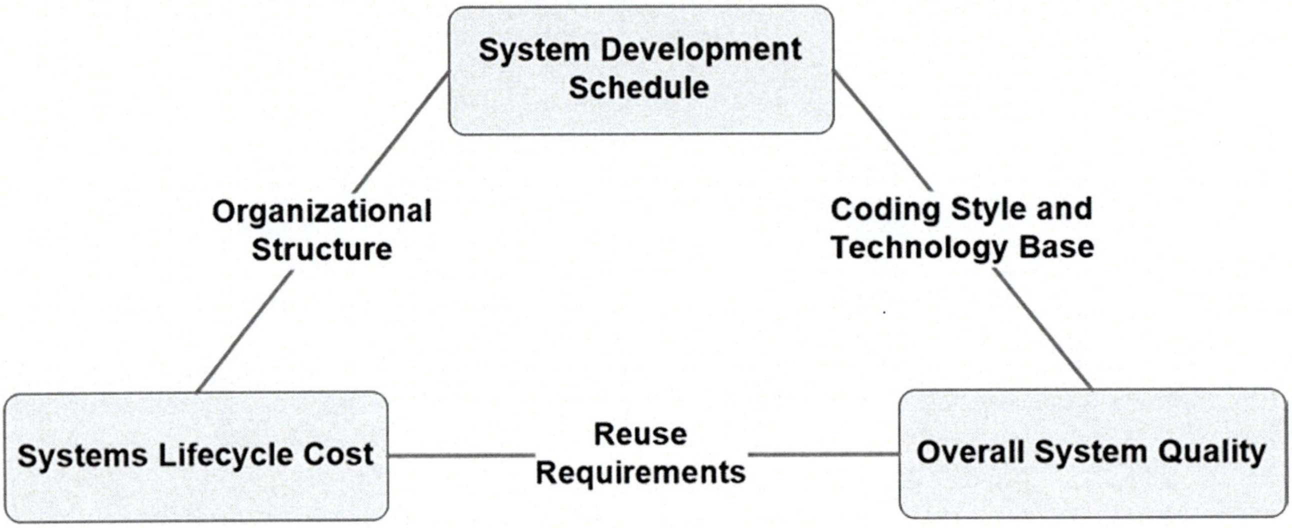

Fig. 3.10 System optimization trade-off analysis parameters

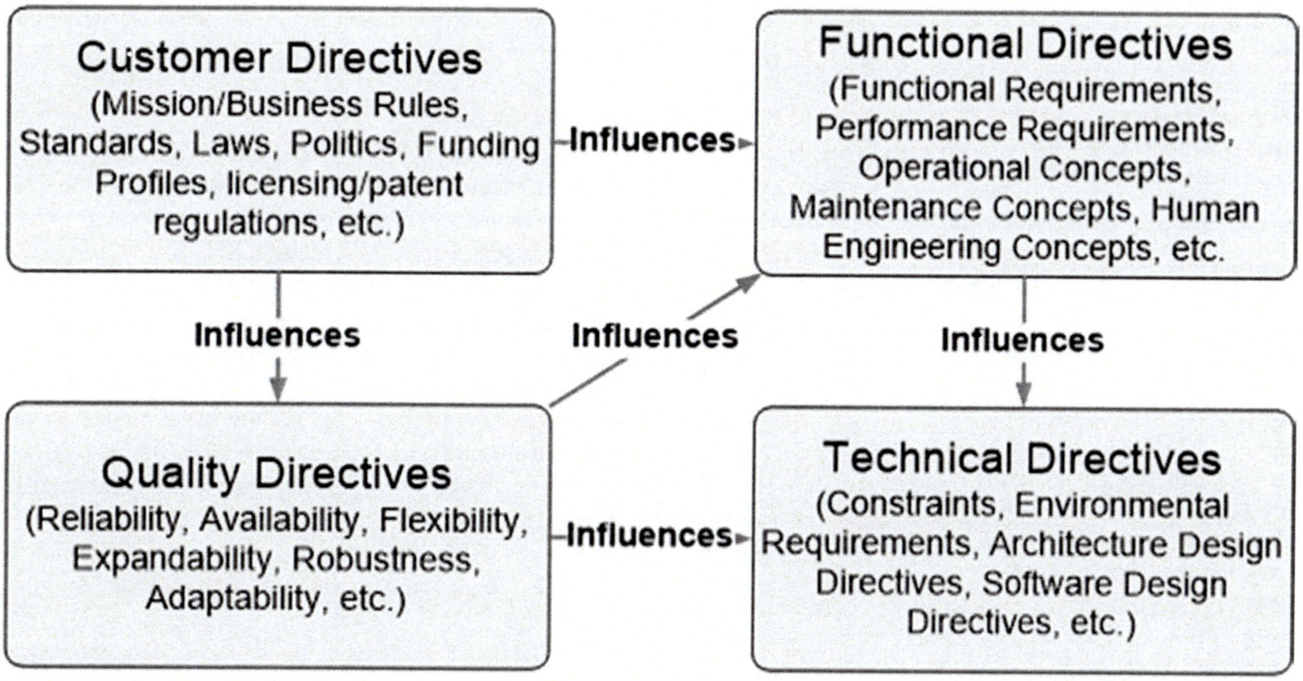

Fig. 3.11 MDRE High-level requirements decomposition criteria

changes in components and configurations will be required, depending on the Navy application/asset, which will affect the overall performance characteristics of the DREX+. Example: using a super-heterodyne receiver vs. direct RF capability. The DREX+ accommodates both configurations, but each come with performance and supportability differences [27].

If you examine the directives shown in Fig. 3.11, you will see directives that come from many engineering disciplines, including systems engineering, software engineering, safety engineering, sustainment engineering, and many others. Each of these must be considered in the initial requirements decomposition to ensure that all of these disciplines are embedded into the initial functional block diagram and requirements derivation/allocation. As the requirements are decomposed utilizing the directives (Fig. 3.11) each decomposition level must be assessed for its viability, given the current and potential functionality of the DREX+ system. Given the overall requirements, environment, constraints, laws, etc. that must be considered, there are multiple, possibly dozens, of ways to adequately decompose a system [28]. To create an objective methodology for the requirements/system decomposition, a set of weightings are created, within each directive category, agreed to by all disciplines, before the decomposition is begun. Table 3.2 below is an example of such a directive-weighting table [29].

As requirements are decomposed and derived, the following set of guidelines should be utilized to assess the adequacy of the derived requirements [27]:

- Correctness.
- Consistency.
- Traceability.
- Disambiguation.
- Testability.
- Atomicity.

3.3.1.1 Completeness

For the DREX+ requirements to be considered complete, we define three characteristics. The first two properties, shown below, can be thought of as "*internal completeness*" and imply a closure, or completeness of available information. The last characteristic involves "*external completeness*" and attempts to ensure that all the problem definition information is contained in the requirements, depending on the context of the application:

- There is no information that is unstated, or "*To-be-Determine*" (TBD).
- The requirement text does not contain undefined entities or objects.
- No information is missing from the requirement.

3.3.1.2 Correctness

Correctness ties completeness, consistency, and traceability are required to understand if the derived DREX+ requirement, as part of the requirements hierarchy, not only reflects the parent requirement from which it is derived, but also traces correctly through the hierarchy, is consistent with

Table 3.2 MDRE requirements directive decomposition weighting example

Directive	Weighting
Functional directives	
Clustering of functions by usage	0.2
Requirements dependencies	0.3
Internal interfaces	0.1
Operational concepts	0.1
Human-machine interface requirements	0.3
Total	1.0
Technical directives	
Technical requirements/constraints	0.5
Design directives	0.3
Communication requirements	0.2
Total	1.0
Customer directives	
Laws and standards	0.1
Licenses, certificates, patents, etc.	0.1
Mission/business rules	0.1
Information requirements	0.1
External interfaces	0.2
Subcontractor relationships	0.1
Political considerations	0.1
Cost profiles	0.1
Schedule profiles	0.1
Total	1.0
Quality directives	
Performance requirements	0.2
Availability requirements	0.2
Reliability requirements	0.2
Maintainability requirements	0.1
Usability requirements	0.1
Security requirements	0.2
Total	1.0

other requirements derived from the same parent requirement, and is complete in its structure and language. There should be no missing components consistent with the functionality of the functional block each requirement is mapped into [30].

3.3.1.3 Consistency

Consistency measures whether any DREX+ functional or nonfunctional requirements conflict with each other. This applies to functional interfaces (both internal and external), as well as performance and quality attributes. If one derived requirement states that two inputs should be compared and a procedure initiated if $A > B$, and another requirement states the procedure should be initiated if $B > A$, these two requirements are not consistent [31]. Consistency of requirements is critical to ensuring a usable system is developed. For each potential application of the DREX+ system, the functional and performance requirements of the system must be compared to the original requirements used for the DREX+ design to ensure no inconsistencies exist within the two sets of requirements [32].

3.3.1.4 Traceability

Ensuring proper requirement traceability is essential to integrating and testing the DREX+ system, throughout the system development. Traceability is concerned with documenting the parent/child relationships between requirements at each level of the architecture. It also documents the relationships between the requirements and other architectural and design artifacts [33]. This ensures that all requirements are addressed at their proper respective architectural levels. Again, for DREX+, the requirements for the Naval application must be vetted against the original design requirements for the DREX+ to ensure complete traceability of requirements. The purposes of traceability are threefold:

- Manage engineering changes across the system development.
- Understand the decomposition of the system into each architecture tier.
- Manage the overall quality of the developed system.

3.3.1.5 Disambiguation

An unambiguous requirement is one that is stated in concise and not esoteric language, without technical jargon, without acronyms that have not been previously defined. An unambiguous requirement contains facts and is written without negative language or compound statements. The disambiguated requirement does not contain opinions and is not subject to interpretation [34]. This is important to ensure the DREX+ functionality is applicable for any Naval application.

3.3.1.6 Testability

For the DREX+ requirements to be testable, the correct implementation of the requirement must be able to be determined through standard methods [35]:

- Demonstration.
- Test—includes instrumentation to validate the tests.
- Inspection.
- Analysis—this may include valid modeling and simulation techniques.

The main point here is that requirements decomposition and derivation is the lifeblood of the DREX+ system development. If the requirement decomposition and derivation is not done correctly, there is little hope of the creating a radar system using the DREX+ that meets the needs of the Navy or other DoD entity's need [36].

Question to ponder

What would be the effect on the overall system sustainability if there were requirements that were not testable?

3.3.1.7 Atomicity

Writing atomic requirements means each requirement represents a single system need and your system either satisfies or fails to satisfy that specific need. It should be independently deliverable and represent in incremental increase in the system's capabilities. Atomic requirements also make supportability and maintainability easier in that a given requirement is independent of other requirements at the same level [37].

3.3.2 MDRE Example: Radar System Requirements

Radar systems transmit a radio signal from a directional antenna that is mounted—for the purposes of this discussion—on a vehicle. The radio signal is reflected off objects that are within the transmitted beam and a portion of the reflected energy returns to the receive antenna. Depending on desired system performance characteristics, transmit and receive antennas can be combined or separate. From this reflection, the distance from the radar unit to the object can be determined. There are two primary waveforms used by radar sensors—pulsing and continuous wave (CW). Pulsed or ultra-wideband (UWB) radar detects obstacles by measuring the time of flight (ToF) of a pulsed signal that is transmitted and then reflected from an object within the radar's beam with distance between transmitter and reflector being proportional to this time. The continuous-wave method works by transmitting a radar signal of a known, stable frequency, and then measuring the Doppler shift of the reflected signal. The Doppler shift indicates the change in frequency of the reflected signal, which is proportional to the speed of the target [38].

Table 3.3 Radar system frequency bands

Radar Band	Frequency (GHz)	Wavelength (cm)
Millimeter	40–100	0.75–0.30
Ka	26.5–40	1.1–0.75
K	18–26.5	1.7–1.1
Ku	12.5–18	2.4–1.7
X	8–12.5	3.75–2.4
C	4–8	7.5–3.75
S	2–4	15–7.5
L	1–2	30–15
UHF	0.3–1	100–30

Typically, these systems operate in the microwave (300 MHz–40 GHz) portion of the radio spectrum. Doppler radar detects the relative motion of an obstacle/target, i.e., detection requires relative movement of the obstacle/target or the vehicle/radar set. Both types of radar (pulsing and continuous wave) are effective for detecting people, other vehicles, large rocks, and buildings. Some obstacles are not good radar reflectors, such as plastics, dry wood, or objects with large flat surfaces that can reflect signals away from the radar antenna. Possible obstacle detection ranges for a radar-based proximity warning system vary from less than a meter to 30 m or more. Radar works well in dirty and dusty environments and does not require any miner wearable component (MWC); however, it is only useful in line-of-sight applications and cannot differentiate between objects and humans.

3.3.2.1 Radar Frequency Bands

Radar systems transmit electromagnetic or radio waves. Most objects reflect radio waves, which can be detected by the radar system. The amount of energy reflected is known as the radar cross section (RCS) and is measured in square meters. The larger the RCS, the more energy is reflected. The frequency of the radio waves used depends on the radar application. Radar systems are often designated by the wavelength or frequency band in which they operate, using the band designations shown in Table 3.3.

The choice of frequency depends on the application requirements. The minimum antenna size is proportional to wavelength and inversely proportional to frequency. Applications often are limited in the size of antenna that can be used. A smaller antenna dictates a higher frequency and lower wavelength choice.

3.3.2.2 Radar Beam Width

Beam width, or the ability of the radar to focus the radiated and received energy in a narrow region, is also dependent on both antenna size and frequency choice. Larger antennas allow the beam to be more tightly focused. Therefore, a higher frequency also allows the beam to be more tightly focused for a given antenna size. The "focusing" ability of the antenna is often described using an antenna lobe diagram, which plots the directional gain of an antenna over the azimuth (side to side) and elevation (up and down).

3.3.2.3 Radar Range

The range of the radar system is also influenced by the choice of frequency. Higher-frequency systems usually are lower power due to electronic circuit limitations, and they experience greater atmospheric attenuation. The ambient electrical noise that can impair operation of analog circuitry also becomes more pronounced at higher frequencies. Most of the radar signal absorption and scattering is due to oxygen and water vapor. Water vapor has high absorption in the "K" band. When this was discovered, the band was divided into Ka for "above 18 GHz" and Ku for "under 18 GHz," the frequencies where radar operation is limited due to water vapor absorption. At higher frequencies in portions of the millimeter band, oxygen causes similar attenuation through absorption and scattering. Another consideration, discussed more fully in the next chapter, is the effect of the radar operating frequency on Doppler frequency measurements. Doppler frequency shifts are proportional to both the relative velocity and the radar frequency. Doppler frequency shifts can provide important information to the radar system. Most airborne radars operate between the L and Ka bands, also known as the microwave region. Many short-range targeting radars, such as on a tank or helicopter, operate in the millimeter band. Many long-range ground-based radars operate at UHF or lower frequencies, due to the ability to use large antennas and minimal atmospheric attenuation and ambient noise. At even lower frequencies, the ionosphere can become reflective, allowing very-long-range over-the-horizon operation.

3.3.2.4 Case Study 1: Navy Long-Range Radar Specification

An example of possible long-range radar specifications is:

- Frequency Band: 850 to 942 MHz.
  - Three selectable 30 MHz bands.
  - 48 discrete frequencies
- Transmitting Power: 360 kW peak 280 kW specified peak power .
  - 12–13 kW average power
- Antenna Parameters: Parabolic Reflector stabilized for roll and pitch.
  - 7.3 m/24 ft. wide, 4.3 m/14.2 ft. high.
- Range 250 nm Minimum Range:
  - 0.5 nmi Frequency Selection: Fixed or frequency agile.
- Range Accuracy: 0.03 nmi.
- Azimuth Accuracy: 0.5 deg.
- PRF 280, 800, 1000 pps.
- Pulse width 125 microsecond.

3.3.2.5 Case Study 2: Border Patrol Surveillance Specification

A sampling of the border patrol surveillance radar specifications is:

- Frequency (GHz) ~9.5 GHz.
- The production system would employ three panels to achieve 360° Az. Beam width at (0°, 0°) Scan –.
- Angle Resolution 1.3° Az; 1.3° El.
- Detection Range > 60 nmi (−20 dBsm RCS, SW1 target; P(d) = 0.9; P(fa) = 10–6); 0 dBsm > 200 nmi;
- Minimum Range 500 ft.
- Range Accuracy Range resolution = 1.5 m (accuracy is a function of target SNR). Range Rate Accuracy <1.6 ft./s.
- Provide unambiguous Doppler measurements at velocities up to 950 knots. Angular Accuracy <0.13 degrees Multiple Target Resolution <1000 ft. spacing w/in 6 nmi Volume.

3.3.2.6 DREX+ Requirements Analysis Discussion

A comparison of the Navy long-range and border patrol surveillance radar specifications shows a wide variance in frequency, accuracy, range, and other specifications, yet all the specs for both radar systems are met with the DREX+ system. The table below illustrates with range of specifications that are accommodated with the DREX+ radar processing environment. The analysis indicates that DREX+ requirements set used to develop the system provides a wide range of applicability across Naval assets and needs [39].

- Frequency: DC to 14 GHz.
- Direct RF ode:
  - Nyquist Zone 1, from DC – 6 GHz.
  - Single Band Use.
- Super-heterodyne down-conversion.
  - From DC-14GHz.
  - Dual Band capable.

Example 1: One long-range radar system is required to run in direct RF mode and has a frequency range from 850–942 MHz DREX+, in direct RF mode operates in a range from DC-6GHz. Some border patrol surveillance type radars utilize a super-heterodyne down-conversion rather than direct RF and operates at 9.5 GHz. DREX+ in a configuration with super-heterodyne conversion can operate from DC-14 GHz. The flexibility of the use of the DREX+ makes it useful across a variety of radar systems, both land, air, and sea. What follows throughout the book are discussions about allocation and derivation of requirements across engineering disciplines within the MDRE process [40].

3.4 DREX+ Maintainability Analysis

3.4.1 The MDRE Maintainability Engineering Process

The Maintenance Engineering Process is an approach to influence the maintenance activities through the design of the system, both hardware and software. The output of this process provides a set of guidelines which serve as a tool to apply Design for Maintenance (DFM) to the system design. The guidelines describe various ways how these future maintenance activities can be influenced. Applying the guidelines can help to reduce the number of maintenance activities, to make them easier to execute or to decrease the logistic support time that is required, as well as the logistic footprint required [41]. DFM is very often overlooked in the overall architecture/design/development activities and is often skipped because of high priorities like schedule-driven objectives. Regardless of the development style (e.g., agile, spiral, waterfall), regardless of the mission/business models used to drive the overall architecture and design, building in the notion of DFM from the beginning of the analysis and design process will influence the entire life cycle process and deliver a system that is maintainable, reducing the overall LCC for the system. When the system does not have maintainability requirements (e.g., Mean Time to Repair <8 h) or system engineering delays, the inclusion of them into the design, the maintainability concepts, requirements, and changes to system architecture to accommodate maintenance typically occurs late in the development life cycle, producing serious system risks, and affecting cost, schedule, and technical decisions to make up for the lack of DFM. DFM influences the architecture, design, and implementation of the system for both the hardware and software capabilities by accounting for the maintenance activities on the system. Taking DFM into account early and often in the design will build maintenance into the overall system implementation, installation, and operations; typically reducing the overall maintenance activities, providing maintenance activities which are easier to execute, and reducing the overall logistics (personnel) support time required for system maintenance. What follows is a set of overriding guidelines when building in DFM from the initial analysis and architectural designs. These overall guidelines were created to help the systems architect/analyst begin the process of Designing for Maintenance [41].

Question to ponder

Why should developers care about maintainability when they are not the ones who will be responsible to maintain the system?

3.4.1.1 Enhancing System Maintainability

Enhancing system maintainability provides for specifying in the designs and materials such that the system does not encounter unnecessarily and prolonged maintenance activities. This includes the selection of materials used in the design, whether it is considered hardware or software/firmware (whether utilizing commercial off the shelf (COTS) or custom). During the systems architecture/analysis, the overall design should consider the amount of time that would be required to service the system, if these hardware and software components were utilized [42].

If, for example, for your radar processing system you have selected a stand-alone computer system that requires manually connecting all the network and peripheral devices, rather than say, a modular blade server rack with slide in blade servers, you impact the availability of the system anytime maintenance is required (see Fig. 3.12 a, b). In the case of software, if you select a COTS software product with a licensing agreement that requires a vendor to be on-site to handle software issues and personnel that operate the system are not allowed to "fix" problems with the COTS software (this is a normal arrangement for COTS SW), either the system would be out of service for a considerable time waiting for a service representative from the COTS SW company, or you would have a considerable cost to have a service representative from the COTS SW company on-site. Such selections should be considered in the Operations and Maintenance, including the ease of performing corrective/preventative actions that can be performed by on-site personnel [43].

3.4.1.2 Enhancing Maintainability through Standardization of Components

When applicable, it is best to utilize standard components; with the advancement of hardware technologies, and standardization across the manufacturing industry, the use of standard components could simplify the design, but could also tie you into a particular technology, possibly violating certain quality attributes that must be met. If the decision is made to custom build either hardware or software, it increases the development cost due to [44]:

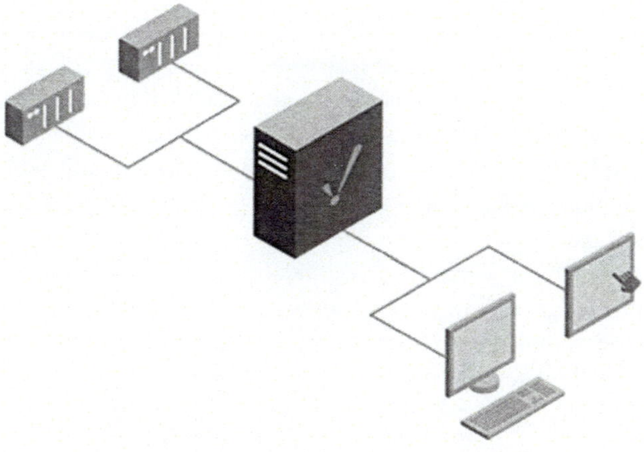

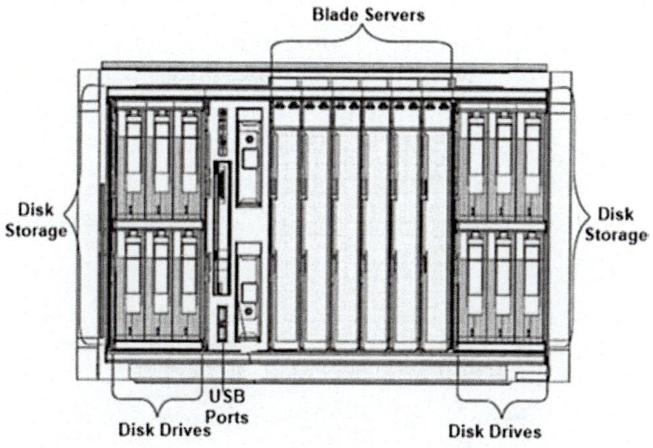

Fig. 3.12 (a) Stand Along Server, (b) Blade Server

- The design and implementation (which may involve customized parts) of the custom solution.
- Maintenance cycle costs since personnel must be trained for maintenance of each custom hardware and/or software component in the system.
- If there are custom parts, you must ensure you specify sufficient customized parts be available or in inventory for the life of the system.

In the last few decades, the use of COTS H/W and S/W has become more prevalent to shorten development cycles and reduce up-front maintenance costs. However, COTS also has certain issues [45]:

Requires lengthy maintenance agreements with the COTS vendors.

Obsolescence of the COTS products.

COTS companies may go out of business.

COTS companies can get acquired and their product lines can be discontinued.

COTS solutions can be less flexible as a system grows and evolves while the operating environment changes over time [46].

Question to ponder

Why is it important to consider handling fixtures and special tooling in the sustainability of a system?

3.4.1.3 Enhancing Maintainability through Standardization of Interfaces

The use of standard interfaces and interface configurations provides similar savings as using standardized components. The Universal Serial Bus (USB) is a standard serial communication interface, which is typically associated with a standard physical component hardware connector. The current version of USB, USB 3.1 specification has a maximum transfer rate of 10Gbps. USB has standardized serial interfaces. Another standard is Transmission Control Protocol/Internet Protocol (TCP/IP), a standard Internet protocol used to define how data is packaged for transfer across the Internet [47]. Again, like standard components, the use of standard interfaces can reduce development and maintenance costs in both labor and materials and can be utilized to enhance system interoperability across the architecture enterprise. However, it depends on the performance requirements, as there are instances where nonstandard interfaces may be required [48].

Question to ponder

Research the new USB 4.0 requirements? How might USB 4.0 capabilities change the maintainability of systems?

3.4.1.4 Enhancing Maintainability through Standardization of Maintenance Manuals/Procedures

Ensure that systems are maintainable through easy-to-understand maintenance manuals and system documentation. The maintenance information should include the required skill levels for the maintenance organization that is responsible for the system. To design a system that requires highly educated personnel to maintain the system, radically increases the overall life cycle cost of the system. If the system is designed such that a PhD is required to understand the maintenance procedures, then the notion of Design for Maintenance has not been a driving force in the overall design of the system (unless, of course, the system under discussion is the Large Hadron Collider (LHC) used by CERN). This is certainly true for Naval radar systems that operate on a variety of Naval assets and environments and must be maintainable continuously by trained Naval personnel [49].

Question to ponder

How can standardization of repair manuals affect the overall maintainability of systems?

3.4.1.5 Enhancing Maintainability through Ease of Accessibility

Ease of accessibility provides for sufficient space around maintenance points to ensure access to the equipment should provide sufficient space for safe access. This can be based on designing the enclosures for ease of access, or access panels. It is also impacted by where and how the system is mounted. For instance, if the system is in a standard rack mount, is there room to remove/replace components in the available space. When maintenance requires the system to be disassembled to perform routine maintenance, the costs are higher to disassemble, and the time of outage is much higher [50].

3.4.1.6 Enhancing Maintainability through Ease of Maintenance Activities

Design the system in such a way that it can only be maintained the right way—as with access points above, care should be taken to design the system so that disassembly and reassembly of the system results in proper performance [51]. Examples:

If the electrical wires [positive, neutral, ground] are important to avoid "*smoking the equipment,*" then the plugs and/or adapters should be designed to ensure that the plug cannot be "reversed" when installed.

If data loss is not allowed in a system, such as in a radar processing system, then redundancy in processing is required such that you can perform "*incremental*" upgrades to the processing services. This then requires the system engineers to address transaction management solutions for data flows in the event the on-line service goes down while the backup service is in maintenance, to support fault recovery [52].

Question to ponder

What can you think of that is easy to maintain due to built-in maintainability requirements? What about something that is very difficult to maintain?

3.4.1.7 Enhancing Maintainability through Designing for Safety

Safety by Design itself is critical in the design for maintenance of a system. Safety, of both the system and the operators/maintainers of the system, is paramount to system operations. This topical area is subject to volumes of books, regulations, and safety standards and alike across various industries [53]. In some electrical power supplies, the electrical circuits are designed to level load the current entering the power supplies to address intermittent connections and currents on the power lines, so that the swings in power do not burn out the electronics receiving power from the power supply. This is critical for Navy radar systems which must be operational continuously to allow detection and processing of threats to Naval assets and personnel.

Question to ponder

How safe is safe enough? Is it possible to build in too much safety into a product?

3.4.1.8 Enhancing Maintainability through Designing for Modularity

Designing modular systems enhances the ability to perform maintenance. Modular designs allow for the software, hardware, or both to be replace/upgrade and like capabilities without large life cycle costs [54]. When a system design has not properly isolated capabilities for maintenance, the entire system is at risk of being down during maintenance, or worse yet, requires a full system retest after a component has received a maintenance update. Modular designs, when properly implemented, allows for maintenance to be performed synchronously to system operations, or reduces the regression testing required on the system before it is placed back into service [55].

3.4.2 The MDRE Maintainability of the DREX+ System

According to the DoD Guide for achieving availability, reliability, and maintainability (RAM), maintainability is the ability of an item to be retained in, or restored to, a specified condition when maintenance is performed by personnel having specified skill levels, using prescribed procedures and resources (i.e., tools), at each prescribed level of maintenance and repair [56]. The maintainability of the DREX+ digital receiver-exciter system depends on having considered the concepts discussed in Sect. 3.1 into the overall design, manufacture, and assembly of the boards and other components that encompass the DREX+ system. We will discuss each of these below.

3.4.2.1 Maintenance Cost as a Part of Overall Life Cycle Cost (LCC)

Here again, we use the DREX+ to illustrate the MDRE process. The DREX+ system's applications for Naval operations requires the DREX+ be utilized in extreme environments over a long period of time (possibly decades) [57]. Under such circumstances, the operational and maintenance cost of the DREX+ production system may account for as much as 80% of the systems LCCs and the operational and maintenance expenses may reach three times the initial cost of the system. However, by employing the design for maintenance process discussed above, we can reduce the overall LCC of the DREX+ while enhancing its availability, reliability, and maintainability (RAM) and make it useful across a variety of Naval applications. Another aspect of the maintainability engineering process that is often ignored it maintainability testing [58]. Maintainability testing is the process of testing the system's ability to update and/or modify the system. This is a crucial part of the overall supportability and manufacturability of a system, since changes and updates to both H/W and S/W are likely throughout the life cycle use of the DREX+ system. Once DREX+ is deployed into an environment, software requires maintenance (updates, patches, etc.) and hardware requires maintenance and updates (tech refresh).

The DoD Guide for Achieving Reliability, Availability, and Maintainability (RAM) reports that maintainability is an issue with many DoD-related projects/products due to:

- System requirements focusing on technical performance with few devoted to RAM [59].
- Using immature technologies to meet performance goals weakens the ability to design systems with high reliability.
- There is often limited collaboration among organizations charged with requirement formation, product-development, and maintenance/sustainment of the product [60].

3.4.2.2 Enhancing the DREX+ Maintainability through Standardization of Components

The strategic partnership between Colorado Engineering Inc. and Intel/Altera provides the opportunity for board design and parts choices consistent with industry standardization. All three boards that comprise the DREX+ system (Crestone, Lightening, and Thunder) utilize the Stratix 5 FPGA processing units [61]. Wherever possible, the same component is utilized throughout the DREX+ system to provide increased maintainability throughout the manufacturing, testing, deployment, and operation of the DREX+.

3.4.2.3 Enhancing the DREX+ Maintainability through Standardization of Maintenance Manuals/Procedures

What is required for the DREX+ system to be available and manufacturable across a variety of Naval programs and for a host of Naval assets is to develop a comprehensive program for design and manufacturing for maintainability [62]. Phase II will lay out a plan for the DREX+ that includes:

Personnel, including reporting responsibility, and a RAM manger.

Develop a model-based conceptual system representation, consisting of DREX+ components, subsystems, manufacturing processes, and performance requirements [63].

Utilize these models throughout the development and manufacturing process to estimate DREX+ performance and RAM metrics.

Identify all critical failure modes and degradations, addressing them in the DREX+ design.

Utilize data from component-level testing to characterize the distribution of Mean Time Between Failure (MTBF) [64].

Conduct analyses to determine if the design can meet the RAM requirements for a given Naval application.

Design diagnostics for fault detection isolation and elimination of false alarms, redundant or degraded system management to provide enhanced mission success.

Ensure designs are modular and can facilitate remove-and-replace maintenance [65].

> **Question to ponder**
>
> Compare the projected production costs of the new F35 fighter ($80 million) to the projected operations and maintenance (O&M) cost ($1.196 trillion). Why are the O&M cost so high?

3.4.2.4 Enhancing the DREX+ Maintainability through Ease of Accessibility

The rack-mountable nature of the DREX+ system makes it easily accessible for removal and replacement should there be an issue with one of the DREX+ units. The DREX+ system is built as a total Line Replaceable Unit (LRU) and the DREX+ should be replaced and sent back for repair [66]. The nature of the DREX+ fault (which will be discussed in the Failure Modes Analysis Report) will indicate which of the three main boards that comprise the DREX+ (the Lightening Analog-to-Digital Converter board, the Crestone Processing board, and the Thunder Digital to Digital-to-Analog Converter board) caused the fault within the DREX+ system.

3.4.2.5 Enhancing the DREX+ Maintainability through Ease of Maintenance Activities

The DREX+ design and operations are well poised for ease of maintenance. The DREX+ is designated as an LRU and fault tree diagnostics provides the repair depot with an understanding, at the board level, of where the problem lies. In addition, a USB port is provided and the DREX+ FPGAs can be programmed for operation through this port, making access to operational parameters and characteristics easy to access and change as needed. This also allows incremental software updates to be uploaded onto the DREX+ system [67]. Each of the three boards within the DREX+ provide ease of programmability, I/O flexibility and a modular, open-systems and standards architecture to provide a scalable and easily configured processing solution for Naval radar, Electronic Warfare (EW), Signal Intelligence (SIGINT) and communications needs. Communication with the DREX+ system is available through an ethernet connection. The on-board health monitoring capabilities that are viewable from the screen of the DREX+ include, but are not limited to:

- Board voltages.
- External temperature sensors.
- Smoke detection.
- Impact detection.
- Field Programable Gate Array (FPGA) status.

Fig. 3.13 DREX+ sample enclosure

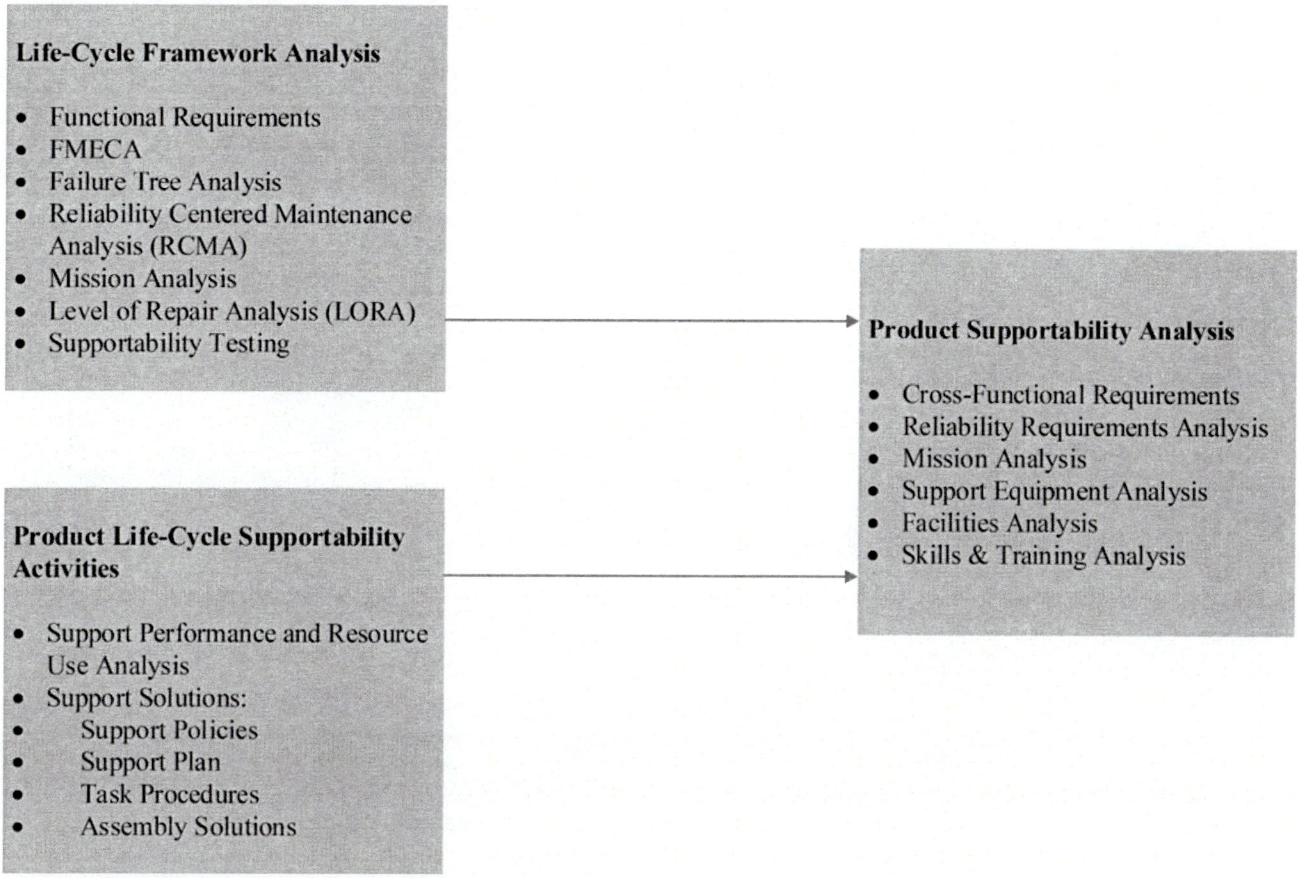

Fig. 3.14 The MDRE design for supportability process

Other characteristics of the DREX+ that enhance maintenance activities is ensuring that all bolts are easily accessible for maintenance personnel [68].

3.4.2.6 Enhancing the DREX+ Maintainability through Designing for Safety

The DREX+ system is rack mountable in a standard 3–6 U server rack. It is designed and built to MIL-STD ruggedization and provides a LED T.V. screen for status control of the unit. This provides health and status without the need to probe the system, enhancing the overall safety of operation of the DREX+. Am NXP K61 Microcontroller provides system health monitoring and protection. Another study that was done is a Damage Mode Analysis (DMA) to identify the types of damage modes each component of the DREX+ may experience (i.e., fire, toxic fumes, smoke), identifying the primary and secondary damage effects to which each component can be exposed to in each of the deployed configurations (e.g., Navy long-range radar, border patrol surveillance, airborne applications) [69].

3.4.2.7 Enhancing the DREX+ Maintainability through Designing for Modularity

The DREX+ system has three main boards, the Lightening ADC board, the Crestone processing board, and the Thunder DAC board. A sample DREX+ enclosure is illustrated in Fig. 3.13, showing the front screen. The DREX+ design was reviewed across multiple Naval applications to isolate and reduce dominate failure modes down to one replaceable module. We analyzed the system to ensure detection and isolation of failures to a single module and lay in that module for depot. This significantly reduced the repair efforts and thus reduce the overall Life Cycle Cost of utilizing the DREX+ system across Naval missions and platforms [70].

3.5 DREX+ Supportability Analysis

Decisions made in the early phases of the DREX+ design greatly affected Life Cycle Cost. Fig. 3.14 below illustrates the "Design for Supportability" process utilized throughout

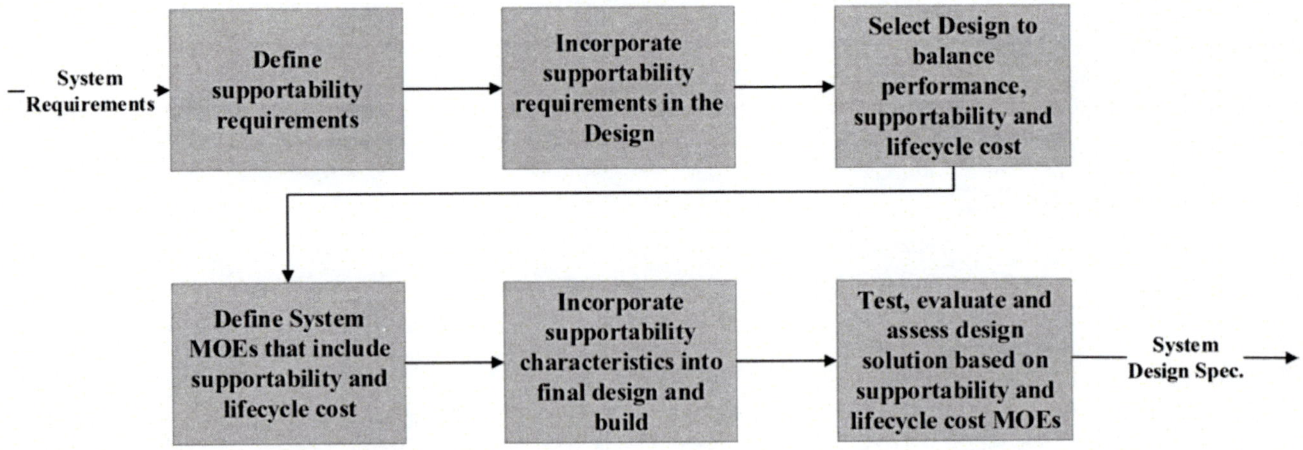

Fig. 3.15 The MDRE supportability design life cycle

the design and implementation phases of the DREX+ system. Design for Supportability (DFS) began at the earliest stages of the DREX+ life cycle to determine how the Function Requirements, Failure Tree Analytics, and RAM requirements affected the overall DREX+ design [71]. In addition, life cycle supportability activities like Resource and Performance analysis and policies factored into the overall supportability plan and subsequent design of DREX+. During the initial design phase, supportability objectives, like mission analysis and skills and training analysis (see Fig. 3.15) and their associated metrics factored into the initial trade studies conducted during the systems engineering architecture design phases and resulted in a system design and sustainment architecture that can be utilized across multiple Navy missions, assets, and platforms [72].

The key to the supportability activities was the development of the maintainability concepts discussed above. Maintainability analysis provided and specified the levels of maintenance for the system and the corrective tasks to be accomplished at every level. This detailed maintainability analysis was performed and included detailed reliability and maintainability modeling, prediction, allocation, and analysis, FMECA and fault tree analyses, and condition-based maintenance (CBM) and reliability centered maintenance (RCM) analyses [73].

3.5.1 Designing for Supportability

Figure 3.15 below illustrates the supportability life cycle. This designing for supportability is based on the supportability process shown in Fig. 3.14. It includes defining and laying out the spares, common and unique tools and facilities and maintenance training characteristics that must be specified and procured throughout the system's life cycle [74]. The initial system design specification must include supportability concepts and requirements; otherwise the system will not be developed with supportability goals, objectives, maintainability thresholds, and/or constraints, and the likelihood of achieving these is greatly diminished. Ultimately these must be verifiable in the final design solution for the system. Measures of Effectiveness (MoEs) directly relating to supportability must be established up front to test, evaluate, and assess the design solution relative to supportability [75].

Question to ponder

Is it possible to build a system that is inherently supportable without specifically deriving supportability requirements?

The supportability life cycle must include the following [76]:

Definition of supportability requirements.
Formulation of supportability design characteristics.
Establishment of supportability goals, constraints, and thresholds.
Implementation of supportability design characteristics.
Identification of support resource requirements across the system life cycle.
Determination of support resources quantities (e.g., spares).

Supportability engineering is not a one-time analysis but has various interrelated analyses throughout the systems life cycle. These must be an integral part of the systems engineering process from the earliest activities through end-of-life of the system [77]. Supportability must be an on-going focus throughout initial design, final design, implementation, operations, and sustainment [78]. As the design matures, focus shifts from supportability concepts to the physical support requirements. Finally, supportability engineering uses operational experience to identify requirements and opportunities for design upgrades and improvement to enhance system supportability. One area often overlooked is the increased sustainability/manufacturability model accuracy. Supportability engineering should be involved in continual analysis, based on an increased accuracy of data and models throughout the systems life cycle. This is illustrated in Fig. 3.16. Example—as the system moves from detailed design to prototype, parts lists

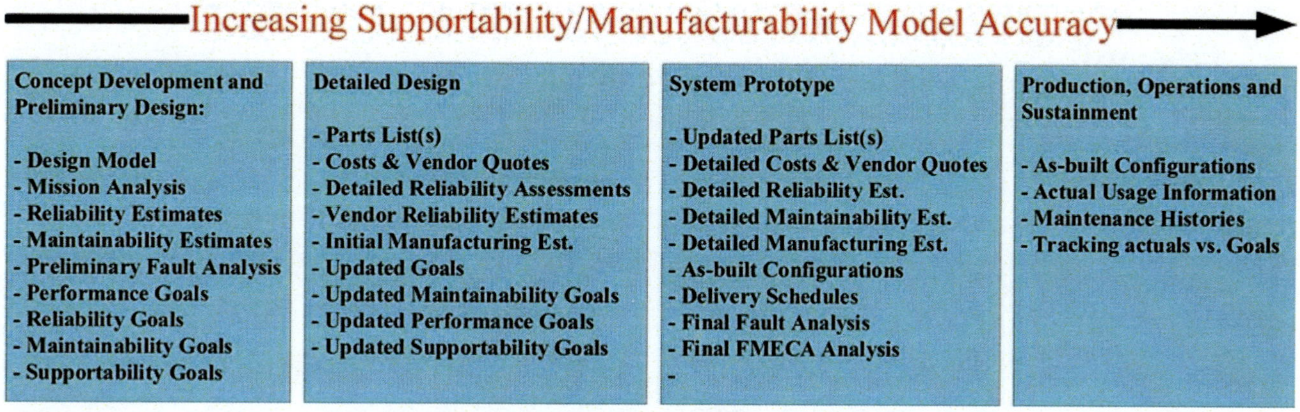

Fig. 3.16 The MDRE supportability/manufacturability model accuracy life cycle

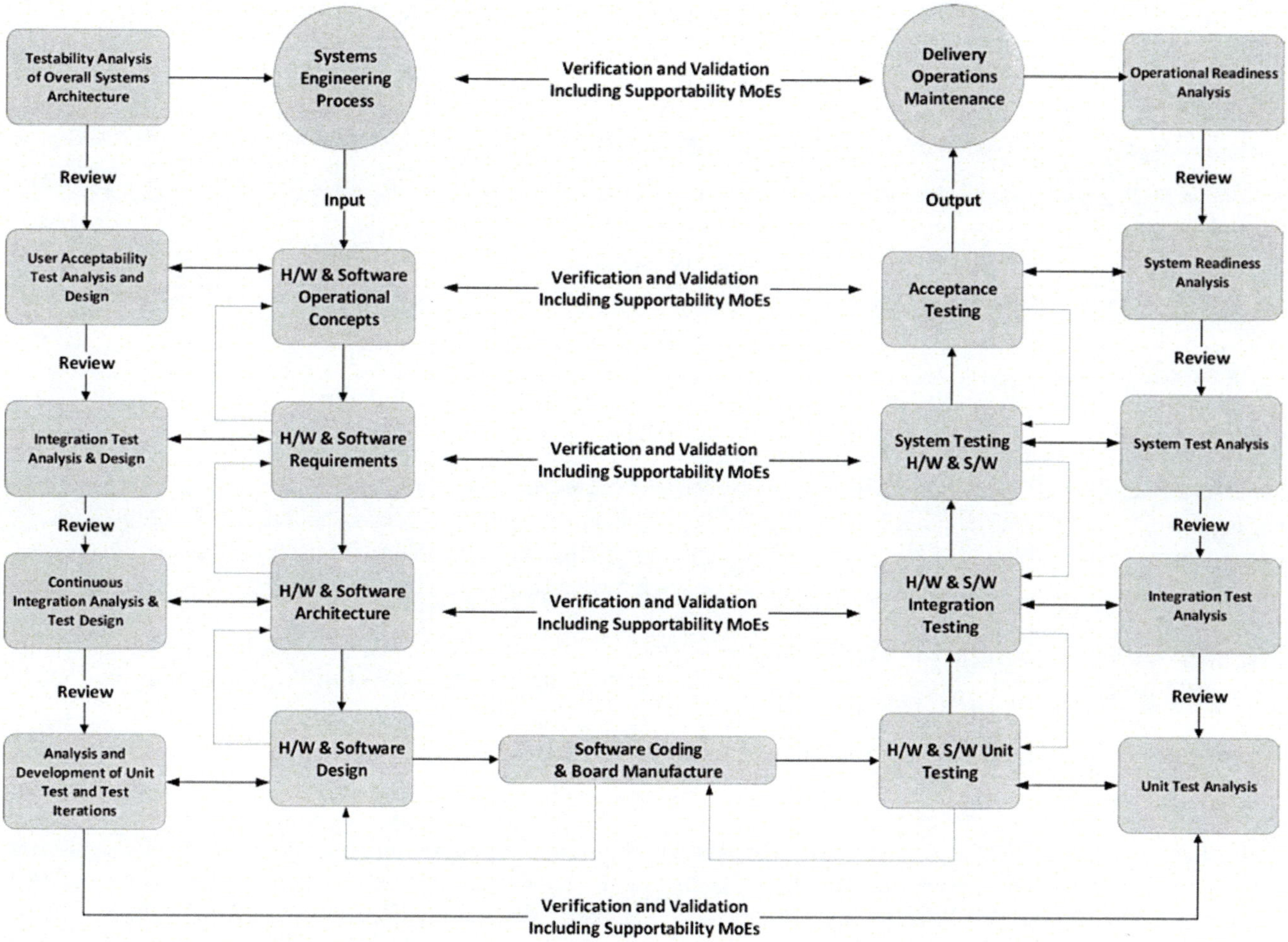

Fig. 3.17 MDRE system test process including supportability

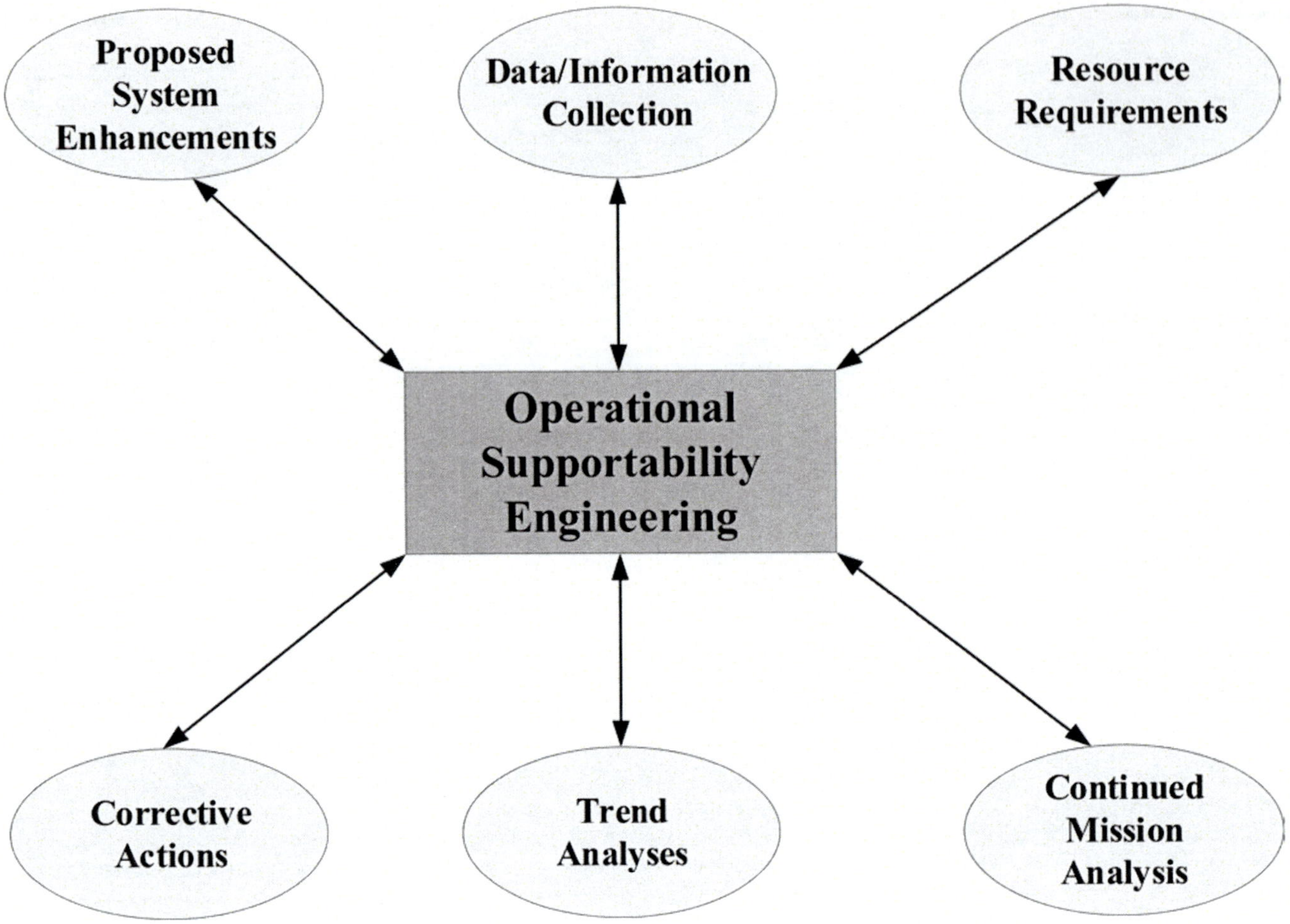

Fig. 3.18 MDRE operational support engineering

are updated, and vendors may come and go as more details about parts are determined. The fault analysis may change the parts required to accommodate reduction of Mean-time-between-failure (MTBF) [79].

3.5.2 Designing for Supportability through Testability

We have discussed the supportability engineering process that provides integration of supportability into the design of a system, which includes incorporation supportability requirements and MoEs into the critical decision-making events throughout the systems development life cycle. Each of the supportability engineering activities must be assessed to ensure that supportability is built in and tested into the system. This includes a feedback-driven testing process that includes testing the supportability MoEs as an integral part of the testing process [80]. Fig. 3.17 illustrates this feedback-driven testing process. This overall testing process shown not only what should be done at each stage in the testing process, but what should be assessed and fed back to previous steps.

3.5.3 MDRE Operational Supportability Engineering

The initial supportability engineering activities discussed lead to delivery of the system. After delivery and acceptance of the system, the focus of supportability is on sustainment of the mission capabilities of the system [81]. Operational supportability must have participation throughout the operational life cycle of the system. This includes continued understanding of resource requirements, continued mission

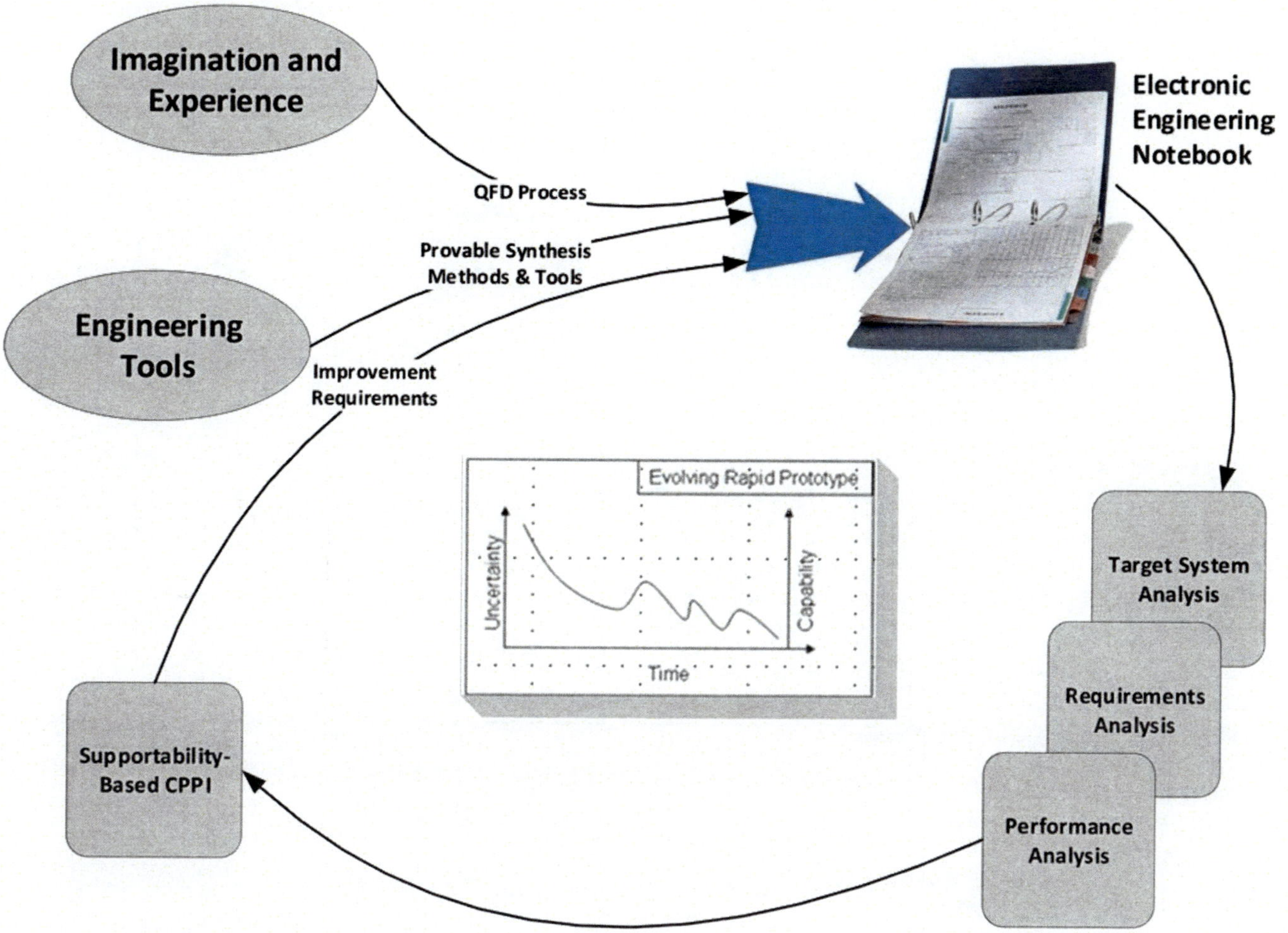

Fig. 3.19 The MDRE supportability continual process improvement process

analyses (since the mission my change over time), and continued analysis and trending of maintenance and test data throughout operations of the system [82]. This is illustrated in Fig. 3.18. All the supportability characteristics and design features developed during the product design must be measurable during operations.

The supportability decisions that were made during the acquisition process, initial design, final design, etc., were based on goals, thresholds, constraints, and design features developed through comparisons with similar systems, historical data on previous systems, and predictions based on future predictive analytics (e.g., FMECA) [83]. Once the system is in operations, continued data collection must take place. All collected data must be accurate, timely, and must be stored in a way that is conducive to data analytics, based on the standards utilized by the organization utilizing the system in operation.

3.5.3.1 MDRE Continued Operational Supportability Engineering

Continued operational supportability includes continuous process/product improvement to ensure the system can provide operational efficiency throughout its life cycle. Supportability reviews should be scheduled consistent with the overall program plans [84]. It should have a set of established procedures which provide consistency of review among all engineering disciplines. The procedures should define the acceptance and rejection criteria for total system supportability. This includes system enhancement to provide continual improvement and operational suitability as mission needs change. Fig. 3.19 illustrates the supportability-based continual process/product improvement (CPPI) process. This includes continual target system analysis (mission analysis), requirements analysis to ensure new/changing requirements are captured in possible enhancements, and performance analysis to ensure the system continually meets operational expectations [85].

Question to ponder

Why do you think continuous product/process improvement is beneficial and necessary for complex systems?

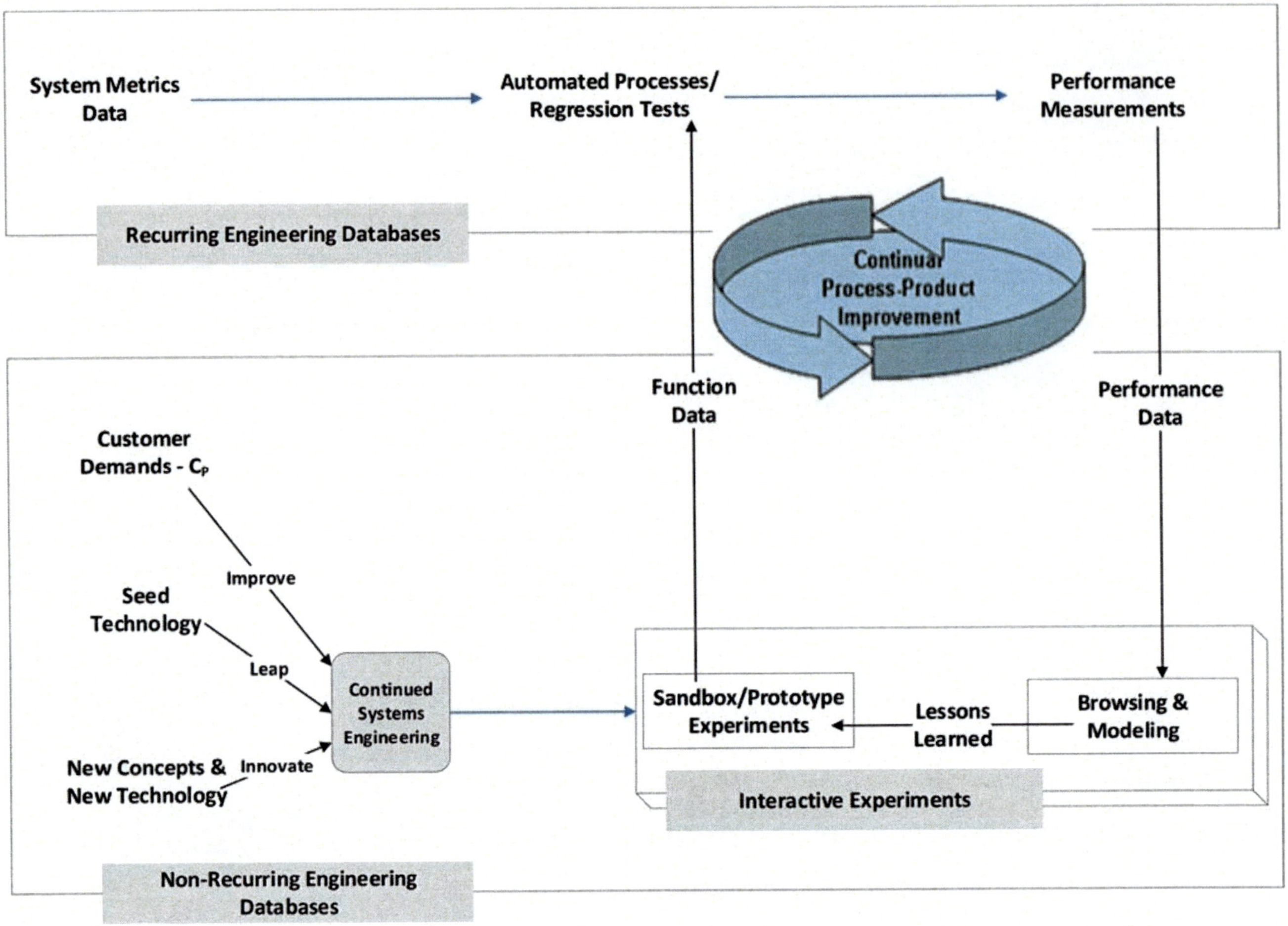

Fig. 3.20 The MDRE system enhancement CPPI process

Once it is determined an enhancement is required for continued operations, the supportability CPPI process is employed. This is illustrated in Fig. 3.20. Enhancements to an existing operational system require a non-recurring engineering effort, along with prototyping the enhancements and testing not only against the new requirements but also automated regression tests to ensure the new enhancement has not interfered with on-going operational requirements [46].

3.6 DREX+ Supportability Conclusions and Discussion

Overall, supportability analyses are intended to produce the procedures and methodologies required to accomplish the planning, development, systems engineering, production, and management for logistics supportability of DoD-related systems and equipment for operational support [86]. Supportability analysis is intended to be conducted as an integral part of the overall systems engineering process for product/system development. It begins at program initiation (Authorization to Proceed (ATP)) and continues throughout system development and operations. Supportability analyses form the basis for logistics design requirements including in the system requirements specification and for decisions concerning how to support the overall system life cycle in the most cost-effective way possible [145]. The supportability analyses in each phase of design, development and operations & sustainment must be scoped to the objectives and level of design anticipated throughout the life cycle. Supportability analysis objectives should be based on:

- The probable hardware and software designs, support concepts, and operational approaches for a new system or enhanced legacy system's reliability and maintainability needs/requirements [87].
- Operations and Support (O&S) costs, logistic support resources and readiness characteristics both from a design and operations perspective [88].
- Ensuring the availability, accuracy, and relevance of data required to perform the proposed support analyses, for both O&S and logistics support [89].
- The potential design impact of performing the supportability analyses, including the estimated supportability,

cost and readiness improvements and reduction of program risks [90].

Supportability analyses should also include an estimate of the cost of each analysis. It should rate the degree of cost effectiveness of performing each analysis, given the projected costs, the anticipated benefits, and the programmatic schedule constraints under which they would be conducted [91]. The supportability analyses should include the integration of many analytical techniques with the goal of designing and developing an effective and efficient logistics support infrastructure. These analyses will include, but are not limited to:

- FMECA.
- Fault Tree Analysis.
- Reliability Centered Maintenance Analysis.
- Level of Repair Analysis.
- Condition-Based Health Analysis.
- Maintenance Task Analysis.

Procedures will be established and integrated into the overall systems engineering process for DREX+. Supportability reviews will be designed, consistent with current commercialization and program plans [92]. These procedures will define the acceptance/rejection criteria pertaining to the total DREX+ supportability requirements.

3.6.1 Sample Supportability Requirements

To illustrate supportability requirements, Table 3.4 provides examples of requirements used for the development of radar systems utilizing the DREX+. They represent requirements that span the range of maintainability, reliability, and availability. This is by no means an exhaustive list but provided here as an example of actual customer requirements levied against the DREX+ system. All the requirements shown in Table 3.4 impact the system supportability, maintainability, and reliability of the system [93]. The first row in the table is a good example of a functional requirements that supports the availability of the system. This is essentially a Built-in-Test (BIT) test requirement for the system. In short, there are times when additional functionality must be specified to support a quality assurance requirement [94].

The transportation requirement (row 3) defines any transportability or material handling requirements for the system. These requirements will help define any special handling fixtures, tooling, etc., that may be required to initially install the system, as well as what is needed to maintain and support the system in the field. This type of requirement is often missed by design engineering. The ability to deploy the system from the development site to the deployable environment site, install the system and get it operational is as important as any other requirements [95]. If the system is difficult to remove from the factory (development site), transport it, install it, and test for operational readiness, does not matter how well it meets other requirements, it is a poorly designed system. Materials requirements like those in rows 4 and 7 deal with construction. They are essentially requirements that are not controlled by performance characteristics, interface requirements, or reference documents. This type of requirement includes appropriate design standards, requirements governing the use or selection of materials, part and processes, interchangeability requirements, safety requirements, etc. This type of requirement impacts the reliability of the system as it specified materials and construction that can withstand the operational environment to the system. In row 6, the second part of this requirement is basically a product marking requirement, even though the entire requirement is a material handling requirement. It implies special handling of ESD sensitive parts. Another important part of MDRE is understanding how the system will be manufactured and ensuring that the requirement specification covers manufacturability. Chapter 4 provides insight into the manufacturability aspects of system requirements and engineering and design [96].

Table 3.4 Sample supportability requirements for DREX+

Category	Requirement
1. Availability	The exciter/synchronizer subsystem shall monitor the transmit pulse output for failure conditions.
2. Reliability	The exciter/synchronizer subsystem shall have a critical failure rate not more than 8 failures per 1000,000 hours of operation
3. Transportability	The LRUs shall be designed to meet interface transportability requirements in accordance with MIL-STD-1366E
4. Materials design	Materials shall not be placed in combinations which accelerate corrosion as defined in MIL-STD-899C
5. Maintainability	The mean time to repair (MTTR) shall be no more than 1 h
6. Material handling	When parts sensitive to electrostatic discharge (ESD), including metal oxide semiconductor (MOS) parts, are utilized in the equipment, the contractor shall affix warning labels to protective packaging.
7. Materials design	Electrical/electronic parts shall not be electrically stressed greater than the derating requirements at maximum environmental stress.

3.6.1.1 Design Problem

The philosophy of the assignment development for this textbook was to create a set of though-provoking questions, problems, and design assignments that help drive home the principles, practices, and methodologies of multidisciplinary requirements engineering (MDRE). We endeavored to provide the opportunity for the student to practice the art and science of requirements engineering, helping the student to understand that there is rarely one answer to a requirements engineering problem, and it is only through study and practice that successful requirements engineers are developed. MDRE is a lifelong journey, and anyone who tells you they have nothing left to learn does not truly understand the discipline or has never been involved in the development of a complex system. With the recent interest in alternative energies and talk of the "Green New Deal," along with the recent issues with ice storms in Texas taking out the power grid, the assignments for this textbook will follow the theme of effectively dealing with both local alternative energy power grids and the national infrastructure required to facilitate large-scale alternative energy production, sustainment, and management.

3.6.1.2 Design Problem Description: Alternative Energy Power Grid

The purpose of this project is to, over the course of the semester, develop a feasibility study, requirements analysis, requirements elicitation, derivation, and allocation, for a system that will facilitate a "local" alternative energy power grid capable of powering a neighborhood of 500 homes and a large biosphere capable of supporting growing year-round fruits and vegetables for the 500-home neighborhood. The student will have to deal with issues of choosing multiple alternative energies that can provide power throughout all seasonal difficulties, from intense summer heat to freezing winter temperatures and possible ice storms. Control, viability, sustainability, and maintainability will be major factors to understand the entire possibly 30-year life cycle of such a system. How to effectively utilize, store, and manage power distribution, along with when to up-convert "extra" energy to the national power grid, will be issues that drive many of the assignments. Simply relying on the current technologies and energy infrastructures will not move the country and world forward. Hopefully, this will provide insight to the students using this textbook to understand some of the issues that face everyone looking out over the next many decades.

3.6.1.3 DC Power Grid Requirements Discussion

In the initial days of power generation and distribution, DC power was far less practical than the AC distribution system that soon supplanted it. However, even after Tesla's AC system became the predominant system (and rightly so), Edison's DC power did not become obsolete for some time. Eventually power plants eliminated their delivery of DC power, but the overall concept of a DC power grid is not dead. Even our computer systems (large server systems) have uninterruptible power supplies which contain batteries that are charged off the grid and used to up-convert to AC to power devices when the AC power grid goes off-line. So, the notion of a DC power grid is not dead, but generally used as a backup system. The point is, the notion of a DC power grid has remained and now that renewable energies are viable, we should re-examine their use for widespread, local, power distribution.

One of the advantages of a locally generated DC station is energy savings as well as allowing a facility to run more easily off various DC sources, such as Solar Arrays. And for power-hungry installations, the notion of generating the power locally is growing in popularity. Part of the reason is that waste heat from the power generation generators (solar, geothermal, etc.) can be used to warm nearby buildings. By combining the production of heat and power, local power grids can squeeze much more useful energy out of the renewable energy they use, so this approach will certainly become more widespread as time goes on. The purpose of this design problem is to look ahead and provide an architecture and specification for a local DC power generation station that can serve a local community.

Currently many systems exist to control and handle alternative energies. There are systems to control and manage solar power, wind power, geothermal power, and others, but very few are designed to control, manage, store, and distribute energy from a power grid consisting of multiple alternative energy sources. As we saw the issues recently caused by the intense ice storms in Texas, one alternative energy source is not the answer to the growing energy needs/wants in our society. Each alternative energy source has limitations and issues associated with its use. By harnessing multiple alternative energies, and managing the storage of the energies produces, communities can be worry free from power generation. In order to facilitate utilizing multiple alternative energy sources, a control center must be included in the design that can manage the energy creation, distribution, storage, and return unused energy to the power grid.

3.6.1.4 High-Level Technical Description of the DC Power Grid System

The purpose of the multiple alternative energies DC power grid is to:

- Supply DC power to a customer community of 500 homes and a biosphere for growing food.
- Provide a storage device to store energy in case of a system failure.
- Determine the amount of surplus energy that is above and beyond what is required to power the grid and charge the backup system, and then deliver that surplus energy to the national power grid systems.

Throughout the semester you will be asked to analyze, design, and architect different components of the overall DC power grid system. This will include research into alternative designs, like Lithium Nanowire Storage systems for backup

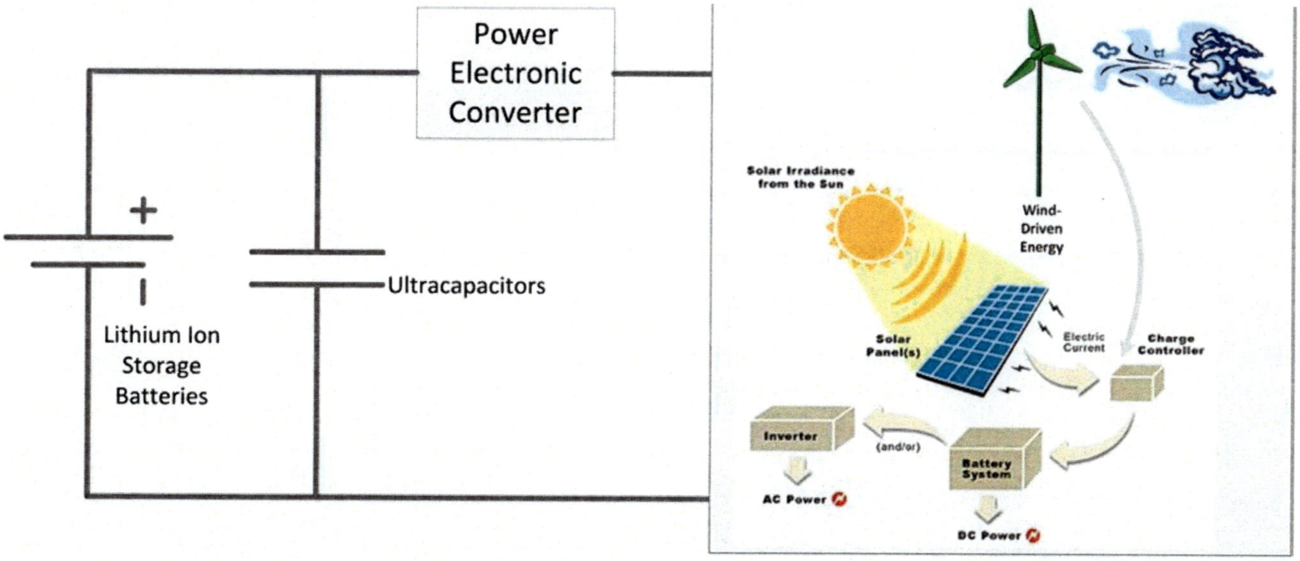

Fig. 3.21 Alternative power management system overview

Fig. 3.22 Example of a whole house power grid

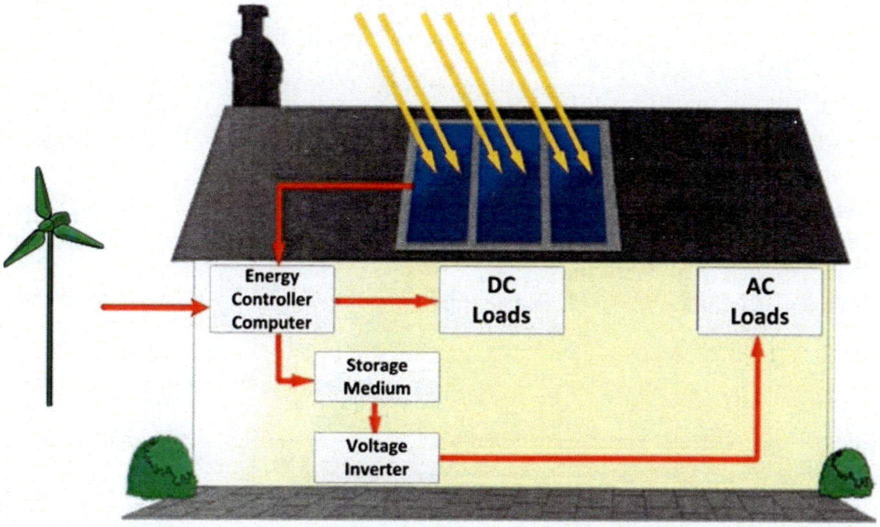

power and/or supercapacitors (also called Ultracapacitors). Fig. 3.21 illustrates these technologies connected to the DC Power Grid Regulator Control Center.

3.6.1.5 High-Level Technical Requirements for the DC Grid System

- Design shall utilize at least two alternative energies, with a maximum of 4.
- Total power generation shall be 3.0 megawatts minimum.
- Each of the alternative energy sources shall be capable of generating at least 60% of minimum wattage.
- DC power grid shall generate DC power, with grid voltage being 48, 24, or 12 volts.
- DC power grid shall be capable of servicing at least 500 single family homes and one 4-acre biosphere with hydroponics capable of growing vegetables year-round. (See Fig. 3.22 for a block diagram of a DC-powered home).
- All lights, appliances, and facilities in each home and the biosphere shall be configured for DC operations, if possible. Things like electric ovens may need DC/AC conversion to accommodate their 220v. requirements.
- Only allowed exceptions are the HVAC system and the Washer/Dryer, although DC is preferred.
- There shall be energy storage capability capable of operating the grid of 500 single family dwellings for 7 days if all alternative forms of energy are not producing.

- Compliance Criteria: Energy storage devices shall be capable of producing 750,000 watts for the entire 7-day period. Nominal energy storage should be batteries or supercapacitors.
- The DC power grid system shall include controlling h/w and s/w to manage the multi-alternative energy strategies and production.
- All energy produced above the required home and energy storage usage shall be upconverted to 220vac for release to the national power grid.
- The DC power grid shall be capable of operating in a temperature range of −40°F to 125°F. without degradation of power generation.
- The DC power grid must meet OSHA regulations.

3.6.1.6 Assignments

- Develop a sustainability study for the whole house DC system and answer the following questions:
- What are five sustainability requirements needed for the alternative energies you have chosen for your design?
- Explain how your requirements include a "design for maintenance" theme.
- How will sustainability differ using batteries vs. supercapacitors to store energy?
- Explain your philosophy for safety within your system design.
- What are two safety requirements you feel are essential for operation of your system?
- What skills will be required for sustainment/maintenance of your system?

References

1. Shaw, M. (1990). Prospects for an engineering discipline of software. *IEEE Software, 7*(6), 15–24.
2. Stevens, R., Brook, P., Jackson, K., & Arnold, S. (1998). *Systems engineering: Coping with complexity*. Prentice Hall Europe.
3. Courage, C., & Baxter, K. (2005). *Understanding your users: A practical guide to user requirements methods, tools, and techniques*. Elsevier Publishing. ISBN 978-1558609358.
4. Huddlestone, J., Sears, R., & Harris, D. (2017). The use of operational event sequence diagrams and work domain analysis techniques for the specification of the crewing configuration of a single-pilot commercial aircraft. *Cognition, Technology & Work, 19*, 289–302.
5. Neighbors, J. (1980). *Software construction using components. Technical report 160*. Department of Information and Computer Sciences, University of California.
6. Neighbors, J. (1984). The Draco approach to constructing software from reusable components. *IEEE Transactions of Software Engineering, SE-10*, 5.
7. Kotonya, G., & Sommervile, I. (2003). *Requirements engineering processes and techniques*. John Wiley & Sons. ISBN 0-471-97208-8.
8. DeMarco, T., & Plauger, P. (1989). *Structured analysis and system specification*. Prentice Hall.
9. Yourdon, E. (1989). *Modern structured analysis*. Prentice Hall.
10. IBM. (2017). IDC, the thinking supply chain.
11. Geriant, J. (2015). The changing face of supply chain risk management (SCM World).
12. O'Marah, K. and Chen, X. (2016). Report: Future of supply chain (SCM World).
13. O'Marah, K. (2017). Your roadmap to a digital supply chain (SCM World).
14. Dittmann, P. (2017). New supply chain technology best practices (white paper, University of Tennessee's Haslam College of Business).
15. Cecere, L. 2017. Driving digital supply chain transformation – A handbook for action.
16. Russell, S., & Norvig, P. (2003). *Artificial intelligence: A modern approach (upper Saddle River)*. Prentice Hall.
17. Taylor, C. (2017). *Artificial intelligence and logistics is transforming business*. Big Data (blog).
18. Kelly, J. (2015). *Computing, cognition and the future of knowing (white paper)*. IBM Research: Cognitive Computing.
19. Siegel, E. (2016). *Predictive analytics: The power to predict who will click, buy, lie, or die*. John Wiley & Sons.
20. James, R., & Lindner, C. (2012). *Business analytics: The next frontier for decision sciences*. Decision Line.
21. Handfield, R. (2017). *The evolution of the transparent and cognitive supply chain, white paper*. IBM and North Carolina State University.
22. Oliver, R., & Webber, M. (1982). Supply chain management: Logistics catches up with strategy. *Outlook, 5*(1), 42–47.
23. Lee, H., Padmanabhan, V., & Whang, S. (1997). The bullwhip effects in supply chains. *MIT Sloan Management Review*.
24. Trent, R. (2004). What everyone needs to know about SCM. Supply chain. *Management Review*.
25. Speier, C., Mollenkopf, D., & Stank, T. (2008). The role of information integration in facilitating 21st century supply chains: A theory-based perspective. *Transportation Journal, 47*(2), 21–38.
26. Dittmann, J. (2013). *Supply chain transformation*. McGraw Hill.
27. Burnette, M. (2015). *Platform life cycle management best practices, white paper*. University of Tennessee's Haslam College of Business.
28. Stank, T., Scott, S., & Hazen, B. (2018). *A savvy guide to the digital supply chain, white paper*. University of Tennessee's Haslam College of Business.
29. Scott, S. (2015). *Supply chain talent—Our Most important resource, white paper*. University of Tennessee's Haslam College of Business.
30. Burnette, M., & Dittmann, J. (2017). *End-to-end supply chain collaboration best practices, white paper*. University of Tennessee's Haslam College of Business.
31. Miyakawa, S. and Ohashi, T. (1986). "The Hitachi Assembly Evaluation Method (AEM)," Proc. International Conference on Product Design for Assembly, Newport, Rhode Island, April 15–17.
32. Boothroyd, G. (1980). *Design for Assembly – A designer's handbook*. Department of Mechanical Engineering, University of Massachusetts, Amherst.
33. Boothroyd, G. (1982). *Design for assembly: The road to higher productivity*. Assembly Engineering.
34. Henchy, L. (1988). American manufacturing fights Back. Business. *Solutions, 1988*, 10.
35. Boothroyd, G., and Dekker, M. (1992). Assembly automation and product.
36. Boothroyd, G., Dewhurts, P., and Knight, W, (1994). Product Design for Manufacture and Assembly. CRC Press, London, ISBN 978-1420089271.

37. Compton, W., (1988), Design and Analysis of Integrated Manufacturing Systems. National Academy Press, Washinton, DC..
38. Lewis, G., and Connelly, H. (2000). Product design for assembly: the methodology applied.
39. Stoll, H., (1988) Design for Manufacture. Manufacture Engineering, pp. 67–73).
40. "IBM Proprinter Case Study". Engineering Systems Research Center. University of California at Berkeley.
41. Hitchens Carl. 2014. Guide to engineering metrology.
42. Franciosa, P. and Ceglarek, D. (2015), Hierarchical synthesis of multi-level design parameters in assembly system. CIRP Ann. - Manuf. Technol., vol. 64, no. 1, pp. 149–152, 2015.
43. Anderson, D. (2004). Design for manufacturability and concurrent engineering: How to design for low cost, design in high quality, design for lean manufacture, and design quickly for fast production. Cambria, CA CIM Press p. 28 ISBN 978–1978072238.
44. Stolt, R., Elgh, F., & Andersson, P. (2017). Design for inspection—Evaluating the inspectability of aerospace components in early states of design. *Procedia Manufacturing, 11*, 1193–1199. Elsevier Science Direct.
45. Crowder, J., Carbone, J., & Demijohn, R. (2016). *Multidisciplinary systems engineering: Architecting the design process*. Springer publishing. ISBN 978-3319223971.
46. Crowder, J., & Friess, S. (2013). *Systems engineering agile design methodologies*. Springer Publishing. ISBN 978-1461466628.
47. Medvidovic, N., Grunbacher, P., Egyed, A., & Boehm, B. (2003). Bridging models across the software lifecycle. *Journal on Systems and Software, 68*, 3.
48. Abadi, M., & Lamport, L. (1991). The existence of refinement mappings. *Theoretical Computer Science, 82*(2), 253–284.
49. Alpern, B., & Schneider, F. (1987). Recognizing safety and liveness. *Distributed Computing, 2*(3), 117–126.
50. Bahrami, A. (1999). *Object-oriented systems development: Using the unified modeling language*. McGraw Hill. ISBN 0-07-234966-2.
51. Bailey, I. and Partridge, C. (2009). Working with extensional ontology for defense applications. Ontology in Intelligence Conference, GMU, Fairfax, VA.
52. Banks, J., Carson, J., Nelson, B., & Nicol, D. (2001). *Discrete event system simulation*. Prentice Hall. ISBN: 0-13-088702-1.
53. Bass, L., Clements, P., & Kazman, R. (2003). *Software architecture in practice* (2nd ed.). Addison Wesley Professional.
54. Beck, K. (1999). *XP explained* (1st ed.). Addison-Wesley Professional. ISBN 0201616416.
55. Beck, K. (2002). *Test driven development*. Addison-Wesley Professional.
56. Beck, K. (2003). *Test-driven development by example*. Addison Wesley – Vaseem.
57. Beck, K., et al. (2001). *Manifesto for agile software development*. Agile Alliance.
58. Bendat, J. (2000). *Non-linear system analysis and identification*. Wiley Interscience.
59. Bettencourt da Cruz, D., & Penzenstadler, B. (2008). *Designing, documenting, and evaluating software architecture technical report TUM-I0818*. Technische Universität München.
60. Boardman, J., DiMario, M., Sauser, B., & Verma, D. (2006). *System of systems characteristics and interoperability in joint command and control*. Defense Acquisition University.
61. Boehm, B. (1984). Verifying and validating software requirements and design specifications. *IEEE Software, 1*(1), 75–88.
62. Booch, G., Rumbaugh, J., & Jacobson, I. (1998). *The unified modeling language users guide*. Addison Wesley Publishing. ISBN: 0-201-57168-4.
63. Books, F. (1975). *The mythical man month*. Addison Wesley.
64. Broy, M., & Stølen, K. (2001). *Specification and development of interactive systems: Focus on streams, interfaces and refinement*. Springer Publishing.
65. Burgin, M. (1982). Generalized Kolmogorov complexity and duality in theory of computations. *Notices of the Russian Academy of Sciences, 25*(3), 19–23.
66. Camarinha-Matos, L., & Afsarmanesh, H. (2008). *Collaborative networks: Reference modeling*. Springer Publishing.
67. Carlock, P., & Fenton, R. (2001). System-of-systems (SoS) enterprise systems for information-intensive organizations. *Systems Engineering, 4*(4), 242–261.
68. Mellon, C. (2013). *Architecture Tradeoff analysis method*. Carnegie Mellon Software Engineering Institute.
69. Checkland, P. (2001). *Systems thinking, systems practice*. John Wiley & Sons.
70. Cheng, B. and Atlee, J. (2009). Current and future research directions in requirements engineering. In Design Requirements Engineering: A Ten-Year Perspective Workshop, Cleveland, OH.
71. Cheng, J., Scharenbroich, L., Baldi, P., & Mjolsness, E. (2005). *Sigmoid: A software infrastructure for pathway bioinformatics and systems biology*. IEEE Intelligent Systems.
72. Chonoles, M., & Schardt, J. (2003). *UML for dummies*. Wiley Publishing. ISBN 0-7645-2614-6.
73. Chung, L., Nixon, B., Yu, E., & Mylopoulos, J. (2000). *Non-functional requirements in software engineering*. Kluwer Academic Publishers.
74. Clausing, D. (2001). Concurrent engineering. Design and Productivity International Conference.
75. Clements, P. (1996). A survey of architecture description languages. Proceedings of the 8th international workshop on software specification and design. IEEE Computer Society.
76. Cockburn, A. (2000). *Writing effective use cases*. Addison-Wesley Publishing.
77. Conway, W. (1995). *The quality secret: The right way to manage*. Conway Quality.
78. Copeland, L. (2001). Extreme programming. Computerworld.
79. Crawford, B., Leon de la Barra, C., Soto, R., and Monfroy, E. (2012). Agile software engineering as creative work. Proceedings of the 5th International Workshop on Co-operative and Human Aspects of Software Engineering, ICSE, Zurick, Switzerland.
80. Creswell, J. (2003). *Research design: Qualitative, quantitative and mixed approached*. Sage.
81. Crowder, J. (1993). *Making change an integral component of advanced design methodologies*. Academic Press.
82. Crowder, J. (2002). Flexible object architecture for the evolving, life-like yielding, symbiotic environment (ELYSE). NSA Technical Paper- CON-SP-0014-2002-09.
83. Crowder, J. (2003a). Using a large linguistic ontology for network-based retrieval of object-oriented components. NSA Technical Paper – CON-SP-0014-2003-03.
84. Crowder, J. (2003b). Ontology-Based Knowledge Management. NSA Technical Paper – CON-SP-0014-2003-05.
85. Crowder, J. (2003c). Agile business rule processing. NSA Technical Paper – CON-SP-0014-2003-06.
86. Crowder, J., & Friess, S. (2014). *The agile manager: Managing for success*. Springer Publishing. ISBN 978-3-319-09017-7.
87. Curcin, V., Ghanem, M., Guo, Y., and Rowe, A. (2004). IT service infrastructure for integrative systems biology. In IEEE International Conference on Services Computing.
88. Curry, E. (2012). System of systems information interoperability using a linked dataspace. In IEEE 7th International Conference on System of Systems Engineering (SOSE 2012), 101–106.
89. Davis, A. (1993). *Software requirements: Objects, functions, and states*. Prentice-Hall.
90. DeLaurentis, D. (2005). Understanding transportation as a system of systems design problem. 43rd AIAA Aerospace Sciences Meeting, Reno, Nevada, AIAA-2005-0123.
91. DeLaurentis, D. (2007). *System of systems definition and vocabulary*. School of Aeronautics and Astronautics, Purdue University.

92. DeLaurentis, D. A., & Callaway, R. (2004). A system of systems perspective for future public policy. *Review of Policy Research, 21*(6), 829–837.
93. Dierolf, D. and Richter, K. (2000). Concurrent engineering teams. Technical Report, Institute for Defense Analysis.
94. Kjokic, S. (2003). The moving optical center in lighting calculations. *Journal of the Illuminating Engineering Society, 32*.
95. Doerr, J., Kerlow, D., Koenig, T., Olson, T., and Suzuki, T. (2005). Non-functional requirements in industry—Three case studies adopting the ASPIRE NFR Method Technical Report 025.05/E, Fraunhofer IESE.
96. Dwyer, M., Avrunin, G., and Corbett, J. (1999). Patterns in property specifications for finite-state verification. In Proceedings of the International Conference on Software Engineering, Los Angeles, CA.

4 MDRE Designing for Manufacturability

Abbreviations

2D	Two dimensional
3D	Three dimensional
AC	Alternating current
ADI	Analog Devices International
AI/ML	Artificial intelligence/machine learning
AOI	Automated optical inspection
BOM	Bill of materials
BSP	Board supply packages
CAD	Computer aided design
CAM	Computer aided modeling
CM	Change management
COC	Certificate of compliance
COTS	Commercial off the shelf
COVID-19	Coronavirus disease 2019
DC	Direct current (or 0 Hz)
DFA	Design for Assembly
DFF	Design for Fabrication
DFI	Design for Inspection
DFM	Design for Manufacturability
DFV	Design for Verification
DoD	Department of Defense
DRC	Design Rule Check
DRC	Design Rule Check
DREX+	Digital receiver/exciter
F	Fahrenheit
FMEA	Failure Modes and Effect Analysis
HVAC	Heating, ventilation and air conditioning
IPC	In process control
KPI	Key Performance Indicators
LCC	Life cycle cost
LCCA	Life cycle cost analysis
LHC	Large Hadron Collider
LRU	Line replaceable unit
MDRE	Multidisciplinary requirements engineering
MTTR	Mean time to repair
NSF	National Science Foundation
O&S	Operations and Support
OEM	Original Equipment Manufacturer
OSHA	Occupational Health and Safety Administration
PCB	Printed Circuit Board
PNP	Pick and place
RAM	Reliability, availability, and maintainability
RF	Radio frequency
SCA	Supply chain analysis
SCM	Supply chain management
SCOR	Supply chain operations reference
SCR	Supply chain roadmap
SW	Software
TOC	Total ownership cost
V	Volts
Vac	Volts AC
WHO	World Health Organization
WIP	Work in progress

4.1 MDRE Supply Chain Management Analysis

4.1.1 MDRE Supply Chain Considerations

Supply chain analysis is the process of evaluating each stage of a product development, implementation, delivery, and sustainment [4]. The analysis requires an evaluation of each step, from the time the company purchases the raw products or supplies (parts) from suppliers, or vendors, to the time the company delivers the product(s) to the customers, to the spare parts required for sustainment of the product. The purpose of the analysis is to determine which stages of the process can be shortened, refined, or made better, to shorten the time it takes to deliver the product to customers without sacrificing the quality of the product or the customer service level of the product and the most efficient way for the customer to sustain the product in normal operations.

The thought process behind supply chain analysis is that the more flexible a company can be, the more likely it

J. A. Crowder, C. W. Hoff, *Requirements Engineering: Laying a Firm Foundation*,
https://doi.org/10.1007/978-3-030-91077-8_4

is to be able to respond to customer needs. Several other benefits may come to a business that undergoes supply chain management analysis as well. These benefits include having just enough inventory rather than having excessive or not enough inventory, improvement of forecasting estimates, and an improvement in scheduling and planning. This includes analysis for the end customer to help them understand the part they must have on hand to effectively sustain the product over the intended product operational life cycle [5].

The selection of hardware, software, and operational procedures, with the inclusion of the Failure Modes and Effect Analysis (FMEA), drives the supply chain support requirements on the system and ultimately the Supply Chain Analysis [6]. This includes sparing for hardware, licensing for software, and safety procedures to be included in the complete set of procedures applied to the human in the loop. In many cases, the logistics support requires formal configuration management to be applied in the support environment to the operational system [7]. Because of this the supply chain requirements and supporting configuration management and subsequent logistics requirements should be treated with the same rigor as the rest of the requirements sets and architectural components. The sell-off supply chain support requirements will vary based on whether it is an inspection requirement, or a functional test requirement. Sparing may have been computed on simple formula such as 10% or 2-deep, where the verification is provided when the correct quantity of LRU spares is provided to the maintenance depot; whereas the ability to build and deliver the software baseline to one or more elements would be tested with the same rigor as functional capabilities within the system [8].

Question to ponder

How might the FMEA results affect the supply chain analysis for a given system?

The supply chain analysis for all systems leads to the success or failure of the deployment of the system. The supply chain analysis and subsequent logistics analysis to support parts sparing and LRU replacement should typically be stood up in advance of the actual deployment of the system. This gives the maintenance personnel time to startup their organizations and supply chains, complete training and like. A system should never go into service without formal maintenance teams in place. In some cases, initially provided by the system developer or an independent organization. In other cases, trained, customer-provided maintenance staff. In short, you should always view supply chain management as a strategic asset to the company and to the customer [9]. This involves:

1. Continually evaluating supply chain activities for efficiency and quality.
2. Detecting and recovering from supply chain issues early before they impact customers.
3. Adopting company policies for the supply chain of materials and establish strong company supply chain management systems.
4. Identify and assess risks related to circumstances in the supply chain (e.g., COVID-19).
5. Devise and adopt supply chain risk management plans in relation to risk assessments, including risk mitigation strategies [10].

Question to ponder

What are the advantages and disadvantages of LRUs that are major systems/subsystems vs. individual component replacement?

4.1.2 MDRE Supply Chain Management Considerations

Supply chain management is defined as the **design, planning, execution, control, and monitoring of supply chain activities with the objective of creating net value**, building a competitive infrastructure, leveraging worldwide logistics, synchronizing supply with demand, and measuring performance across the complete product life cycle. There is a definite life cycle to supply chain awareness, policy creation, and continued supply chain management control across each product/production line within a given company, which includes the vendor partners involved in each product [11]. Table 4.1 below illustrates some of the steps in the supply chain life cycle from initial awareness of supply chain management to full on supply chain management process control. For instance, it is reported that 94% of *Fortune 1000 companies* are experiencing disruption to their supply chains as a result of COVID-19. Yet that disruption is probably just the tip of the iceberg in terms of the virus' overall economic impact. In a recent news article, research firm *Oxford Economics* has warned that global economic growth could shrink by more than US$1 trillion now as a result of the World Health Organization declaring COVID-19 a pandemic.

Question to ponder

Investigate Covid-19's effect on supply chain in the food industry. Why is supply chain management so important?

Table 4.1 MDRE company supply chain management awareness life cycle

Category	Details
Supply chain management	Basic supply chain awareness by management team
	Need to eliminate fire-fighting and ad-hoc supply chain decision-making
	Basic supply chain information available
	Check 3 W rule for these differential, high-speed data or other sensitive signals
	Focussing on bringing the company's cost and procedures in line with good supply chain practices
Supply chain infrastructure	Management team(s)' awareness of supply chain management's potential
	A vision and plan for supply chain management created and reviewed
	Functional supply chain-focused change and establishment of policies
Supply chain cost reduction	Plans to work with supply chain suppliers and partners established
	Key goal becomes time and cost reduction along the supply chain
	IT investments targeted to improve supply chain efficiency
	Development of a culture of continuous process/product improvement
Supply chain collaboration	Supply chain suppliers and partners identified for collaboration
	Collaboration upstream and downstream established
	Business rules and metrics established between vendors and partners to improve supply chain performance
	Mutually beneficial supply chain improvements established
Supply chain strategic contribution to reduction of life cycle costs	Utilization of supplier/vendor inputs into company strategy
	Multi-program/projects processes implemented and managed using IT to improve communications and information exchange between company, suppliers, and customers
	Competitive advantage established due to efficiencies and innovative supply chain practices

4.1.3 MDRE Disruption Is the New Normal

Deeply worrying as COVID-19 is, epidemics are just one of the many disruptive events that can hit supply chains unexpectedly, to the detriment of business performance. Supply chain disruption is a continued risk to multiple systems, especially in our highly globalized parts supply chains where there commonly exists long, complex, supply chains [11]. Typically, classical supply chain structures are fundamentally ill-equipped to cope with increasing numbers of unplanned disruptions. It is imperative that organizations and programs like DREX+ have supply chain management plans and procedures in place, including a robust risk assessment and fault analysis system. Table 4.2 below illustrates an assessment of risks and the stages of design through development during which phases the risks are applicable. Fig. 4.1 shows a sample supply chain root cause analysis for problems within a risk management system for a given product [12].

No one could have predicted the full social and economic impact of the COVID-19 outbreak—or, indeed, of any event with global impact. Nevertheless, it has served as a reminder to the DREX+ team that the risk of an unexpected disruptive event is ever-present and if we want to continue to serve our DoD customers during a period of disruption, we need to be proactive in their planning [13]. These are steps we can take to help ensure the DREX+ supply chain is transformed in ways that help the process of building and delivering the DREX+ system functions effectively, even when stressed and stretched by unexpected global events. It is not simply about protecting profits. The resilience of the supply chains is critical to securing the health and well-being of commercial, military, and civil assets across the world.

4.1.4 MDRE Supply Chain Management System Implementation

So, how can the DREX+ team or any other development team boost the resilience of its supply chain? Fundamentally, they must move away from having rigid, linear supply chains to operating within agile, networked ecosystems by focusing on five key areas:

1. Assessment and strategy. Conduct an end-to-end supply chain risk assessment to stress test the supply chain, identify critical risk scenarios, and define potential responses [14].
2. Capability build-out. Invest in key supply chain capabilities, including visibility and monitoring, alternative business operating models, alternative supplier sourcing strategies, network flexibility, and agile planning [15].
3. Intelligence monitoring. Implement risk monitoring and reporting tools, as well as an early warning system that enables a rapid early response to risks or disruptions. Undertake new product risk assessments and look for changes in demand and supply. Conduct ongoing risk and controls assessments, including systems and facility risk and cyber reviews [16].
4. Operating procedures. Put in place a Plan B for disruptive events, covering operating procedures and responses to predefined supply disruption triggers, such as a natural disaster or terrorist attack. Work to ensure there is clear delegation of authority and decision-making and that external and internal communication protocols are in place [17].

Table 4.2 MDRE supply chain risk assessment across design/development/sustainment phases

Supply chain risk	Design	Plan	Source	Implement	Deliver	Sustain
Increase complexity	x	x	x	x	x	x
Variant diversity	x	x	x	x	x	x
Technology dependence	x		x			x
Long-term availability (end-of-life)	x		x			x
Transfer of expert knowledge	x		x	x		x
Delayed of insufficient information flows	x	x	x	x	x	x
Insufficient forecast accuracy/planning safety		x	x	x		x
Demand fluctuations		x	x	x		
Insufficient system supported processes/IT processes	x	x	x	x	x	x
Geographical location risks				x	x	
Political risks in procurement markets			x		x	x
Missing certifications/expert knowledge			x			x
Single parts sourcing	x		x			x
Procurement costs	x		x			
Flexibility		x	x			
Capacities			x	x		
Qualification		x	x			x
Long replenishment Lead times/delivery times		x	x			x
Allocation		x	x	x		
Process disruptions					x	x
Quality	x		x	x		x
Wastage				x	x	
Packaging			x	x	x	
Transport Damae/company emergencies			x	x	x	x

5. Major crisis management. Put in place a crisis management framework for major events where predefined responses will be inadequate. This should be accompanied by governance procedures, a desired operating model, and standard ways of working [18].

4.1.4.1 MDRE Supply Chain Management System Development Process

Figure 4.2 below illustrates the steps for development of a comprehensive supply chain management system for an ongoing program like the DREX+ system. Since we are aiming for deployment across a wide variety of applications, the DREX+ implementation, enhancement, and sustainment could be in place for 30+ years. A continually improving process is required for this type of supply chain life cycle. A checklist of supply chain steps that must be continually monitored during the entire supply chain management life cycle are:

Supply Chain Checklist:

1. Adapt product supply chain based on the needs of the customer [19].
2. Customize the product's logistic network for each customer.
3. Align demand planning across the product supply chain, so that all parties (customers and company) benefit by not keeping too much stock in hand.
4. Outsource strategically while keeping core competencies in-house.
5. Develop information technology that supports multi-level decision-making for product support and tie it to product business practices to fully understand any deficiencies and customer needs [20].
6. Adopt product service metrics using "Time-Driven Activity-Based Costing" to determine customer's overall supportability and affordability.

This is important because control of the configuration of components and subsystems—both hardware and software revision—is needed to ensure that when installed into the product, the product continues to perform as designed. This is particularly a problem with commercial off-the-shelf items—where the supplier has changed the Board Support Packages (BSP) that changed the system (e.g., processors) such that it is now possible the SW will not run. If the system contains COTS hardware and/or software, or a custom designed component or subsystem—changes in the suppliers' process for building the part—resulted in a high initial failure rate during product test. This can be especially problematic when you are purchasing relatively few numbers of parts from a supplier relative to the number of parts they are building—as a small customer, you will have little influence over your supplier.

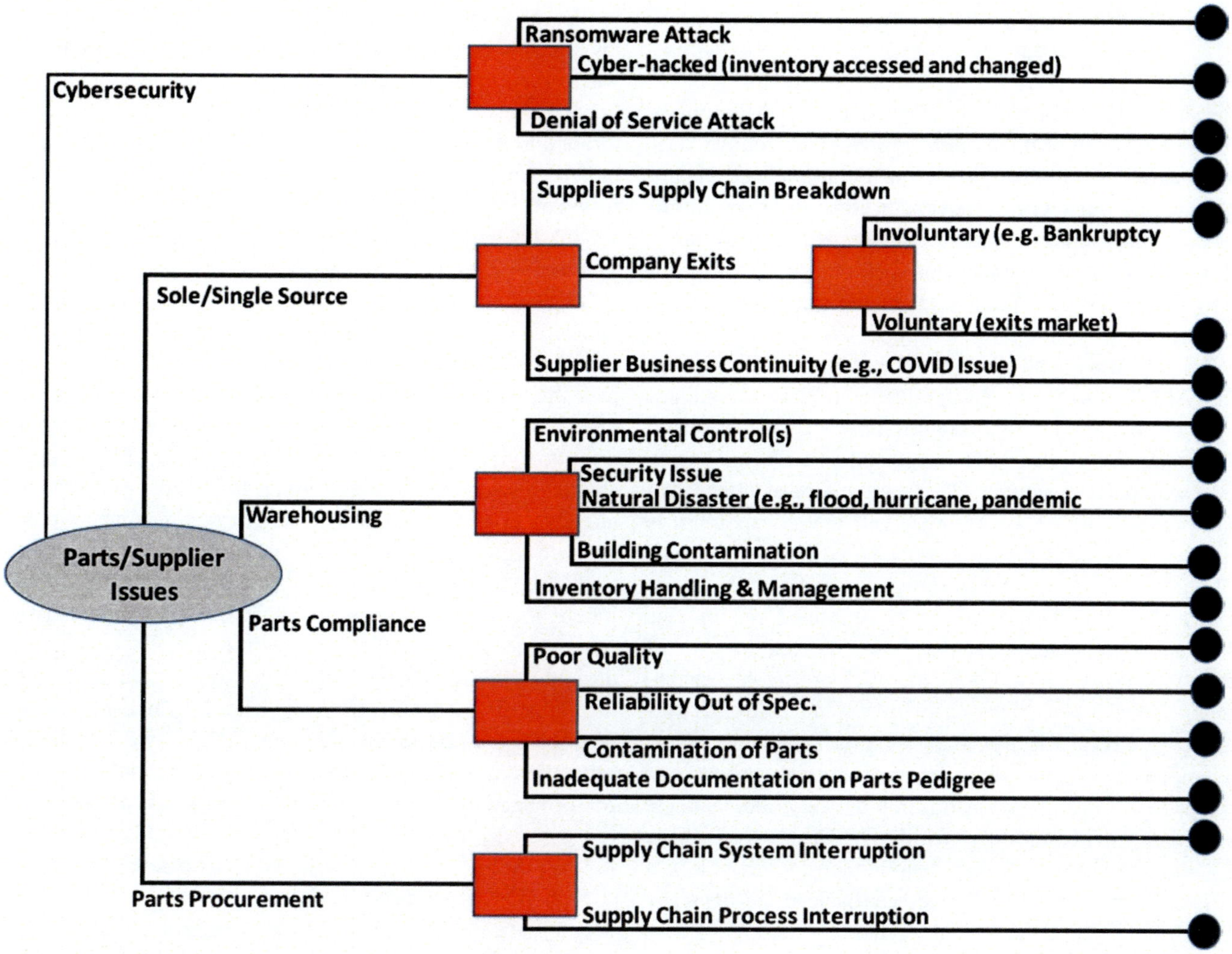

Fig. 4.1 Supply chain issues root cause analysis

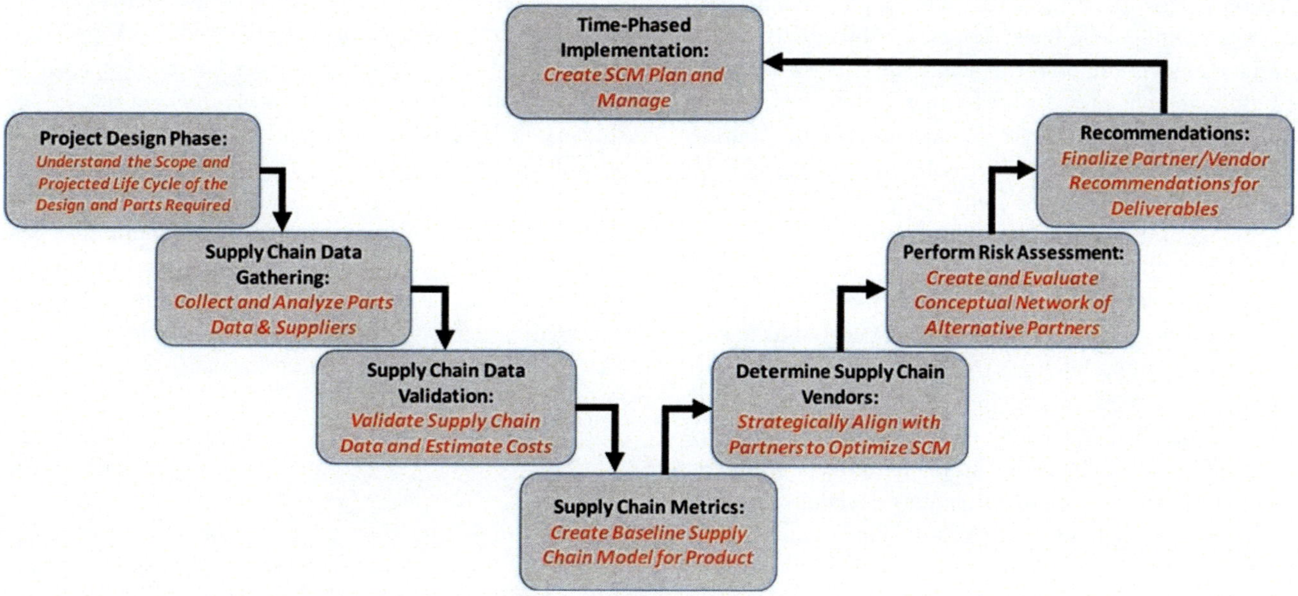

Fig. 4.2 MDRE supply chain management system development process

4.1.5 MDRE Supply Chain Management Critical Steps

One of the keys to improved performance which should be explored in for the MDRE DFM process is to reduce variability and forecast error through artificial intelligence and machine learning (AI/ML) advanced algorithms for autonomous supply chain forecast management. These algorithms will support multiple methods of short-term forecasting, as well as predictive analytics and prescriptive analytics [21].

The role of demand sensing and autonomous forecasting using AI/ML. AI/ML and the application of these techniques to prescriptive problem-solving provides only half the capability required to turn your supply chain network upside down with decisions driven by desired outcomes rather than stale process data [22]. The other half of the capability is provided by a set of robust network-based multi-party software/hardware/parts services across demand, supply, logistics, and fulfillment which can be tuned to operate automatically and autonomously, based on creating a set of desired process outcomes. Keeping track of time-based analytics is crucial for supply chain management (SCM) systems [23]. ML algorithms to autonomously monitor, track, and make predictions about future deliveries, issues, and forecasts can make or break on-time delivery and seamless sustainment for a complex system like DREX+. Unavailability of a single resistor or IC can delay either deliver of a new DREX+ system or repair of one sent back to the company depot from a commercial, civil, or military asset can cause severe difficulties in the operational system(s).

Once DREX+ has a functional supply chain management (SCM) system, a company can realize a real-time demand and supply network across all trading/supplier/vendor partners, which will enable the systems, like the DREX+, team to surface many more problems and issues than would have been discoverable previously. In fact, given the scale of today's supply chains and the number of monthly transactions, in parts procurement for multiple systems, like the DREX+, across a variety of application can exceed human capacity, requiring the SCM to take full advantage of automation technology using AI/ML algorithms. Once the AI/ML algorithms are deployed throughout the SCM, to make quick decisions, you must define decision constraints to keep snap decisions from de-railing the SCM. An example might be to not change parts suppliers/vendors without vetting them to ensure the parts quality meets standards or to ensure the parts are not coming from unreliable sources. Within the confines of SCM, constraints automating decision-making to resolve routine, repetitive decision-making reduces time, reduces cost, and enhances repeatability, reliability, and sustainability across the system life cycle [24].

The original prescriptive analytic optimization method, the simplex method, was limited to solving linear programs where the constraints and objectives could be modeled as linear expressions of real value variables. While a breakthrough at the time, this is still very limiting as many common business problems cannot be expressed using this formulation. We believe the SCM optimization problem should be approached using "meta-heuristics." Methods/techniques should be investigated like Simulated Annealing, Tabu Search, Ant Colony, Particle Swarm, and Genetic Algorithms. While these algorithms can model many types of optimization problems, much testing would be required to pick the appropriate one. In addition, machine learning has proven highly effective at predictive (as opposed to prescriptive) analytics [25]. The supply chain management team should investigate ML algorithms for simple prescriptive analytics or as an input into the more traditional prescriptive analytic algorithms. The SCM team should strive to establish an SCM solution architecture where the independent variable is the outcome you want to achieve, then adjust every other variable in the network (within reason) to give the desired result. Rather than adjusting variables like lead time, inventory, and capacity to the old levels which has led to poor performance in the past, the AI/ML overlay should drive the entire SCM network to achieve optimal operating performance while also achieving the targeted outcomes.

4.1.5.1 MDRE Supply Chain Management Metrics Development

Metrics are a way to measure performance, and, in turn, communicate that information to supply chain management team [26]. Their value is how supply chain management is supporting the direction and strategy of the business. It is important that they present a strategic and tactical understanding of what is happening and how well it is happening. Supply chain management is a process that flows across the SCM team and from parts/component suppliers through to customers and sustainment personnel. The challenge of a supply chain is the length; scope; geographic reach; number of internal and external stakeholders and participants; and overall complexity. One major factor to keep in mind as you look at the SCM is that supply chains are nonlinear, or at least not as linear as some project them. There are supply chains within supply chains. Viewing nonlinear supply chains as linear contributes to performance issues and to measuring operations [27].

Question to ponder

Why is it important to keep track of the health and status of suppliers in your system's supply chain?

Metrics, or KPIs (key performance indicators), must be measurable. As a result, numerous metrics are about the logistics components. While some of these are valuable, you must remember they do not present a complete view of the total supply chain. Assessing logistics parts is a node-link approach and does not recognize the supply chain process. Supply chain management is a process that is often measured on its logistics costs [28]. That approach is a root cause of issues. Such computations do not measure the supply chain or its performance and can include factors outside of the supply chain. In addition, the way accounting treats supply chain costs is typically dated. They go before supply chain management and global activity were recognized. For instance, for the DREX+ to be a useful system across multiple Navy missions and assets, the DREX+ supply chains consider useful metrics throughout the DREX+ life cycle. This means improved performance and new metrics [29].

Supply chain complexity raises the question of whether measuring the overall supply chain is sufficient [30, 31]. Segmenting can reflect similar supply chain characteristics, business unit, or other ways. Supplementing the macro performance with segmented KPIs provides understanding and insight to what is happening on both centralized (within the SCM team) and decentralized (across a complex and varied applications) view. It will enable seeing underlying factors to measure that are important to multiple system operations. At the same time, too many metrics become measures for measures sake. Below outlines four of the SCM metrics than one should consider.

4.1.5.2 MDRE SCM Inventory Velocity

Inventory is a hot button with the dual challenge of having capital tied up and while being able to service customer orders. Aligning inventory and the supply chain network is an additional challenge. Moving inventory more quickly through the supply chain has become a requirement. This is also important with inventory planning and with being able to respond to positioning products [32]. This will become increasingly important for systems like the DREX+ as more and varied uses for the system across various applications become apparent. It is important for lean and the waste of excess inventory. In many of these cases, inventory is stationary, not moving dynamically across the supply chain. Speed increases the value of inventory, while reducing working capital. It is critical with aligning networks, positioning inventory, and satisfying overall sustainability and manufacturability requirements.

If inventory control is outdated, decision-making will not likely meet the customers' expectations. For the DREX+ system, SCM is pivotal in achieving the many forms of system, board-level, and part-level velocity [33]. Also, inventory management, as traditionally understood, has been replaced by alignment and by velocity. Velocity can be measured as turns or days of inventory. It can be applied to finished systems, e.g., a complete DREX+ system; WIP (work in process), especially that is transferred to another location, e.g., to a third-party vendor for board assembly and checkout; and raw material, e.g., individual resistors or capacitors. Segmenting as to division, product category, or other relevant ways for SCM is important.

4.1.5.3 MDRE SCM Time Compression

This KPI ties to inventory velocity and is an integral part of the providing the immediacy the customer wants and requires. Time is a hidden waste both inside and outside the company and is an enemy of supply chains. Reducing it is important with the new business reality. Remember, time is important with building inventory velocity and minimizing inventory waste [34]. There are two points here. One is that the largest time sector is with the inbound supply chain, especially if there are multiple sourcing companies involved (e.g., DREX+ has multiple vendors). The other, and often overlooked, is the non-movement part. That is a fixed block of time that is not compressed, unless it is recognized. An example would be the time to "bake-in" a circuit board. You cannot "accelerate" the bake time and still produce a high-quality printed circuit board.

4.1.5.4 MDRE SCM Perfect Order: The Customer

This is an outstanding metric. This KPI comes from the SCOR (Supply Chain Operations Reference) model, measuring accuracy (performance measures), completeness (meets all the requirements), and timeliness (schedule adherence) of the delivery of the system. This metric is about the customer (e.g., in the DREX+ case, the Navy). It validates the customer mantra. This is what customer service is—delivery of systems, complete, accurate, and on time. It is implicit in doing business with companies for the products like the DREX+ system. But, as simple as that sounds, compliance is difficult; the issues are in the details [35].

4.1.5.5 MDRE SCM Perfect Order: The Supplier

The supplier (company) is at the opposite end of the supply chain from the Navy and the perfect customer order. This metric then creates a yin and yang. Supplier (both the company for complete systems and vendors for parts to make the system, e.g., DREX+) performance—purchase orders delivered complete, accurate, and on time—is very important to supply chain success. This is a fundamental metric [36]. Supply chain performance begins on the inbound side with suppliers. The impact of weak supplier accomplishment ripples throughout the supply chain and impacts the actuality and fundamentals, both operational and financial, of the company as well as mission success for the applications utilizing systems like the DREX+. We will provide examples of SCM-related activities associated with the DREX+ system to provide how these techniques play out in real-world products.

4.1.5.6 MDRE SCM Metrics Discussion Example: DREX+

The four metrics described above have been instantiated within the DREX+ SCM system because they encompass the entire DREX+ supply chain and its process. Success with them mitigates DREX+ failures and problems with procurement, growth, profitability, and sustainment success. They are inter-related, connected, and bring cohesion to the DREX+ SCM, what it does, and how it does it. These define it and provide a way to see the intricacies and convolutions which can be lost in the daily happenings. They recognize what can be viewed as unrelated parts of the supply chain—when they are not [37]. The challenge to further improving DREX+ SCM performance is ongoing adherence to these metrics, throughout the DREX+ life cycle. Success lies at the macro and granular levels. Some of the work ahead for the DREX+ SCM that should be explored in the future and is useful for any SCM system includes:

- Focusing and improve the DREX+ SCM process.
- Increasing the visibility across the supply chain.
- Integrate the financial supply chain with the product supply chain.
- Align the inventory network.
- Extending the supply chain upstream as far as possible.
- Implement advanced integration of process and technology like AI/ML techniques.

4.1.6 MDRE Supply Chain Management Applications

To ensure quality and repeatable processes within the company's SCM process, many steps are taken to ensure manufacturability throughout the design, implementation, testing, and delivery process [38]. Here we discuss specific processes and procedures that should be implemented within a company's SCM system.

4.1.6.1 MDRE SCM Traceability

For all parts procured for systems, including the DREX+ system, components should be sourced from vetted Tier 1 distributors and OEMs with CoCs (Certificate of Compliance) and Data/Lot code information to ensure counterfeit parts and components are not utilized in the printed circuit board (PCB) assembly and systems implementations throughout the SCM process. Component traceability can be generated down to the component reference designator on each PCB assembly.

Question to ponder

Why would part/component traceability be important, say in a heart pacemaker?

4.1.6.2 MDRE SCM Parts Compatibility

To increase the sustainability/maintainability of systems like the DREX+ and reduce the overall life cycle cost of manufacturing, delivering, and servicing each system, each board, component, or subsystem within the deployable system should use similar parts for similar functions. For the DREX+ system, this entailed using similar parts across the three boards that make up the DREX+ system, the Lightening II, Thunder II, and Crestone boards. All three boards use identical parts for similar functions. An example is shown in Fig. 4.3. Figure 4.3 illustrates the parts for all three boards that are procured from Analog Devices, Inc. (ADI). Even though all three boards were designed separately, similar functions utilize the same ADI part. An example is the Multiplexer 1X1632. As you can see from the Bill of Materials (BOM) from each board, they all three utilize the ADI ADG1606BCPZ-REEL7 component. Similar results can be seen by scanning the information shown in Fig. 4.3.

Similar results are illustrated in Fig. 4.4 for the Texas Instruments components utilized across the three boards. From Fig. 4.4 you can see that similar functions utilize the same part from the same vendor, reducing the overall life cycle cost from the initial purchase of components to the spares that must be kept on-hand for maintenance and sustainability.

4.1.6.3 MDRE SCM PCB Manufacturing and Assembly

To increase the overall quality and manufacturability of the systems like the DREX+ system, companies should consider utilizing a certified vendor for board fabrication and assembly. Such companies utilize traceable processes that increase the overall repeatability of the systems like the DREX+ manufacturing process. These are outlined below and discussed in detail in Sect. 4.2.

4.1.6.4 MDRE SCM Front-End Production Processes

To ensure repeatability of the manufacturing process, upon receipt of the required PCB component data, board manufacturing, assembly and testing companies conduct the following checks [39]:

- Verifies that all necessary data is present to ensure accurate and repeatable production results.

Thunder II – Analog Devices Inc. Components

ceipn	Description	Designator	Quantity	mfr1	mfr1_pn
400-000078-01	IC - MULTIPLEXER 1x16 32 LFCSP	U110	1	Analog Devices	ADG1606BCPZ-REEL7
400-000159-01	IC - 3.2 GHz GENERAT+DISTRI	U76	1	Analog Devices	HMC7044LP10BE
405-000013-01	IC - VREF 2.048V 0.24% SOT-23	U23	1	Analog Devices	ADR380ARTZ-REEL7
405-000035-01	IC - REG 2A ULTRALOW NOISE	U79, U81, U88	3	Analog Devices	ADP7159ACPZ-04-R7
405-000037-01	IC - REG LINAR 3A 16LFSCP	U80, U82, U83, U84, U85, U86, U89, U91	8	Analog Devices	ADP1763ACPZ-R7
407-000005-01	IC - DAC 16 BIT 12.6 GSPS 2CH	U59, U60, U61, U62	4	Analog Devices	AD9172BBPZ
451-000019-01	IC - Sniff Sense PQFN24	U120	1	Analog Devices	ADPD188BI

Lightning II – Analog Devices Inc. Components

ceipn	Description	Designator	Quantity	mfr1	mfr1_pn
400-000078-01	IC - MULTIPLEXER 1x16 32 LFCSP	U115	1	Analog Devices	ADG1606BCPZ-REEL7
400-000120-01	IC - 3.2 GHz 14 Output Fanout Buffer	U74	1	Analog Devices	HMC7043LP7FE
400-000159-01	IC - 3.2 GHz GENERAT+DISTRI	U79	1	Analog Devices	HMC7044LP10BE
405-000013-01	IC - VREF 2.048V 0.24% SOT-23	U23	1	Analog Devices	ADR380ARTZ-REEL7
405-000035-01	IC - REG 2A ULTRALOW NOISE	U76, U80, U81, U88, U90, U91, U93, U95	8	Analog Devices	ADP7159ACPZ-04-R7
405-000037-01	IC - REG LINAR 3A 16LFSCP	U82, U83, U84, U85, U86, U87, U92, U96	8	Analog Devices	ADP1763ACPZ-R7
406-000003-01	IC - 14 BIT 3 GSPS 2 CH ADC	U58, U59, U60, U61	4	Analog Devices	AD9208BBPZ-3000
451-000019-01	IC - Sniff Sense PQFN24	U120	1	Analog Devices	ADPD188BI

Crestone – Analog Devices Inc. Components

ceipn	Description	Designator	Quantity	mfr1	mfr1_pn
400-000078-01	IC - MULTIPLEXER 1x16 32 LFCSP	U115	1	Analog Devices	ADG1606BCPZ-REEL7
400-000120-01	IC - 3.2 GHz 14 Output Fanout Buffer	U74	1	Analog Devices	HMC7043LP7FE
400-000159-01	IC - 3.2 GHz GENERAT+DISTRI	U79	1	Analog Devices	HMC7044LP10BE
405-000013-01	IC - VREF 2.048V 0.24% SOT-23	U23	1	Analog Devices	ADR380ARTZ-REEL7
405-000035-01	IC - REG 2A ULTRALOW NOISE	U76, U80, U81, U88, U90, U91, U93, U95	8	Analog Devices	ADP7159ACPZ-04-R7
405-000037-01	IC - REG LINAR 3A 16LFSCP	U82, U83, U84, U85, U86, U87, U92, U96	8	Analog Devices	ADP1763ACPZ-R7
406-000003-01	IC - 14 BIT 3 GSPS 2 CH ADC	U58, U59, U60, U61	4	Analog Devices	AD9208BBPZ-3000
451-000019-01	IC - Sniff Sense PQFN24	U120	1	Analog Devices	ADPD188BI

Fig. 4.3 Comparison of ADI Components utilized across the Lightning II, Thunder II, and Crestone PCBs

Thunder II – Texas Instruments Components

211-000005-01	CR - 2 CHANNEL PROTECTION USB 3.0	U75, U77	2	Texas Instruments	TPD2EUSB30DRTR
400-000077-01	IC - CONFIG MULT-FUNC GATE SC70-6	U103	1	Texas Instruments	SN74AUP1G57DCKR
402-000001-01	IC - OPAMP R-R 3MHZ SOT23-5	U112	1	Texas Instruments	TLV271IDBVR
405-000018-01	IC - DDR TERMINATION REGULATOR	U117	1	Texas Instruments	TPS51200DRCR

Lightning II – Texas Instruments Components

ceipn	Description	Designator	Quantity	mfr1	mfr1_pn
211-000005-01	CR - 2 CHANNEL PROTECTION USB 3.0	U73, U77	2	Texas Instruments	TPD2EUSB30DRTR
400-000077-01	IC - CONFIG MULT-FUNC GATE SC70-6	U110	1	Texas Instruments	SN74AUP1G57DCKR
402-000001-01	IC - OPAMP R-R 3MHZ SOT23-5	U117	1	Texas Instruments	TLV271IDBVR
405-000018-01	IC - DDR TERMINATION REGULATOR	U123	1	Texas Instruments	TPS51200DRCR

Crestone – Texas Instruments Components

ceipn	Description	Designator	Quantity	mfr1	mfr1_pn
211-000005-01	CR - 2 CHANNEL PROTECTION USB 3.0	U6, U72, U77, U78	4	Texas Instruments	TPD2EUSB30DRTR
400-000077-01	IC - CONFIG MULT-FUNC GATE SC70-6	U91	1	Texas Instruments	SN74AUP1G57DCKR
400-000079-01	IC - BUS TRANSCVR TRI-ST 2B US8	U39, U41	2	Texas Instruments	SN74AVC2T45DCUR
402-000001-01	IC - OPAMP R-R 3MHZ SOT23-5	U89	1	Texas Instruments	TLV271IDBVR
402-000008-01	IC - HIGH SPD DUAL WIDEBAND VOLTAGE FDBCK	U88	1	Texas Instruments	OPA2690IDG4
405-000018-01	IC - DDR TERMINATION REGULATOR	U73, U100, U126	3	Texas Instruments	TPS51200DRCR
450-000004-01	IC - COMPARATOR HS LVDS	U5, U128	2	Texas Instruments	LMH7220MK/NOPB

Fig. 4.4 Comparison of Texas Instruments Components utilized across the Lightning II, Thunder II, and Crestone PCBs

- Runs Design Rule Check (DRC) on the company-provided Gerber Data.[1]
- Performs a Bill of Materials (BOM) scrub for the following:
  - Part number to part description discrepancies.
  - Part availability and/or long lead times.
  - Part listed in the Bill of Materials (BOM), but not present in the Gerber data or vice versa.
- Analysis of component size to Printed Circuit Board (PCB) footprint in accordance with the Institute of Printed Circuits (IPC) and Original Equipment Manufacturer (OEM) requirements.
- Validation of component polarity based on provided schematics.

[1]The Gerber data format enables data exchange between CAD development and CAM production and is used primarily to output the layout data used for Printed Circuit Boards (PCBs).

- Concerns regarding assembly or compliance with the system requirements.

4.1.6.5 MDRE SCM Sustainability

Overall, the sustainability of the SCM process will affect the overall logistics, maintainability, and overall supportability of a system across applications. Material (LRU) logistics planning will be crucial to the sustainability across systems, especially those with diverse geographical or environmental considerations [40]. Each system and its application must be evaluated, based on the criticality, environment, use, and priority within operational usage for the specific application. This will require a Supply Chain Roadmap (SCR) be created that illustrates the use of system and the logistics that must be put in place to sustain and maintain the system for the anticipated life cycle of each application.

4.1.7 MDRE PCB Assembly Analysis

Since most modern systems have hardware and software as part of their creation, we will spend some time discussing manufacturability in terms of assembly of the PCBs. The PCB design, assembly, and test procedures are critical for the overall manufacturability success of a product like DREX+. Design features that make the boards difficult to assemble and test add to the overall life cycle cost and impact the supportability of the product, both from a production cost and a life cycle cost due to scrap rate of the PCBs [41]. Design for assembly and the overall design for manufacturability is necessary for complex boards like those utilized in the DREX+ system. A thorough review process is necessary to identify and eliminate assembly and test difficulties ahead of the assembly and manufacturing to reduce part cost and minimize risk during PCB assembly.

4.1.7.1 MDRE Design for Assembly (DFA) Analysis

DFA should go beyond the simple Design Rule Check (DRC) process by adding levels of interpretation of the design's effect on the assembly process. Items to consider when putting a DFA process in place are fabrication notes from the board manufacturer, machining views, and possible customer requirements that fall outside normal processing and testing procedures. Studying such requirements and procedures that fall outside the scope of a standard automated DRC enables changes that can reduce the complexity and increase the efficiency of the PCB assembly and test process [42]. Improving assembly efficiency and accuracy increases first pass test yield, repeatability, decreases failure rates, and increases the overall supportability and manufacturability of the PCBs, and thus, the overall manufacturability of the systems like the DREX+.

> **Question to ponder**
>
> Think back to the early days of Mass Production. Where do you think the notion of Design for Assembly came from? Compare the concept to Henry Ford's assembly line concept.

DFA is becoming increasingly important for the PCB design. The scope of the DFA is significant and involves both the bare PCB fabrication/testing and subsequent component assembly and test. As a result, DFA for PCB design is an intricate interaction between Design for Fabrication (DFF), Design for Assembly (DFA), and Design for Manufacturability (DFM) requirements. Examples of DFF checks include an "acid etch trap" where an area on the PCB may retain etchant longer than intended, resulting in open connections on the board. Another example of a DFF issue is shown in Fig. 4.5 and illustrates what is called a design "silver," which is susceptible to detaching during the overall assembly flow, with the possibility of re-depositing and shorting distinct connections [43]. Other examples include a trace routed through a solder mask opening created for a test point, or an analysis of the board stack-up thickness to via size aspect ratio to evaluate the drilling requirements and costs. These layout topologies are DRC clean, but represent additional risks during volume PCB manufacturing and assembly.

PCB manufacturers utilize a "bed of nails" tester to test the etch traces on multi-layer boards to ensure internal trace continuity and to check for short circuits between etches. For high volume PCBs, this bed of nails tester is also utilized after the passive components—resistors, capacitors, inductors—have been installed to check that proper values have been installed. This provides an "early defect" detection

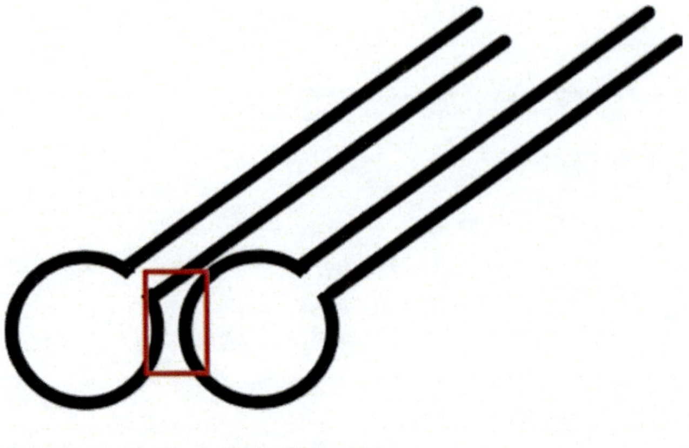

Fig. 4.5 Example of a PCB "Silver" issue

capability which will increase first pass yield on completed circuit boards, as well as reduce troubleshooting time of fully assembled boards.

In summary, the complexity of modern PCB design, manufacturing, and assembly processes means there are numerous opportunities for PCB failure. Some of these are the result of design oversite, such as insufficient clearances or incorrect measurements, which negatively affect the functionality of the final PCB, and ultimately the system (e.g., DREX+). Other issues may result from manufacturing processes such as drilling errors or over-etching.

4.1.8 MDRE Design for Assembly (DFA) Implementation

4.1.8.1 MDRE Design for Assembly (DFA) Background

Geoff Boothroyd, supported by an NSF grant at the University of Massachusetts Amherst, developed the Design for Assembly method (DFA) in 1977, which could be used to estimate the time for manual assembly of a product and the cost of assembling the product on an automatic assembly machine [44]. Recognizing that the most important factor in reducing assembly costs was the minimization of the number of separate parts in a product, he introduced three simple criteria which could be used to determine theoretically whether any of the parts in the product could be eliminated or combined with other parts. These criteria, together with tables relating assembly time to various design factors influencing part grasping, orientation, and insertion, could be used to estimate total assembly time and to rate the quality of a product design from an assembly viewpoint. For automatic assembly, tables of factors could be used to estimate the cost of automatic feeding and orienting and automatic insertion of the parts on an assembly machine. In the 1980s and 1990s, variations of the DFA methods have been proposed, namely: the GE Hitachi method which is based on the DFA; the Lucas method, the Westinghouse method and several others which were based on the original DFA method. All methods are now referred to as *design for assembly* methods [45]. Most products are assembled manually and the original DFA method for manual assembly is the most widely used method and has had the greatest industrial impact throughout the world.

The DFA method was originally made available in the form of a handbook where the user would enter data on worksheets to obtain a rating for the ease of assembly of a product. Starting in 1981, Geoffrey Boothroyd and Peter Dewhurst developed a computerized version of the DFA method which allowed its implementation in a broad range of companies. There are many published examples of significant savings obtained through the application of DFA. For example, in 1981, Sidney Liebson, manager of manufacturing engineering for Xerox, estimated that his company would save hundreds of millions of dollars through the application of DFA [46]. In many companies DFA is a corporate requirement and DFA software is continually being adopted by companies attempting to obtain greater control over their manufacturing costs. There are many key principles in design for assembly [47]. To illustrate its use on real systems, the DFA techniques applied to the DREX+ assembly process are outlined below.

4.1.8.2 MDRE Design for Assembly (DFA) Applied to the DREX+

The PCB assembly production process for the DREX+ system includes the following:

- A single production unit for each of the three DREX+ boards is run manually prior to running a full-scale production to verify the pick and place[2] (PNP) software programming and reflow profiles for each DREX+ board. This applied to both the top and bottom sides of each DREX+ PCB.
- Solder paste is inspected both 2D and 3D to ensure consistency.
- Reflow profiles are generated for each specific assembly for each DREX+ PCB and the profile is stored for future DREX+ PCB assemblies.
- PnP machines are configured to electrically test passive components to ensure the parts used for the DREX+ PCB match the defined Bill of Materials (BOM) for each DREX+ PCB.
- DREX+ assembly documentation is created for each PCB, distributed, and then electronically stored to ensure the assembly instructions are up to date and easily accessible for future DREX+ builds.

4.1.8.3 MDRE Design for PCB Inspection and Verification

Design for Inspection (DFI) is an engineering principle that proposes that inspection methods and measurement instruments used to certify manufacturing conformity should be considered early in the design of products. Production processes should be designed in such a way that features of the product are easy to inspect with readily available measurement instruments, and so that measurement uncertainty is considered in the tolerance that are applied. The concept can be applied in almost all engineering disciplines. DFI describes the process of designing or engineering a product to facilitate the measurement to reduce the overall costs of

[2] Pick and place machines are a key element of any PCB assembly line enabling components to be automatically placed on a printed circuit board quickly and accurately.

manufacturing and delivering products that satisfy customers [48].

Question to ponder

If a system is not designed for verification, how might it affect the ability to guarantee a system's reliability?

The role of inspection in the manufacturing process is to ensure that the manufacturing process is producing components that meet the specification requirements. Inspection does not assure the quality of the product, only a robust and repeatable manufacturing process can achieve this. Therefore, inspection is often considered as an overhead although an extremely important one. Like Design for Manufacture (DFM) and Design for Assembly (DFA) (which seek to avoid designs which are difficult to make), the concept of DFI considers measurement capabilities at an early stage in the product development life cycle and uses knowledge of the fundamental principles of metrology to achieve cost reduction [49]. If the inspection method and instruments are considered and selected at the design stage, the likelihood that a tolerance feature cannot be inspected or requires a specialized instrument is substantially reduced. High precision features require specialized manufacturing and metrology; these can have limited availability in the supply chain and therefore often have increased cost [50]. The concept of DFI should complement and work in collaboration with DFM and DFA. There are three key areas when considering DFI, datum selection, tolerances and accessibility, plus general metrology considerations. Getting the most from inspection techniques will help improve quality. It is still difficult for systems designers to build machines that allow finished products to be inspected easily. To do so requires an understanding of the product being manufactured and how inspection tasks can improve the quality control process [51].

Inspection can represent a significant percentage of an existing product's manufacturing cost. DFI may naturally be called for in redesign of a product to reduce that cost component. However, DFI will not always reduce inspection costs: it can also lead to increased rate of inspection, because more convenient or higher quality measurement may justify increasing measurements, say from a sampling rate satisfactory to support a basic level of tolerance to a higher rate (e.g., to 100%). Or DFI may make it economical for 100% inspections to measure more features or to make repeated measures of the same feature at different points within the manufacturing process. This would be justified if it would reduce internal failure costs (such as costs of rework or scrap) or external failure costs (such as customer returns) within the cost of quality framework [52].

Design for Verification (DFV) is a set of engineering guidelines to aid designers, assemblers, and testers in ensuring right first-time manufacturing and assembly of complex PCB assemblies and components. The guidelines were developed as a tool to inform and direct designers during early-stage design phases to trade off estimated measurement uncertainty against tolerance, cost, assembly, measurability, and production requirements [53]. The DFV framework was designed and developed specifically for low rate, high-value, complex manufacturing processes, like the DREX+, to enable system design engineers to design and plan for effective verification procedures for high-value[3] systems to reduce failure rates and life cycle costs, improve manufacturability integrity and efficiency, and increase the overall quality and conformance to customer requirements [54].

4.1.8.4 MDRE DFA Quality Inspection/Verification Process

- The single production unit assembled as part of the production process (described above) is inspected for each side of the assembled board.
- Full 2D and 3D automated optical inspection (AOI) is performed on all assembled PCBs.
- A sampling of PCBs is taken with parts containing hidden solder joints for 2D and 3D X-ray inspection to ensure quality of all solder joints.
- Visual inspection of each PCB is performed by an IPC trained and certified inspection at in-process and final quality control gates for each PCB.

4.1.8.5 MDRE DFA Quality on-Site Inspection/Verification Traceability

PCB Traceability: Each PCB is given a unique serial number to allow traceability throughout the PCB assembly and test process. All records are stored and retained for inspection by customer personnel if required.

Component Traceability: Verify that all sources for components are sources from Tier 1 distributors and OEMs with CoC (certificate of compliance) and date/lot code information. This is to ensure counterfeit part mitigation throughout the DREX+ production process. Component traceability can be provided down to the reference designator on each PCB, if required by customer personnel. Reports (AS9102 format) should be generated if required.

[3] Here – high-value refers not only to the original cost of the unit (e.g., DREX+) but also for the high-cost to the customer if the unit fails during critical operations (e.g., aircraft carrier long-range radar system).

4.2 MDRE Manufacturability Analysis

4.2.1 Design for Manufacturability (DFM)

Design for Manufacturability (DFM) is the engineering practice of designing products to decrease the complexity of the manufacturing process. The concept is widely utilized, but the implementation differs widely depending on the domain of the product and manufacturing process and technologies. DFM describes the process of designing products to facilitate the manufacturing process to reduce manufacturing costs and reduce the overall test and verification of the product. DFM allows potential manufacturing problems to be identified and fixed in the design phases of a product where it is most cost effective to solve issues rather than in the middle of production [55].

Fig. 4.6 Design for manufacturability process flow

4.2.1.1 MDRE DFM and System Design Process

Ultimately, DFM is about designing a system or process to achieve the goals of cost effectiveness, high quality, reduced production schedule, and, in the end, reduced overall life cycle costs for a fielded system. Utilizing proven and the development of more advanced DFM processes and practices can go a long way to ensure quality, reduced delivery lead-times, and reduced long-term life cycle costs. DFM analysis is becoming increasingly important in the printed circuit board (PCB) design domain and is crucial for systems like the DREX+. The scope of DFM analysis is significantly different for PCB design than for other systems. PCB manufacture involves both bare board fabrication/test and subsequent component assembly, verification, and test. As a result, the DFM analysis for the PCB designs is an intricate interaction between Design for Fabrication (DFF) and Design for Assembly (DFA) requirements [56]. Unlike the pass/fail nature of DRC rules, manufacturability processes to support PCB DFM requires additional features. In addition to simply fixing DFM rule violations, PCB layout designers should flag DFM results, assign a severity, and generate reports for broader review by the engineering design team [57].

4.2.2 Design for Manufacturability (DFM) Example: DREX+

All the previous discussions, from Design for Assembly, Design for Inspection, Design for Verification, to Supply Chain Management and Design for Supportability all play into Design for Manufacturability (DFM) and play into the overall life cycle cost of a product. The engineering processes to enable Design for Manufacturability required a multidisciplinary approach of active participant early in the design process to build in (during design) rather than bolt on (late in the manufacturing process) manufacturability for a system like the DREX+. We looked at the structure of the team representing board designers, RF designers, industrial designers, logistics experts, manufacturing engineers, quality control professionals, procurement specialists, and compliance experts (a given person may meet more than one of these) who all bring their collective subject matter expertise and experience to the table. This is required to support the creation of functional designs that optimize quality, reliability, maintainability, and manufacturability at the lowest life cycle cost possible [58]. The overall DFM process flow is shown in Fig. 4.6.

DFM should be a feedback-driven process to accommodate issues discovered at every level in the design, implementation, deployment, and support phases throughout the life cycle of a system. Modeling and Simulation is an important step in the DFM life cycle, for it provides a valuable look into the design and how it meets requirements, even if it is an idealized environment. Modeling and Simulation can provide valuable insights into how the overall system will respond to inputs, but no simulation can cover all situations that a system will encounter in the actual operational conditions. This is particularly true for Naval applications where the environment can radically change many times throughout a systems life cycle, over and over (heat, humidity, vibrational characteristics, etc.). The feedback-driven DFM life cycle is crucial to the successful deployment and operation of Naval systems. When requirements change, as they often do across a large development program, the Life Cycle Cost Analysis (LCCA) should be revisited to see how the changes in requirements affect the overall life cycle of the system. Changes in requirements can radically change the supportability of a system from the original design and affect the DFA, DFI, DFF, and RAM considerations [59].

4.2.2.1 MDRE DFM Best Practices

As part of the overall DFM process, besides the right engineering disciplines, DFM best practices are required to guide the overall design, implementation, assembly, test, deployment, and sustainment of any system, including the DREX+ system. What follows are important, but not all-inclusive best practices [60].

Minimum Parts Count: Fewer parts mean a reduction in the intensity of the board design, implementation, and test of the DREX+ PCBs. In addition to less complexity, it also means a reduction in inventory and a reduction in handling and assembly [61].

Designing Parts for Multiple Purposes: As described in above, similar functions required across PCBs within the systems like the DREX+ utilize similar parts. Understanding the requirements across the entire system during initial design of the system allows designers to incorporate features that facilitate commonality across the PCBs and reduce inventory, procurement costs, sustainability, and maintainability; thereby reducing overall life cycle cost [62].

Developing a Modular Design: A modular design approach simplifies activities such as inspection, test, assembly, and verification, but also adds versatility to easily create product variations. An example is discussed above with the differences between the Navy long-range and border patrol surveillance radar designs, the Navy's design utilized a direct RF configuration and the Sentry design utilizing a superheterodyne approach without a redesign of the DREX+ boards [63].

Use Standard Components and Materials when Feasible: Utilizing standard components, where feasible, simplifies the supply chain process and simplifies the reliability calculations, given that standard components normally have established reliability factors.

Design for Ease of Fabrication: Selection of the best combination of material and fabrication processes results in a reduction of overall manufacturing costs. Tight scrutiny of tolerance requirements and identification when a looser tolerance is feasible on certain features in the design allows a reduction in overall manufacturing costs while still meeting overall requirements. Excessively tight tolerances may lead to higher setup costs and increased rejection rates as well as longer lead times. Evaluation of final operational requirements are necessary for functionality and compliance [64].

Minimize and Simplify Assembly: PCB parts should be designed such that they can be positioned, oriented, and assembled only one way. This avoids errors during assembly operations (Pick and Place) due to variations in part dimensions or the accuracy of the positioning device. It may also reduce the programming required to get accurate and repeatable assembly of the PCBs. The more symmetrical the parts, the easier to program and install. Conversely, if a part cannot be symmetrical, it is better to design it to be very asymmetric to make the orientation obvious and easy to get correct [65].

Question to ponder

What could be an adverse effect on sustainability if minimizing parts count were not part of the overall design decisions for a system?

4.2.2.2 MDRE DFM Frame of Reference

Understanding the overall operational system environment is essential to understanding how to decompose the system requirements, need, and goals into a system architecture. Based on the environment, the requirements, the performance needs, the external interfaces, the concepts of operations, constraints, and other applicable design considerations, a basic set of reference frames should be developed that guide the architecture design, system development, transition, and operations of the system [66]:

4.2.2.3 MDRE DFM Logical Analysis

During Logical Analysis, the required system functionality and overall behavior is analyzed. This analysis defines, selects, and designs the logical architecture required to create a Logical Framework from which design can be assessed for compliance with all system requirements for all operational scenarios required and for long-term operations. The Logical Analysis will include Trade-off analyses between requirements and proposed solutions. This Logical Analysis will include creation of Functional Models (functional decomposition) of the system requirements, Semantic Models that describe the relationships between information and data required within the system (e.g., Data Flow Diagrams, Entity-Relation Diagrams), and Dynamic Models that describe state transition diagrams and activity required to describe and meet the overall system requirements. The Logical Architecture that flows from this analysis is a set of technical concepts and principles that support the logical operations of the systems [67]. Improper handling of the Logical Analysis and subsequent Logical Architecture is the reason for many system design and implementation failures. Some of the common problems with improper Logical Analysis are outlined in Sect. 4.2.3 below and issues that drive redesign are highlighted.

4.2.3 Design for Manufacturability (DFM) Root Causes of Redesign

The discussion below outlines some of the root causes of redesign throughout the life cycle of system development. However, one major factor that plays into the overall redesign process is how the organization handles Change Management (CM). CM is a collective term for all approaches to prepare, support, and organizations in making changes to systems throughout the development and deployment stages. Drivers of change may include the ongoing evolution of technology, internal reviews of processes, crisis response, customer change demands, and competitive pressure. It includes methods that redirect or redefine the use of resources, engineering process, budget allocations, or other modes of operation that significantly change a company or organization. Change management refers to how people and teams are affected by such design changes within a system development. It deals with many different disciplines, including information technology and engineering solutions [68]. The feedback-driven DFM process illustrated in Fig. 4.6 defines this process where success in seen in terms of:

- An overall design process, not just a product.
- The interrelationship between each step in the DFM process.
- The balance between each engineering discipline, making sure all aspects of the design are equally considered throughout the system life cycle (e.g., security engineering, systems engineering).
- The overall level of teamwork and engineering skills, including communication.

4.2.3.1 Insufficient Performance Margin(s)

Insufficient performance margin can result in either failure in the operational environment or failure during stress testing and verification. Insufficient performance margin arises when design requirements are not matched to existing supplier or manufacturing capabilities. This may result with the design is too sensitive to manufacture variations or the design is difficult to fabricate and/or the design is difficult to assemble. This can cause costly redesign or a change in requirements which could drive utilization of expensive manufacturing processes to achieve performance goals.

Question to ponder

Why would insufficient performance margins be an issue? Can you think of examples?

4.2.3.2 Projected Life Cycle Cost Too High

A life cycle cost analysis (LCCA) should be performed on the original design specifications to assess the projected total ownership cost (TOC) to the customer for the system being designed. LCCA is useful when there are alternative designs for the system that fulfill the same performance requirements but may differ in initial cost, ease or difficulty in board manufacture and ease or difficulty in assembly and test. The LCCA should be performed early in the design phase while there is still the possibility of refining the design to ensure the lowest life cycle cost to the customer [69].

4.2.3.3 Improper Handling of Mission Needs

The overall system design should consider the customers' operational concepts. If these are not well understood or folded into the overall architecture, the resulting developed system may not be useable for the purposes for which it was originally intended.

4.2.3.4 Poor Communication and/or Inadequate Handling of Requirements

There are times design engineers will pay attention to some requirements more than others because they are well versed in designs using some requirements and not well-versed in others. The consequences are that the ensuing architecture may define a system the designers are comfortable with, not one that meets the requirements and needs of the customer and system operators.

4.2.3.5 Improper Functional Decomposition of the System

If the design engineer creates too many functions within a given level of the architecture, or too many levels of architecture, the number of internal interfaces and data/information flows increases, overly complicating the overall design, implementation, operations, and maintenance of the system.

4.2.3.6 Improper Consideration of Inputs and Outputs

Many designers consider only the functionality and subsequent actions needed to support the functionality while not considering the system inputs and outputs as part of the overall, integrated design, deciding, instead, to deal with the system inputs/outputs as an afterthought later. The designers must realize that inputs and outputs are an integral part of the overall design and must be included at every step in the architectural design process. Improper handling of inputs/outputs can cause major redesigns later when there is a mismatch between the designed system and the environment it must operate within [70].

Question to ponder

What would happen if a system was designed leaving out one of the required output interfaces?

4.2.3.7 Improper Emphasis on Static Functional Decomposition

Static decomposition of the system is important but is a small part of the overall architectural design and should only be performed after the functional scenarios have been created and folded into the design. The static decomposition should be one of the last operations that the design deals within the architecture design [71].

4.2.3.8 Improper Handling of System Management

Monitoring of the system (System Management) is often handled as a set of system functionality. Behavioral management as well as temporal management of the system are related but must be managed within the design individually. These are very distinct functions and mixing them will result in improper use of both within the overall functional design [72].

4.3 Design for Manufacturability (DFM) Discussion

DFM is very often overlooked in the overall architecture/design/development activities and is often skipped because of high priorities like schedule-driven objectives. Regardless of the development style (e.g., agile, spiral, waterfall), regardless of the mission/business models used to drive the overall architecture and design, building in the notion of DFM from the beginning of the analysis and design process will influence the entire life cycle process and deliver a system that is maintainable, reducing the overall LCC for the system. When the system does not have manufacturability requirements, which should include maintenance requirements (e.g., Mean Time to Repair <8 h), or the system designers delay the inclusion of them into the design, the manufacturability concepts, requirements, and changes to system architecture to accommodate manufacturability typically occur late in the development life cycle, producing serious system risks, and affecting development cost, schedule, and technical decisions to make up for the lack of DFM [73]. DFM influences the architecture, design, and implementation of the system for both the hardware and software capabilities by accounting for the manufacture activities on the system. Taking DFM into account early and often in the design will build maintenance into the overall system implementation, installation, and operations; typically reducing the overall manufacturability activities, providing manufacturability activities which are easier to execute, and reducing the overall logistics (personnel) support time required for system operations and sustainment. For proper treatment of DFM, the activities and functions should be required for the overall life cycle manufacturability of the any system across platforms, applications, and environments.

So far, we have dealt with the basics of the need for MDRE and a few of the special aspects of MDRE that are often missed in the process (i.e., manufacturability and sustainability). Now we delve into the heart of MDRE and discuss the overall MDRE requirements process in Chap. 5.

4.3.1 Design Problem

The philosophy of the assignment development for this textbook was to create a set of thought-provoking questions, problems, and design assignments that help drive home the principles, practices, and methodologies of multidisciplinary requirements engineering (MDRE). We endeavored to provide the opportunity for the student to practice the art and science of requirements engineering, helping the student to understand that there is rarely one answer to a requirements engineering problem, and it is only through study and practice that successful requirements engineers are developed. MDRE is a life-long journey, and anyone who tells you they have nothing left to learn does not truly understand the discipline or has never been involved in the development of a complex system. With the recent interest in alternative energies and talk of the "Green New Deal," along with the recent issues with ice storms in Texas taking out the power grid, the assignments for this textbook will follow the theme of effectively dealing with both local alternative energy power grids and the national infrastructure required to facilitate large-scale alternative energy production, sustainment, and management.

4.3.2 Design Problem Description: Alternative Energy Power Grid

The purpose of this project is to, over the course of the semester, develop a feasibility study, requirements analysis, requirements elicitation, derivation, and allocation, for a system that will facilitate a "local" alternative energy power grid capable of powering a neighborhood of 500 homes and a large biosphere capable of supporting growing year-round fruits and vegetables for the 500-home neighborhood. The student will have to deal with issues of choosing multiple alternative energies that can provide power throughout all seasonal difficulties, from intense summer heat to freezing winter temperatures and possible ice storms. Control, viability, sustainability, and maintainability will be major factors to understand the entire possibly 30-year life cycle of such a system. How to effectively utilize, store, and manage power distribution, along with when to upconvert "extra" energy to the national power grid will be issues that drive many of the assignments. Simply relying on the current technologies and energy infrastructures will not move the country and world forward. Hopefully, this will provide insight to the students using this textbook to understand some of the issues that face everyone looking out over the next many decades.

4.3.3 DC Power Grid Requirements Discussion

In the initial days of power generation and distribution, DC power was far less practical than the AC distribution system that soon supplanted it. However, even after Tesla's AC system became the predominant system (and rightly so), Edison's DC power did not become obsolete for some time. Eventually power plants eliminated their delivery of DC power, but the overall concept of a DC power grid is not dead. Even our computer systems (large server systems) have uninterruptible power supplies which contain batteries that are charged off the grid and used to upconvert to AC to power devices when the AC power grid goes off-line. So, the notion of a DC power grid is not dead, but generally used as a backup system. The point is, the notion of a DC power grid has remained and now that renewable energies are viable, we should re-examine their use for widespread, local, power distribution.

One of the advantages of a locally generated DC station is energy savings as well as allowing a facility to run more easily off various DC sources, such as Solar Arrays. And for power-hungry installations, the notion of generating the power locally is growing in popularity. Part of the reason is that waste heat from the power generation generators (solar, geothermal, etc.) can be used to warm nearby buildings. By combining the production of heat and power, local power grids can squeeze much more useful energy out of the renewable energy they use, so this approach will certainly become more widespread as time goes on. The purpose of this design problem is to look ahead and provide an architecture and specification for a local DC power generation station that can serve a local community.

Currently many systems exist to control and handle alternative energies. There are systems to control and manage solar power, wind power, geothermal power, and others, but very few are designed to control, manage, store, and distribute energy from a power grid consisting of multiple alternative energy sources. As we saw the issues recently caused by the intense ice storms in Texas, one alternative energy source is not the answer to the growing energy needs/wants in our society. Each alternative energy source has limitations and issues associated with its use. By harnessing multiple alternative energies, and managing the storage of the energies produces, communities can be worry free from power generation. To facilitate utilizing multiple alternative energy sources, a control center must be included in the design that can manage the energy creation, distribution, storage, and return unused energy to the power grid.

4.3.4 High-Level Technical Description of the DC Power Grid System

The purpose of the multiple alternative energies DC power grid is to:

- Supply DC power to a customer community of 500 homes and a biosphere for growing food.
- Provide a storage device to store energy in case of a system failure.
- Determine the amount of surplus energy that is above and beyond what is required to power the grid and charge the backup system, and then deliver that surplus energy to the national power grid systems.

Throughout the semester you will be asked to analyze, design, and architect different components of the overall DC power grid system. This will include research into alternative designs, like lithium nanowire storage systems for backup power and/or supercapacitors (also called ultracapacitors). Fig. 4.7 illustrates these technologies connected to the DC Power Grid Regulator Control Center.

4.3.5 High-Level Technical Requirements for the DC Grid System

- Design shall utilize at least two alternative energies, with a maximum of 4.
- Total power generation shall be 3.0 megawatts minimum.
- Each of the alternative energy sources shall be capable of generating at least 60% of minimum wattage.
- DC power grid shall generate DC power, with grid voltage being 48, 24, or 12 volts.
- DC power grid shall be capable of servicing at least 500 single family homes and one 4-acre biosphere with hydroponics capable of growing vegetables year-round. (See Fig. 4.8 for a block diagram of a DC-powered home).
- All lights, appliances, and facilities in each home and the biosphere shall be configured for DC operations, if possible.

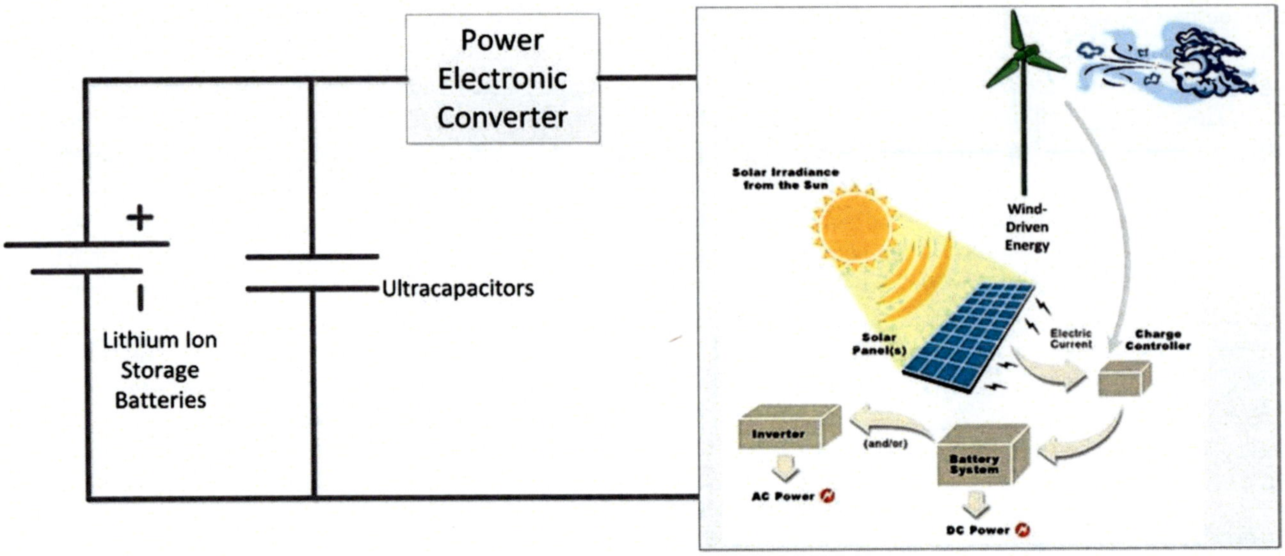

Fig. 4.7 Alternative power management system overview

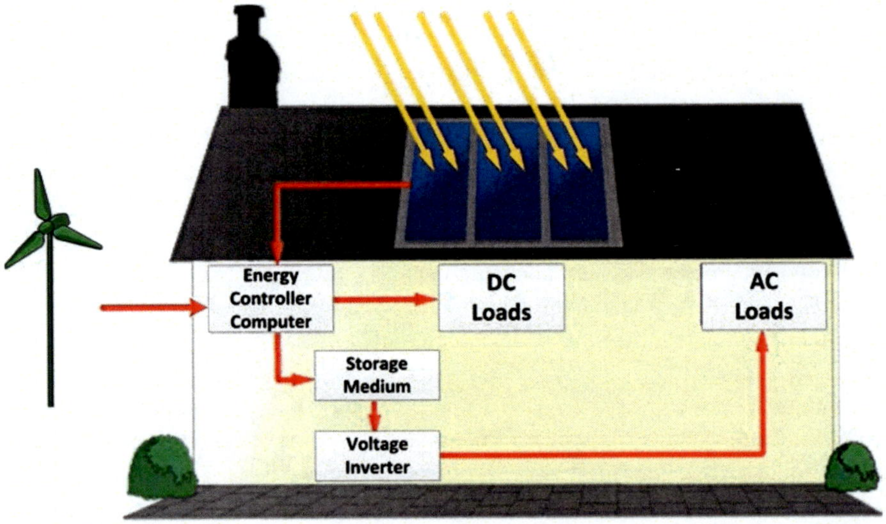

Fig. 4.8 Example of a whole house power grid

Things like electric ovens may need DC/AC conversion to accommodate their 220v. requirements.

- Only allowed exceptions are the HVAC system and the Washer/Dryer, although DC is preferred.
- There shall be energy storage capability capable of operating the grid of 500 single family dwellings for 7 days if all alternative forms of energy are not producing.
  - Compliance Criteria: Energy storage devices shall be capable of producing 750,000 watts for the entire 7-day period. Nominal energy storage should be batteries or supercapacitors.
- The DC power grid system shall include controlling h/w and s/w to manage the multi-alternative energy strategies and production.
- All energy produced above the required home and energy storage usage shall be upconverted to 220vac for release to the national power grid.
- The DC power grid shall be capable of operating in a temperature range of −40 degrees F. to 125 degrees F. without degradation of power generation.
- The DC power grid must meet OSHA regulations.

4.3.6 Assignments

- Develop a manufacturability analysis for your system and include:
- Include a discussion of the supply chain needed for your system.

- This includes vendors for the parts required to build the system.
- Explain how manufacturability differs between each alternative energy source you have chosen.
- Include a discussion of what happened in Texas in February 2021 and how your system requirements and design will ensure this does not happen to your system. Note—it is important to understand what capabilities/processes were not implemented in the Texas power system compared to comparable systems in the northern United States—e.g., dehumidifiers in natural gas lines.

References

1. Dwyer, M., Avrunin, G., and Corbett, J. (1999). Patterns in property specifications for finite-state verification. In Proceedings of the International Conference on Software Engineering, Los Angeles, CA.
2. Edvardsson, J. (1999). A survey on automatic test data generation. Proceedings of the Second Conference on Computer Science and Engineering in Linkoping. CiteSeerX: 10.1.1.20.963.
3. Egyed, A., Grunbacher, P., and Medvidovic, N. (2001). Refinement and evolution issues in bridging requirements and architecture-the CBSP approach. In International Software Requirements to Architecture Workshop, Toronto, CAN.
4. Egyed, A., Grunbacher, P., Heindl, M., and Biffl, S. (2009). Value-based requirements traceability: Lessons learned. Design Requirements Engineering: A Ten-Year Perspective. Springer-Verlag, Berlin Heidelberg, GER, ISBN 978-3-540-92966-6.
5. Erdobmus, H., & Torchiano, M. (2005). On the effectiveness of test-first approach to programming. *Proceedings of the IEEE Transactions on Software Engineering, 31*(1), NRC 47445.
6. Feathers, M. (2004). *Working effectively with legacy code*. Prentice Hall.
7. Ferguson, R., & Korel, G. (1996). The chaining approach for software test data generation. *ACM Transactions of Software Engineering and Methodology, 5*(1), 63–86.
8. Fiadeiro, J., Lopes, A., & Bocchi, L. (2006). A formal approach to service component architecture. *Web Services and Formal Methods, 4148*, 193–213.
9. Fowler, M. (1999). *Refactoring—improving the design of existing code*. Addison Wesley Longman. ISBN 0-201-48567-2.
10. Glinz, M. (2007). On non-functional requirements. International Conference on Requirements Engineering, New Delhi, India.
11. Gold-Bernstein, B., & Ruh, W. (2005). *Enterprise integration: The essential guide to integration solutions*. Addison Wesley.
12. Gotel, O. and Finkelstein, C. (1994). An analysis of the requirements traceability problem. International Conference on Requirements Engineering, Colorado Springs, CO.
13. Hamilton, M. and Hackler, W. (2000). A rapid development approach for rapid prototyping based on a system that supports its own life cycle. Proceedings, 10th International Workshop on Rapid System Prototyping.
14. Hartman, J. (2006). *A systems approach to understanding gene interaction networks using the yeast model system*. SDPS Systems Biology Workshop.
15. Heidrick, J., Munch, J., Riddle, W., & Rombach, D. (2006). New trends in software process modelling. *World Scientific, 18*. ISBN: 978-981-256-619-5.
16. Heinz, J. (2000). What went wrong. Aerospace & Defense Science.
17. Held, J. (2008). *The modelling of systems of systems*. PhD Thesis. University of Sydney.
18. Helmreich, R., & Wilhelm, J. (1991). Outcomes of crew resource management training. *International Journal of Aviation Psychology, 1*(4), 287–300.
19. Herrmann, A., Paech, B., & Plaza, D. (2006). An automated process for the solution of requirements conflicts and architecture design. *International Journal and Knowledge Engineering, 16*, 1–34.
20. Hershey, P., & Blanchard, K. (1989). *Management of organizational behavior: Utilizing human resources*. Prentice-Hall.
21. Jaakola, H., & Thalheim, B. (2011). Architecture-driven modeling methodologies. In A. Heimbürger et al. (Eds.), *Proceedings of the 2011 Conference on Information Modelling and Knowledge Bases XXII*. IOS Press.
22. Jamshidi, M. (2005). System-of-systems engineering—A definition. *IEEE SMC, 2005*, 10–12.
23. Korel, G. (1990). Automated software test data generation. IEEE Transactions on Software Engineering, CiteSeerX: 10.1.1.121.8803.
24. Koskela, L. (2007). *Test driven: TDD and acceptance TDD for Java developers*. Manning Publications.
25. Kossiakoff, A., & Sweet, W. (2003). *Systems engineering: Principles and practice*. John Wiley & Sons.
26. Kotonya, G., & Sommerville, I. (1998). *Requirements engineering: Processes and techniques*. John Wiley & Sons Ltd. ISBN: 978-0-471-97208-2.
27. Kotov, V. (1997). Systems-of-systems as communicating structures. Hewlett Packard Computer Systems Laboratory Paper HPL-97-124, pp. 1–15.
28. Kruchten, P. (2000). *The rational unified process: An introduction*. Addison-Wesley Longman Publishing.
29. Lamsweerde, A. (2001). Goal-oriented requirements engineering. In Proceedings of the Fifth IEEE International Symposium on Requirements Engineering, Toronto, CAN.
30. Lamsweerde, A. (2009). *Requirements engineering: From system goals to UML models to software specifications*. Wiley & Sons.
31. Liu, L. and Yu, E. (2001). From requirements to architecture design—Using goals and scenarios. In From Software Requirements to Architectures Workshop (STRAW), Toronto, CAN.
32. Lormans, M. and Van Deursen, A. (2006). Can Isi help reconstructing Requirements Traceability in Design and Test? Conference on Software Maintenance and Reengineering, IEEE, Piscataway, NJ, ISBN 0-7695-2536-9.
33. Luckey, M., Fernandez, D., Baumann, A., and Wagner, S. (2010). Reusing security requirements using an extended quality model. In Workshop SESS at the IEEE International Conference on Software Engineering, Cape Town, South Africa.
34. Luskasik, S. J. (1998). Systems, systems-of-systems, and the education of engineers. *Artificial Intelligence for Engineering Design, Analysis, and Manufacturing, 12*(1), 55–60.
35. Luzeaux, D., & Rualt, J. (2010). *Systems of systems*. ISTE Ltd and John Wiley & Sons.
36. Madeyski, L. (2010). *Test-driven development—an empirical evaluation of agile practice*. Springer Publishing. ISBN 978-3-642-04287-4.
37. Maier, M. (1998). Architecting principles for system of systems. *Systems Engineering, 1*(4), 267–284.
38. Manfred, B. (2007). Model-driven architecture-centric engineering of (embedded) software intensive systems: Modeling theories and architectural milestones. *Innovations in Systems and Software Engineering, 3*(1), 75–102.
39. Manthorpe, W. (1996). The emerging joint system-of-systems: A systems engineering challenge and opportunity for APL. *Johns Hopkins APL Technical Digest, 17*(3), 305–310. [2].
40. McHugh, O. Conboy, K., and Lang, M. (2012). Agile practices: The impact on trust in software project teams. IEEE Software May/June 2012.

41. Medvidovic, N., & Taylor, R. (2000). A classification and comparison framework for software architecture description languages. *Software Engineering, IEEE Transactions, 26*(1), 70–93.
42. Medvidovic, N., Grunbacher, P., Egyed, A., & Boehm, B. (2003). Bridging models across the software lifecycle. *Journal on Systems and Software, 68*, 3.
43. Menker, L. (2000). Results of the Aeronautical systems division critical process team on integrated product development. Technical Report, Wright Patterson AFB.
44. Meyers, R. (2009). *Encyclopedia of complexity and systems science*. Springer Publishing. ISBN 978-0-387-75888-6.
45. Mittal, S., & Martin, J. (2013). *Netcentric system of systems engineering with DEVS unified process*. CRC Press.
46. Morris, P., Hough, G., & Morris, W. (1987). *The anatomy of major projects: A study of the reality of project management*. Wiley & Sons.
47. Navarro, E., Ramos, I., and Perez, J. (2003). Software requirements for architectured systems. In Proceedings of the 11th IEEE International Requirements Engineering Conference, Kyoto, Japan.
48. Offutt, J. and J. Hayes. (1996). A semantic model of program faults. International Symposium on Software Testing and Analysis. CiteSeerX: 10.1.1.134.9338.
49. Paech, B., et al. (2003). An experience-based approach for integrating architecture and requirements engineering. International Software Requirements to Architecture Workshop, STRAW 2003, Portland Oregon.
50. Papoulis, A. (1965). *Probability, random variables, and stochastic processes*. McGraw Hill.
51. Parker, J. (2010). Applying a system of systems approach for improved transportation. *S.A.P.I.EN.S., 3*, 2.
52. Parnas, D. (1972). On the criteria to be used in decomposing systems into modules. *Communications of the ACM, 15*(12), 1053–1058.
53. Parnas, D. (1979). Designing software for ease of extension and contraction. *IEEE Transactions of Software Engineering, 5*(2), 128–138.
54. Pavlich-Mariscal, J. (2005). *A framework for composable security definition, assurance, and enforcement*. MODELS Conference.
55. Pei, R. S. (2000). Systems-of-systems integration (SoSI)—A smart way of acquiring Army C4I2WS systems. *Proceedings of the Summer Computer Simulation Conference*, 574–579.
56. Pender, T. (2003). *UML Bible*. Google Books. ISBN 0-7645-2604-9.
57. Penzenstadler, B. and Koss, D. (2008). High confidence subsystem modeling for reuse. International Conference on Software Reuse, Beijing, China.
58. Perks, C., & Beveridge, T. (2003). *Guide to enterprise IT architecture*. Springer Publishing. ISBN: 0-387-95132-6.
59. Pohl, K., & Sikora, E. (2007). Structuring the co-design of requirements and architecture. In *International conference on requirements engineering: Foundations for software quality*. Springer Publishing.
60. Popper, S., Bankes, S., Callaway, R., & DeLaurentis, D. (2004). *System-of-systems symposium: Report on a summer conversation*. Potomac Institute for Policy Studies.
61. Ramesh, B. (1998). Factors influencing requirements traceability practice. *Communications of the ACM, 41*(12), 37–44.
62. Ramesh, B., & Jarke, M. (2001). Toward reference models for requirements traceability. *IEEE Transactions on Software Engineering, 27*(1), 58–93.
63. Reynard, W., Billings, C., Cheaney, E., & Hardy, R (1986). The development of the NASA aviation safety reporting system. *NASA Reference Publication, 1114*.
64. Robertson, J., & Robertson, S. (2007). *Mastering the requirements process*. Addison-Wesley. ISBN-13: 978-0321815743.
65. Ross, J., Weill, P., & Robertson, D. (2006). *Enterprise architecture as strategy: Creating a foundation for business execution*. Harvard Business Review Press, Harvard University.
66. Russell, S. and Zilverstein, S. (1991). Composing real-time systems. In Proceedings of the Twelfth International Joint Conference on Artificial Intelligence, Sydney, Australia.
67. Safonov, M. and Laub, A. (2001). The role and use of the QFD matrix. IEEE Transactions in Automatic Control.
68. Sage, A., & Cuppan, C. (2001). On the systems engineering and management of systems of systems and federations of systems. *Information, Knowledge, Systems Management, 2*(4), 325–345.
69. Schenk, T. (2005). Introduction to photogrammetry. Department of Civil and Environmental Engineering and Geodetic Science, Ohio State University, Columbus, OH.
70. Shah, N., Laws, J., Wardman, B., Zhao, P., & Hartman, L. (2007). Accurate, precise modeling of cell proliferation kinetics from time-lapse imaging and automated image analysis of agar yeast culture arrays. *BMC Systems Biology, 1*.
71. Shan, T. and Hua, W. (2006). Solution architecting mechanism. Proceedings of the 10th IEEE International EDOC Enterprise Computing Conference (EDOC 200), 23–32.
72. Staszewski, J. (2004). Models of expertise as blueprints cognitive engineering: Applications to land mine detection. Proceedings of the 48th Annual Meeting of the Human Factors and Ergonomics Society, 48, 458–462.
73. Taylor, F. (1911). *Shop management*. The Principles of Scientific Management. Harper & Row.

5 MDRE: The Requirements Engineering Process

Abbreviations

AC	Alternating current
DC	Direct current (or 0 Hz)
F	Fahrenheit
HVAC	Heating, ventilation, and air conditioning
MBSE	Model-based systems engineering
MDRE	Multidisciplinary requirements engineering
OSHA	Occupational Safety and Health Administration

Every aspect, from a functional assessment of the system, to manufacturing and sustainment considerations, as well as factors like human engineering and assessment of complexity must be considered throughout the MDRE process [2]. This process is illustrated in Fig. 5.1. As depicted in Fig. 5.1, all aspects of the system life cycle must be integrated together, modeled, and folded into the set of requirements that will ultimately lead to the system design, implementation, deployment, and sustainment/maintenance. If any aspect is left out (e.g., security or maintainability) it can greatly affect the overall efficacy and longevity of system performance and may drive total life cycle cost higher than was expected for the system [3]. MDRE requirements management must start with the initial investigation and elicitation of the system-level requirements. The lead MDRE systems engineer must embrace the MDRE process, as depicted in Fig. 5.2.

Modern requirements management is a feedback-driven continual process that must endure throughout the life cycle of the system. And while the original MDRE team may hand-off the system to the system integrators and/or operations staff, the process continues with the team of sustainment/maintenance/operations personnel [4]. This drives-home the need for completely documenting not only the requirements, but also the process and decisions made during MDRE requirements analysis to ensure the operations team understands the decisions and why they were made, including how the requirements decisions drove the overall design of the system.

5.1 The MDRE Requirements Process Flow

Requirements management is the foundation of the MDRE process and must be put in place from the very beginning of the system requirements process [5]. Fig. 5.3 illustrates the high-level view of the MDRE requirements process flow.

5.1.1 Elicitation

In general, requirements elicitation involves collecting the known set of requirements that exists for the system to be designed and implemented. This is a critical first step in MDRE and can be difficult, depending on the system and the customer(s) desiring the system. Often, the high-level requirements are handed to you by the customer(s) and it is then the MDRE team's responsibility to interpret the requirements and ensure that the development team and the customer are on the same page as to the meaning of each requirement. This may be a daunting task and, in some cases, stretches out for months as differing opinions, even among the customer base, can create issues for the MDRE team. The result of elicitation must be a common understanding among the development team, the customer team, and the various stakeholders before design and development ensues.

5.1.2 Requirements Analysis and Validation

The high-level purpose of requirements analysis is to determine needed requirements that are specifically provided by the customer. This entails understanding the needs of the customer and system users that were not explicitly called out in the original requirements set provided to the design team

J. A. Crowder, C. W. Hoff, *Requirements Engineering: Laying a Firm Foundation*,
https://doi.org/10.1007/978-3-030-91077-8_5

Fig. 5.1 The MDRE approach to requirements engineering

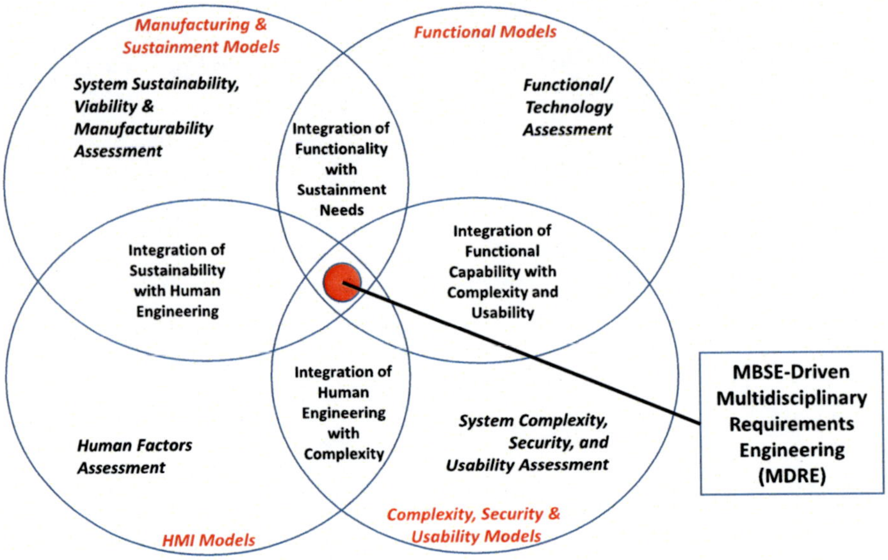

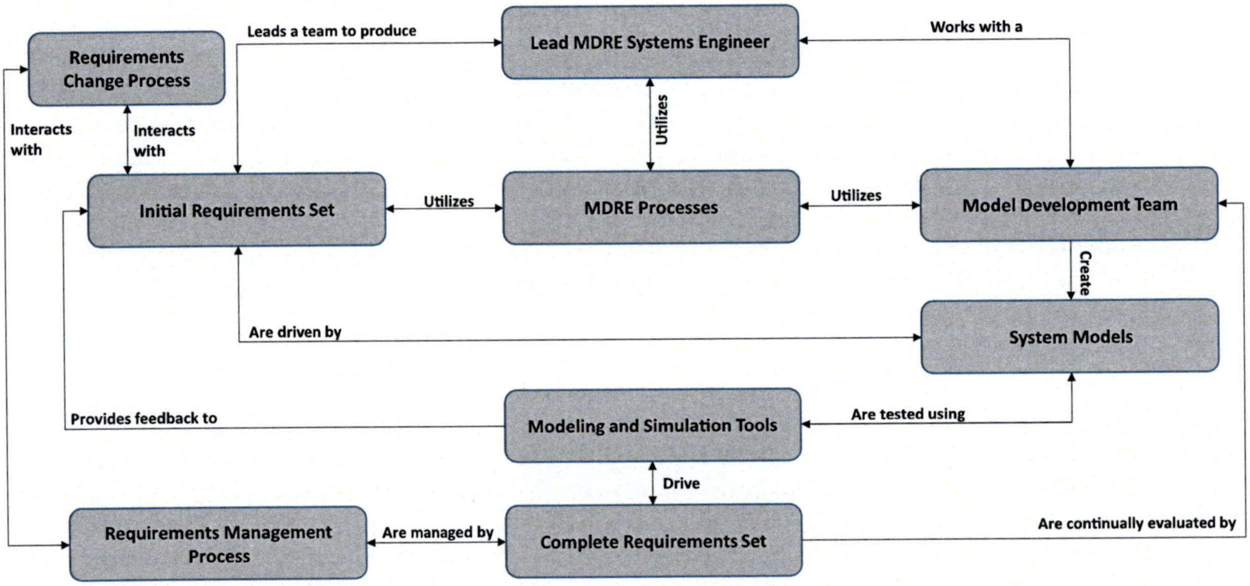

Fig. 5.2 MDRE requirements management dynamics

and were not expressed during the requirements elicitation effort. Every project of any size usually has requirements that were not expressed but were assumed. It is beneficial to understand these requirements as early as possible in the design/development life cycle. Finding needed requirements later in the project will drive overall costs and can cause major rework if they drive changes in the overall design [6]. In the MDRE requirements analysis process, software and hardware designs are not moved forward until the requirements analysis process is completed. Often there is a tendency and push to start coding before the analysis is completed. This most often leads to major changes to software once new/undiscovered requirements are known. One major use of requirements analysis is to ensure all parties (designers, customers, coders, etc.) are using the same vocabulary and do not have different interpretations and definitions for terms and descriptions. Not taking this step often leads to final products that do not meet customer expectations because their definitions and interpretations of requirements were different than the design engineers, which were different than the software engineering [7]. Also, requirements analysis is important to ensure that the requirements are not being forced to adhere to an existing system specification or existing models.

> **Question to ponder**
>
> What could happen if the software development starts before the requirements are established and agreed to?

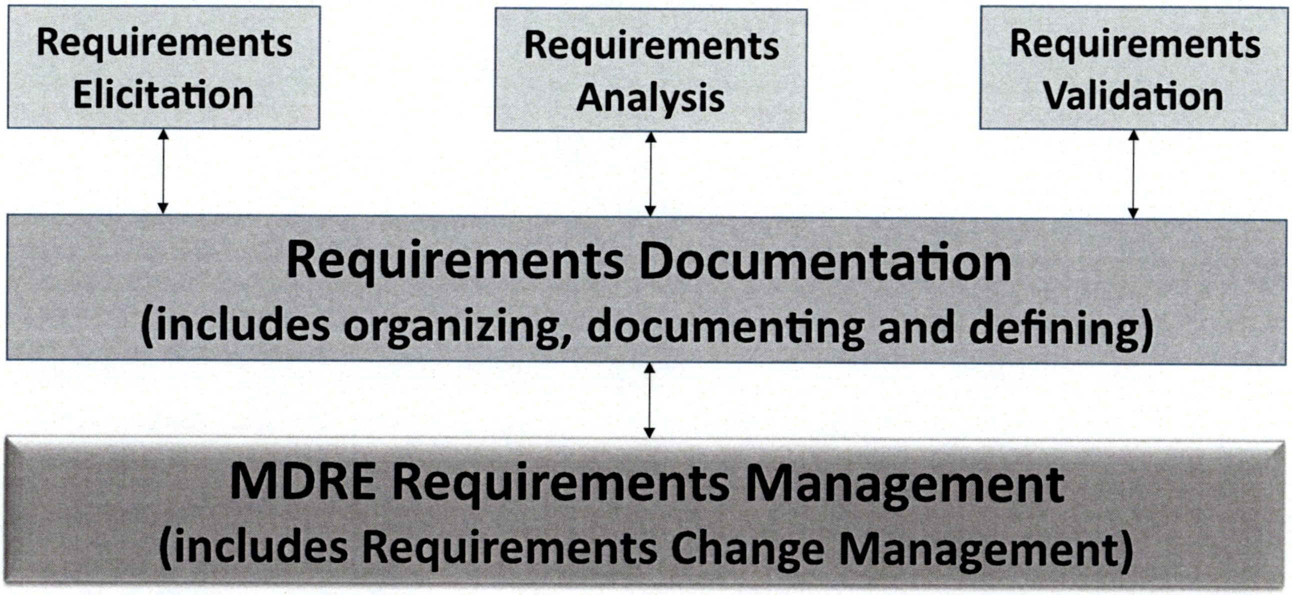

Fig. 5.3 The MDRE high-level requirements process flow

It is important to develop requirement specifications and system models that adhere to the current system being designed and implemented, even if it is "close" to another system design. An example would be a system where a technology refresh is required because the hardware is no longer supported. But, because the current hardware is not supported, a hardware redesign is required while the overall system design remains the same or very close to the same. Figure 5.4 illustrates the MDRE requirements analysis process. The process begins with a complete understanding of the domain or environment the system will be utilized within. Without a complete understanding for the system's application domain, it is not possible to understand and analyze what requirements are needed that have not already been specified. Each requirement that is collected or "discovered" through the requirements analysis process must be classified as to its requirements type (e.g., functional, performance). Each requirement must be correlated and reviewed against the current requirements to ensure that there are not conflicts between requirements, and, if there are, these conflicts resolved. This may entail rewriting current requirements or adjudicated them with the customer [8]. Part of the conflict resolution process involved prioritization of requirements to understand the importance placed on certain requirements. Each requirement must then be validated to ensure it aligns with the overall statement of objectives for the system under development. This is an on-going process to ensure that the design (solution), the customer(s), the stakeholders, and the envisioned transition of the system align with the objectives. The requirements analysis process is a feedback-driven iterative process that continues throughout the life cycle of the system [9]. The final step is creation of the requirements specification with each requirement defined, including how each requirement will be tested. We will discuss all of these in much more detail throughout the book.

The requirements collection process includes derivation of requirements that the requirement analysis indicates are required but are not already part of the requirements set. Requirement derivation is the process of writing new requirements, based on a parent requirement that specializes the requirement for a given subsystem or component of the overall system design. An example would be adapting a system architecture level requirement to a software module or component. Requirement derivation provides context to a specific subsystem, component, or module within an overall systems architecture [10]. An example would be a high-level requirement that the system batteries shall be capable of sustaining operations for 36 hours on a full charge. This might be broken into two separate lower-level requirements:

1. The storage batteries shall provide 36 hours of operational power when fully charged.
2. The system shall detect and provide an alarm when the battery level is below 25% of full charge.

5.1.3 Requirements Classification

Requirement classification, also called allocation, is the process of assigning requirements to specific requirement classes (e.g., performance) [11]. Each collected and derived

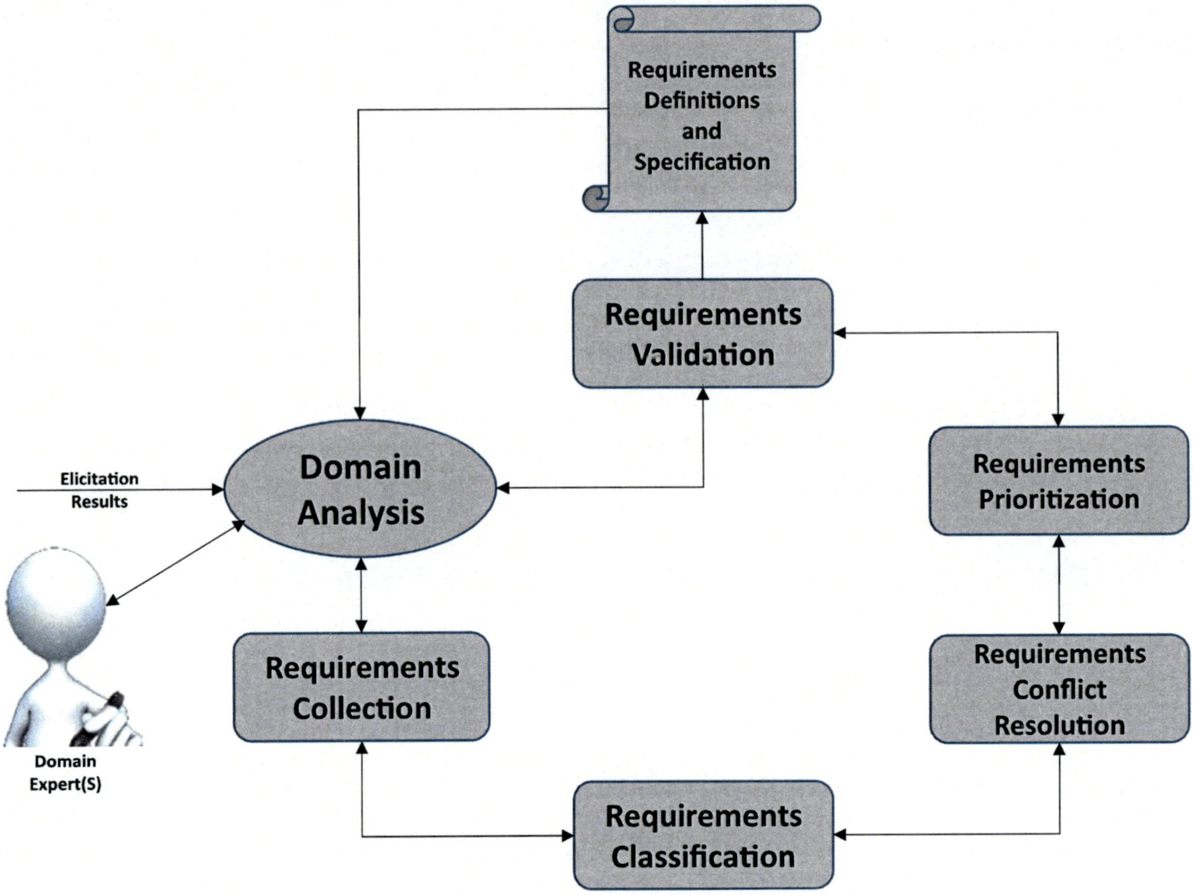

Fig. 5.4 The MDRE requirements analysis process

requirement must be classified/ allocated to a specific requirement class. All requirements must be allocated appropriately to ensure the system design meets all standards (e.g., safety) [12]. As discussed in Chaps. 3 and 4, three of the major areas that get missed in the requirements elicitation, collection, and derivation process are sustainability, maintainability, and manufacturability. These requirement categories often require the most derivation and analyses to ensure the overall system life cycle is considered up front in the initial design phase(s) of the project [13].

Question to ponder
What do you think the results of misclassifying requirements might be?

5.1.4 Requirements Documentation

Documentation of the system requirements is essential for MDRE. The MDRE engineers must ensure that the system is designed, developed, deployed, and maintained according to the agreed upon set of requirements. System requirements documentation, as well as maintenance manuals, should be provided with all delivered systems. The requirements document should be easily understandable by the target operator(s)/maintainer(s) of the system in the environment it is intended to operate within. Generally, requirements are documented and maintained in a requirements management software tool that provides the capabilities to link requirements from the lowest level to high-level system architecture requirements. It is crucial to maintain the correct links between requirements to ensure the system meets its requirements at all levels of operation. We will discuss requirements management tools in detail in later chapters. Here we want to expound on the need for adequate requirements documentation.

5.1.4.1 Requirements Vs. Features

During the elicitation process, many engineers and stakeholders will vie for features they desire to be in the system design that are not part of the requirements set provided to the design team. The goal of the MDRE process is to ensure that requirements decisions are based on the objectives and needs and not the "this would be great to have" wants of people not providing the funding and not associated with approving the final design. This can be especially true in agile development projects where the requirements process is more fluid. It is easy for the requirements to wander into ambiguous or vague desires. Having requirements change is an integral part of an agile development process, and control and management of the requirements becomes even more important in this type of development and requires more discipline on the part of the MDRE.

Question to ponder

Why do you think features are considered separate from requirements? How might a feature turn into a requirement?

Whether agile, waterfall, or other development strategies, you must eliminate the "designed by whining" syndrome where the engineer that complains the most gets their requirements into system specification. It requires training for engineers to understand how to specify requirements. The goal of the MDRE is to ensure that requirements decisions are made regarding necessary features and objectives, those things that will generate the system operational needs, including hidden needs like sustainability, reliability, availability, and maintainability.

5.1.4.2 Ongoing Requirements Documentation

It is essential that requirements are adequately maintained and documented as the system is developed, deployed, and maintained. Sometimes different groups within the development team will push for an application or feature that was not specified in the original requirement specification. By maintaining continuity of the requirements documentation, you can see why or why not a given feature was approved or rejected. Continued requirements documentation management requires software tools for creating and managing requirements specifications. Descriptions of requirements documentation tools will be provided in Chap. 13 on requirements traceability.

5.1.5 Requirements Prioritization

It is often the case that during the design and development phases of a project things change. Budgets get cut, timelines get shortened, any number of things can happen that result in the reality that not everything that was desired for the system can be implemented within the budget and/or timeline provided by the customer. It is important that part of the requirements process is the prioritization of all requirements to understand what is most important to the customer and to the operational viability of the system. Fig. 5.5 provides a fun example of what happens when all parties do not agree on the interpretation of requirements.

Question to ponder

Why is it important to prioritize requirements? Do you think functionality should be higher priority than performance?

5.1.6 Requirements Management

Requirements engineering, and therefore requirements management, is crucial for all levels of industry, aerospace, transportation, aviation, automotive, utilities (e.g., nuclear power generation), and many others. Unfortunately, there are no industry standards for what constitutes a requirement across all these industries. Almost all industries have the notion of requirements traceability (which will be covered in later in this chapter), but how the traceability is done, and the tools used vary greatly across industries [14]. This drives home the need for standardization of requirements engineering and requirements management. MDRE is a discipline that establishes and manages the entire life cycle of requirements, from elicitation and requirements specification, to analysis, verification, and validation as well as operational management of requirements across the system life cycle. While most development efforts start with requirements definitions and analysis, not all do, many "start building/coding" and hope the requirements eventually catch up with them. This creates instability into the design, implementation, and particularly the sustainment process, since without a complete and coherent set of requirements, the sustainability and maintainability personnel have very little idea what is in the system they must maintain [15]. At the highest level, MDRE requirements management is a process that aims to:

- Have an organization in place to manage the requirements of a project during every stage of design, development, test, deployment, operations, and, ultimately, system retirement.
- Manages the requirements, including syntax, principles, attributes, etc. Every aspect of each requirement should be managed by the MDRE requirements management engineer/team [16].
- Manage the documentation and traceability of requirements to guide and direct the production of the system.
- Manage the requirements change process to ensure any changes make their way through every part of the requirements that are affected by the change [17]. This includes:
  - Identification of the requirement(s) change(s).

Fig. 5.5 Example of misinterpreted requirements

- Performing impact analyses of the change(s).
- Determine the effectivity of the change(s), i.e., when the change will be implemented.
- And ensuring system tests are adequate to test not only the new functionality but also potential issues that arise from the change(s).

MDRE requirements management includes management of all the data, architecture, processes, as well as requirements, all traced back to each other to ensure a complete design, development, deployment, and sustainability life cycle. This includes not only tracing the requirements from the lowest level to the top architecture requirements, but also tying the requirements to architecture, both hardware and software, to understand which requirements are instantiated in which part of the system. Fig. 5.6 illustrates the MDRE requirements and data management process. This includes information analysis to understand the relationships between organizations within the development team, technology utilized for the systems, processes used throughout the life cycle of the system, as well as how data are utilized, shared, stored, etc. throughout the system. The MDRE requirements

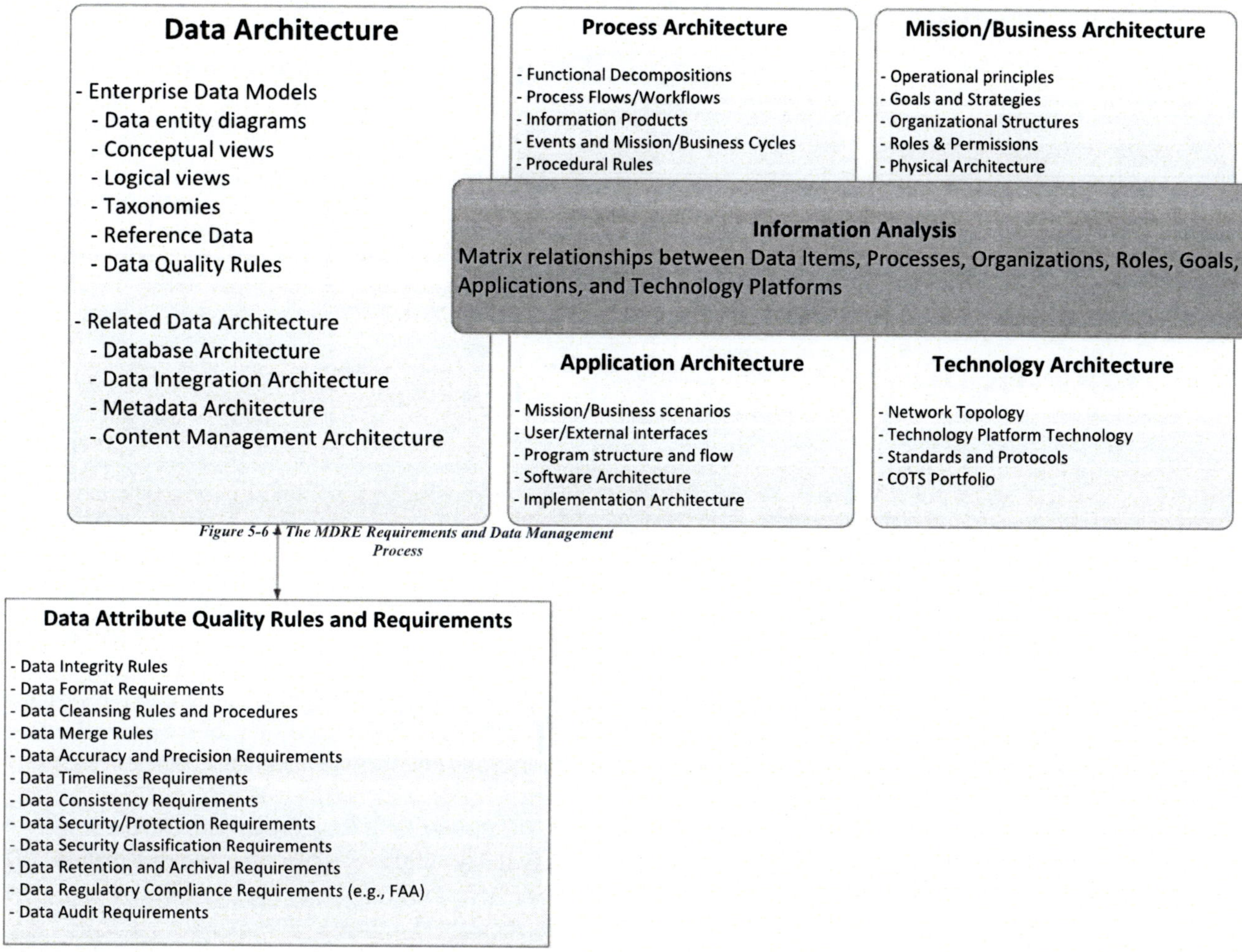

Fig. 5.6 The MDRE requirements and data management process

data management process is the backbone of MDRE documentation and management. In the end, MDRE requirements management is a continual process that constantly must re-examine:

- System objectives to ensure the requirements span the range of all needs of the system.
- The user/operator community through User Scenarios.
- Understanding all functional and nonfunctional requirements.
- Knowledge of regulations and constraints that the system must consider.
- The overall system design and development process.

Figure 5.7 illustrates this overall MDRE requirements management process. This will be discussed in more detail in Chaps. 14 and 17.

Question to ponder

What skills do you think would be important for requirements management?

5.1.6.1 Design Problem

The philosophy of the assignment development for this textbook was to create a set of thought-provoking questions, problems, and design assignments that help drive home the principles, practices, and methodologies of multidisciplinary requirements engineering (MDRE). We endeavored to provide the opportunity for the student to practice the art and science of requirements engineering, helping the student to understand that there is rarely one answer to a requirements engineering problem, and it is only through study and practice that successful requirements engineers are developed.

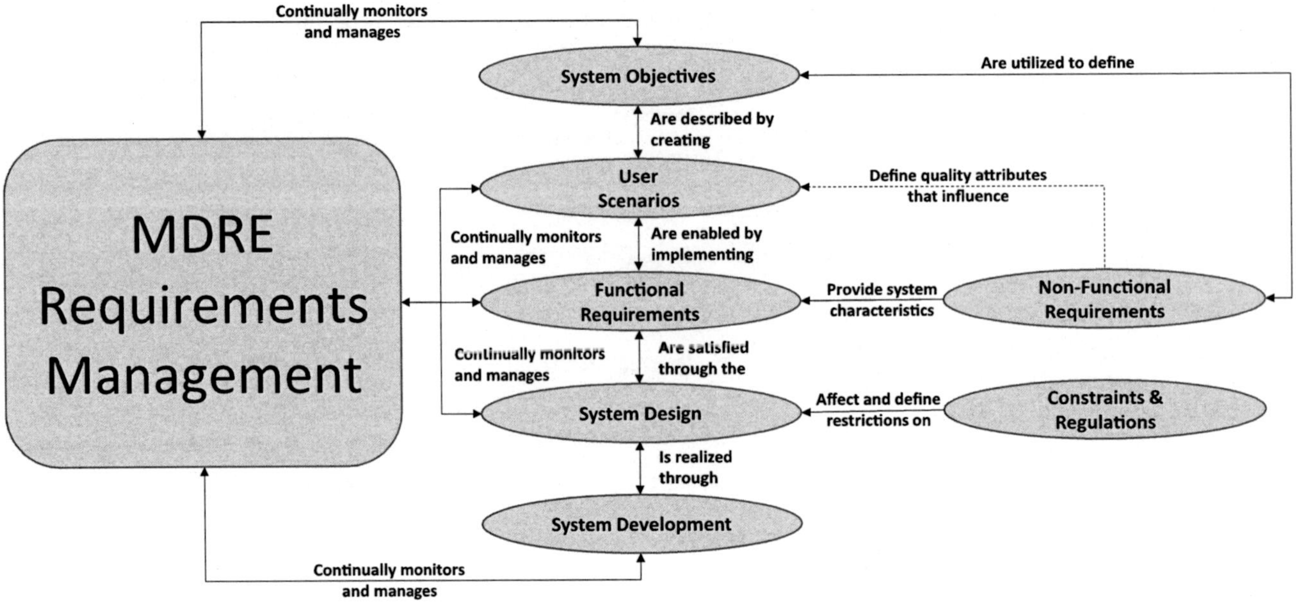

Fig. 5.7 MDRE requirements management context diagram

MDRE is a life-long journey, and anyone who tells you they have nothing left to learn does not truly understand the discipline or has never been involved in the development of a complex system. With the recent interest in alternative energies and talk of the "Green New Deal," along with the recent issues with ice storms in Texas taking out the power grid, the assignments for this textbook will follow the theme of effectively dealing with both local alternative energy power grids and the national infrastructure required to facilitate large-scale alternative energy production, sustainment, and management.

5.1.6.2 Design Problem Description: Alternative Energy Power Grid

The purpose of this project is to, over the course of the semester, develop a feasibility study, requirements analysis, requirements elicitation, derivation, and allocation, for a system that will facilitate an "local" alternative energy power grid capable of powering a neighborhood of 500 homes and a large biosphere capable of supporting growing year-round fruits and vegetables for the 500-home neighborhood. The student will have to deal with issues of choosing multiple alternative energies that can provide power throughout all seasonal difficulties, from intense summer heat to freezing winter temperatures and possible ice storms. Control, viability, sustainability, and maintainability will be major factors to understand the entire possibly 30-year life cycle of such a system. How to effectively utilize, store, and manage power distribution, along with when to upconvert "extra" energy to the national power grid will be issues that drive many of the assignments. Simply relying on the current technologies and energy infrastructures will not move the country and world forward. Hopefully, this will provide insight to the students using this textbook to understand some of the issues that face everyone looking out over the next many decades.

5.1.6.3 DC Power Grid Requirements Discussion

In the initial days of power generation and distribution, DC power was far less practical than the AC distribution system that soon supplanted it. However, even after Tesla's AC system became the predominant system (and rightly so), Edison's DC power did not become obsolete for some time. Eventually power plants eliminated their delivery of DC power, but the overall concept of a DC power grid is not dead. Even our computer systems (large server systems) have uninterruptible power supplies which contain batteries that are charged off the grid and used to upconvert to AC to power devices when the AC power grid goes off-line. So, the notion of a DC power grid is not dead, but generally used as a backup system. The point is, the notion of a DC power grid has remained and now that renewable energies are viable, we should re-examine their use for widespread, local, power distribution.

One of the advantages of a locally generated DC station is energy savings as well as allowing a facility to run more easily off various DC sources, such as Solar Arrays. And for power-hungry installations, the notion of generating the power locally is growing in popularity. Part of the reason is that waste heat from the power generation generators (solar, geothermal, etc.) can be used to warm nearby buildings. By combining the production of heat and power, local power grids can squeeze much more useful energy out of the renewable energy they use, so this approach will certainly become more widespread as time goes on. The purpose of this design problem is to look ahead and provide an architecture and specification for a local DC power generation station that can serve a local community.

Currently many systems exist to control and handle alternative energies. There are systems to control and manage solar power, wind power, geothermal power, and others, but very few are designed to control, manage, store, and distribute energy from a power grid consisting of multiple alternative energy sources. As we saw the issues recently caused by the intense ice storms in Texas, one alternative energy source is not the answer to the growing energy needs/wants in our society. Each alternative energy source has limitations and issues associated with its use. By harnessing multiple alternative energies, and managing the storage of the energies produces, communities can be worry free from power generation. In order to facilitate utilizing multiple alternative energy sources, a control center must be included in the design that can manage the energy creation, distribution, storage, and return unused energy to the power grid.

5.1.6.4 High-Level Technical Description of the DC Power Grid System

The purpose of the multiple alternative energies DC power grid is to:

- Supply DC power to a customer community of 500 homes and a biosphere for growing food.
- Provide a storage device to store energy in case of a system failure.
- Determine the amount of surplus energy that is above and beyond what is required to power the grid and charge the backup system, and then deliver that surplus energy to the national power grid systems.

Throughout the semester you will be asked to analyze, design, and architect different components of the overall DC power grid system. This will include research into alternative designs, like lithium nanowire storage systems for backup power and/or supercapacitors (also called ultracapacitors). Fig. 5.8 illustrates these technologies connected to the DC Power Grid Regulator Control Center.

5.1.6.5 High-Level Technical Requirements for the DC Grid System

- Design shall utilize at least 2 alternative energies, with a maximum of 4.
- Total power generation shall be 3.0 megawatts minimum.
- Each of the alternative energy sources shall be capable of generating at least 60% of minimum wattage.
- DC power grid shall generate DC power, with grid voltage being 48, 24, or 12 volts.
- DC power grid shall be capable of servicing at least 500 single family homes and one 4-acre biosphere with hydroponics capable of growing vegetables year-round. (See Fig. 5.9 for a block diagram of a DC-powered home).
- All lights, appliances, and facilities in each home and the biosphere shall be configured for DC operations, if possible. Things like electric ovens may need DC/AC conversion to accommodate their 220v. requirements.
- Only allowed exceptions are the HVAC system and the Washer/Dryer, although DC is preferred.
- There shall be energy storage capability capable of operating the grid of 500 single family dwellings for 7 days if all alternative forms of energy are not producing.
  - Compliance Criteria: Energy storage devices shall be capable of producing 750,000 watts for the entire

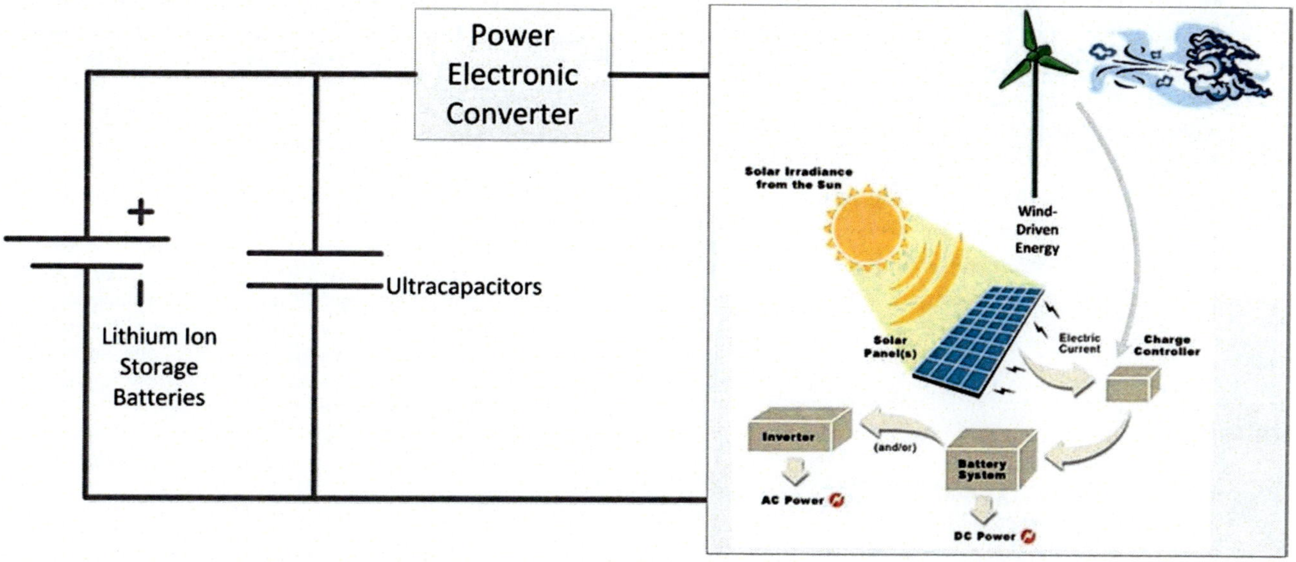

Fig. 5.8 Alternative power management system overview

Fig. 5.9 Example of a whole house power grid

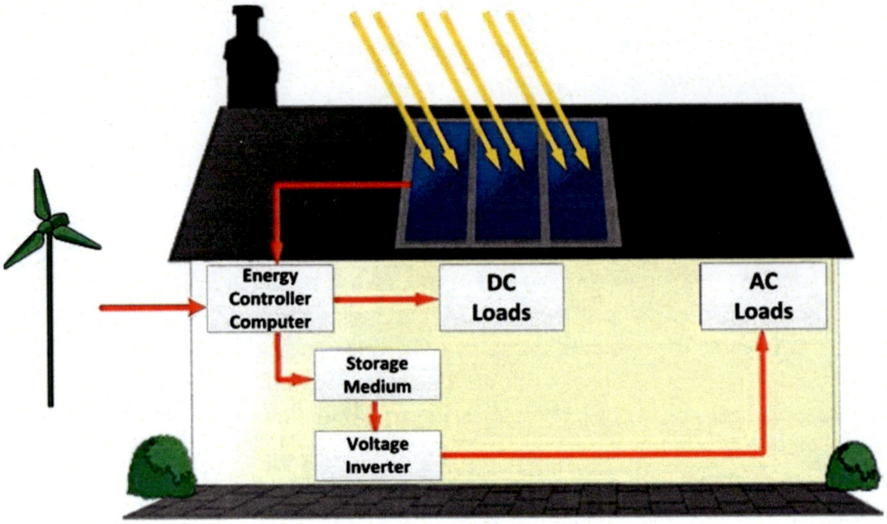

7-day period. Nominal energy storage should be batteries or supercapacitors.

- The DC power grid system shall include controlling h/w and s/w to manage the multi-alternative energy strategies and production.
- All energy produced above the required home and energy storage usage shall be upconverted to 220vac for release to the national power grid.
- The DC power grid shall be capable of operating in a temperature range of −40 degrees F. to 125 degrees F. without degradation of power generation.
- The DC power grid must meet OSHA regulations.

5.1.6.6 Assignments

- Write a report on each of the following, answering the questions asked in each section:
  - The elicitation process for your system.
    - Describe the major functional requirements.
    - Describe the major nonfunctional requirements.
  - The requirements analysis for your system.
    - Describe three requirements that came out of your requirements analysis.
    - Describe how you validated that your requirements set will meet the objectives of the system.
    - Describe your prioritization process and describe your top 5 prioritized requirements and why they are the highest priority requirements.

References

1. Wagner, S. and Deissenboeck, F. (2007). In 5th Workshop on Software Quality, IEEE Computer Society Press.
2. Yaneer, B. (2004). The characteristics and emerging behaviors of system-of-systems. In: NECSI: Complex Physical, Biological and Social Systems Project.
3. Zachman, J. (1987). A framework for information systems architecture. *IBM Systems Journal, 26*, 3.
4. Zachman, J. (2003). *The Zachman framework for enterprise architecture: Primer for enterprise engineering and manufacturing*. Zachman International.
5. Ertas, A., & Tanik, M. (2000). Transdisciplinary engineering education and research model. *Transactions of the SDPS, 4*(4), 1–11.
6. Crowder, J. and Carbone, J. (2012). Reasoning frameworks for autonomous systems. Proceedings of the AIAA Infotech@ Aerospace 2012 Conference, Garden Grove, CA.
7. Simon, D. (2000). Design and rule base reduction of a fuzzy filter for the estimation of motor currents. *International Journal of Approximate Reasoning, 25*, 2.
8. Lucena, J. (2013). *Engineering education for social justice*. Springer Publishing. ISBN 978-94-007-6350-0.
9. Azim, S., Gale, A., Lawlor-Wright, T., Kirkham, R., Khan, A., & Mehmood, A. (2010). The importance of soft skills in complex projects. *International Journal of Managing Projects in Business, 3*, 387–401.
10. Antonsson, E. (1987). Development and testing of hypotheses in engineering design research. *ASME Journal of Mechanisms, Transmissions, and Automation in Design, 109*(2), 153–154.
11. Petroski, H. (1992). *To engineer is human: The role of failure in successful design*. Vintage Books. ISBN-10: 0679734163.
12. Saunders, T. 2005. System-of-systems engineering for air force capability development: Executive summary and annotated brief. US Air Force Publication, Scientific Advisor Board (Air Force), SAB-TR-05-04, Washington, DC.
13. Joshi, C., Trani, A., & Baik, H. (2005). *Development of a decision support tool for planning rail systems: An implementation in TSAM*. MS Thesis. Virginia Polytechnic Institute and State University.
14. IEEE Std. 830-1998 (1998) Recommended practice for software requirements specifications. Los Alamitos:IEEE Computer Society.
15. Kerzner, H. (2003). *Project management—A systems approach to planning, scheduling, and controlling*. John Wiley.
16. McConnell, S. (1998). *Software project survival guide*. Microsoft.
17. Phillips, D. (2000). *The software Manager's handbook*. IEEE Computer Society.

6 MDRE: Systems Engineering and Requirements

Abbreviations

AC	Alternating current
CDP	Concept definition phase
CONOPS	Concept of Operations
COTS	Commercial off-the-shelf
CPP	Concept prototype phase
DC	Direct current (or 0 Hz)
DoD	Department of Defense
EIA	Engineering Industries Alliance
F	Fahrenheit
HVAC	Heating, ventilation and air conditioning
ICD	Interface control document
IEC	International Electrotechnical Commission
IEEE	Institute for Electrical and Electronic Engineers
ISO	International Organization for Standardization
MBSE	Model-based systems engineering
MDRE	Multidisciplinary requirements engineering
MDSE	Multidisciplinary systems engineering
NASA	National Aeronautics and Space Administration
OCE	Office of Chief Engineer
OPSCON	Operations Concept
OSHA	Occupational Safety and Health Administration
PDR	Preliminary design review
SoO	Statement of objectives
SoW	Statement of work
SRR	System Requirements Review
UL	Underwriters Laboratories
V	Volts
Vac	Volts AC

Requirements process standards have been developed within the multidisciplinary systems engineering discipline using establish baselines for requirements management and ultimately for system design. One example is the IEEE 29148-2011–Systems and Software Engineering Lifecycle Processes–Requirements Engineering Standards. Others include the IEEE 12207–Systems and Software Engineering–Software Lifecycle Processes [1]. These are international standards for establishing a common requirements framework for both systems and software development. The EIA 632 Processes for Systems/Requirement Engineering was established by the Engineering Industries Alliance (EIA) to focus on standards for conceptualization and implementation of complex systems through comprehensive and common requirements management practices. While a formal MDRE process helps to ensure all the requirements necessary for the life cycle of the system are gathered and maintained, in terms of capturing, recording, and managing requirements, this often begins with informal requirements. The notion of formal vs. informal requirements may seem confusing, as formal requirements are necessary to "sell-off" a system to the customer, but often the entire process of creating a set of high-level system requirements, a Statement of Objectives, and Statement of Work starts with an informal notion of what is needed by the customer [2].

Question to ponder

Think of two informal requirements that would be applicable to deciding on a car? Turn one into a formal requirement.

6.1 Formal Vs. Informal Requirements

Many times, informal methods are utilized as a basis for customer/stakeholder informational notions of the high-level objectives of the system. Plain English statements and hand-drawn diagrams may often become an initial basis for concept description and can be discussed and validated before a formal set of customer Statement of Objectives is created. As discussed above, it is extremely difficult and risky to begin development without a complete set of

J. A. Crowder, C. W. Hoff, *Requirements Engineering: Laying a Firm Foundation*,
https://doi.org/10.1007/978-3-030-91077-8_6

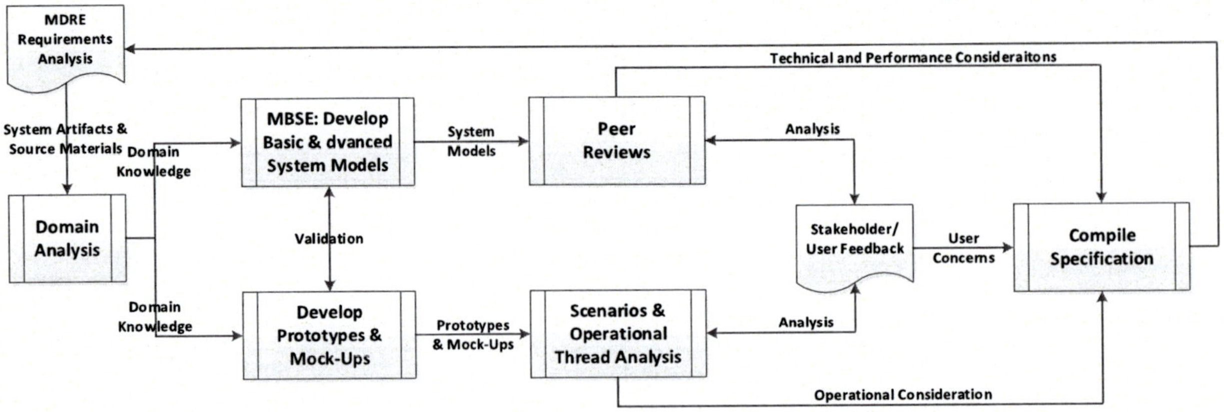

Fig. 6.1 High-level MDRE requirements engineering process

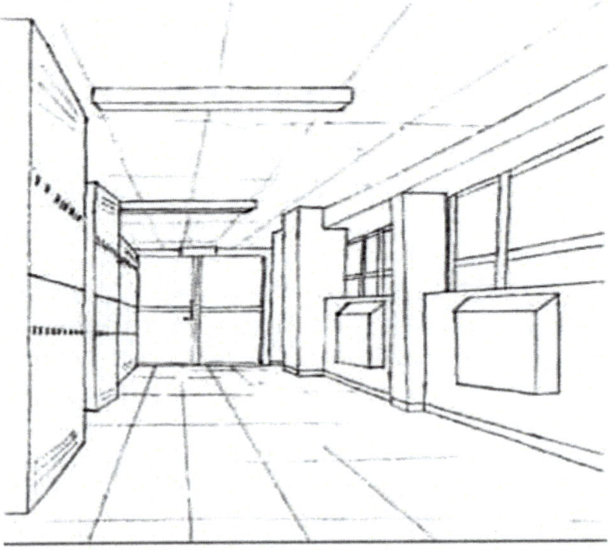

Fig. 6.2 Example of an informal concept drawing

requirements [3]. A good quality requirements specification starts with the MDRE process illustrated in Fig. 6.1, which is an expansion of the requirements analysis process described in Fig. 5.4.

The formal use of prototypes and mock-ups (especially for user displays) helps the customer and end user understand the "operational" concept for the design. Formal or informal user scenarios and an operational thread analysis ensures that the final system operates as expected both from a functional perspective as well as a "user experience" perspective. After evaluating the prototypes, mock-ups, and operational scenario information, the customer and stakeholders can provide feedback to the MDRE team—allowing adjustments to the requirements (whether formal or informal) before major development of the system begins. Next, we discuss the nature of informal vs. formal requirements.

6.1.1 Informal Requirements

Most development projects require formal methods as early as possible in the MDRE process [4]. However, from the customer/client perspective it is often the case that the original future system concepts are described using informal methods. It is important to capture these initial informal notes, sketches, and requirements (written as plain sentences), as they are often the basis for the first set of formalized requirements [5]. Fig. 6.2 provides an example of an informal sketch that might be used to provide insight into a project that would be formalized later with formal drawing(s) and exact dimensions. However, this would be useful to relay overall objectives from customer to contractor.

> **Question to ponder**
>
> What could happen if the customer and the contractor do not agree on the interpretation of the requirements?

Informal requirements text and/or drawings may be used to kick-start the requirements elicitation and analysis process if the customer has not provided a formal requirements document. Table 6.1 illustrates the difference between informal and formal requirements [6].

In the early stages of requirements analysis, informal discussions between the MDRE engineers and stakeholders/customers are often necessary to take informal, vague, ambiguous, and incomplete requirements, objectives, and operational notions and produce a set of understood and agreed upon objectives that will drive creation of a formal requirements document [7]. This is needed to understand the implicit information expressed in the informal requirements and objectives (i.e., what the customer was thinking when

Table 6.1 Informal vs. formal requirements presentation

Informal requirements	Formal requirements
Customer/user community	MDRE engineer(s)
Natural language	Formal requirements language
Imprecise	Precise
Non-technical	Technical
Application terminology	Engineering technology (systems, hardware, software)

they wrote them) to create explicitly expressed components, objectives, and high-level requirements. In the early stages, the requirements analysis research will provide the MDRE engineers with guiding principles to improve the natural language requirements without necessarily introducing formalism into the requirements before understanding and agreement between the MDRE and customer has been achieved [8]. If informal requirements/objectives are provided instead of a formal requirements specification, it is often beneficial to involve stakeholders/customers in a joint development effort to produce a formal specification. This has the advantage of reducing ambiguities between how the MDRE engineer interprets informal documents and how the customer interprets them. This may draw out the requirements process but provides instantaneous feedback between customer and the MDRE team [9]. It also provides a common point of reference/agreement of how (and when) these requirements are verified.

6.1.2 Formal Requirements

As we have discussed, one of the first MDRE processes is the elicitation phase. The effectiveness and efficiency with which elicitation is performed set up the development project either for success or failure and drives the entire system life cycle. Elicitation of formal system requirements is built on several successive factors that must be adhered to throughout the elicitation and analysis processes [10].

6.1.2.1 Requirements Elicitation Process

The MDRE elicitation process is very dependent on the contractor organization's business processes, organization procedures, and lines of management control (responsibilities). If the MDRE team is not allowed to control the elicitation process and the lines of responsibility are not clearly delineated, it will be difficult to perform an effective requirements elicitation [11].

6.1.2.2 Communication between the MDRE Team and the Customer

Communication within the MDRE team and between the MDRE team and the customer/stakeholder base is critical to a successful requirements elicitation process. One of the main sources of miscommunication is terminology differences between the customer and the MDRE team [12]. Differences in terminology lead to requirements ambiguity since there is not a clear understanding and/or agreement on terms used within the requirements. Lack of formality and rigor can easily derail a development. And even in an agile development where formality has been suspended for software functionality, the inherent complexity of systems still requires rigor, even with informal requirements and user stories [13]. Successful project development never happens by "luck."

6.1.2.3 Requirements Traceability

The complexity of systems makes the ability to trace requirements from top-level system requirements to low-level hardware or software requirements crucial to the overall success of a project. Given the frequency of software versions throughout the life cycle of a system, formalization of hardware and software requirements is essential. Lack of formalization can result in the loss of the original intent of the requirements and makes software maintenance difficult, if not impossible [14]. Also, making sure the hardware, systems, and software requirements are in sync across multiple changes is crucial to sustainability and maintainability of a system, not to mention crucial to maintaining a consistent requirement set throughout the development phase. Many times, the system architecture documentation is out of sync with the software documentation due to multiple changes in software requirements that were never reflected in the overall system architecture documentation. This can lead to excessive down-time when issues arise trying to determine which documentation of the system is correct [15].

6.1.2.4 Documentation Organization and Management

Many system development projects start with a formal set of requirements. However, some projects start with requirements in a variety of formats and media. It can be photographs, sketches, natural language text files, videos—all can be uploaded onto a computer system and sorted. No matter how the information comes to the MDRE team, which may even be in email or scanned in handwritten notes, these media objects, which may constitute objectives and/or requirements, must be formally tagged, documented, and organized. This will make the elicitation and analysis processes possible and much more efficient [16].

6.1.2.5 Requirement Language Formality

Even though informal requirements may initially be transmitted to the MDRE team with natural language, such notation can suffer from omissions, over-specified requirements, redundancy, contradictions, incompatibility between requirements, and a host of other issues that must be adjudicated by the MDRE team before formal elicitation and analysis can proceed; and formalism is necessary to reduce/eliminate ambiguity and establish methods and test cases to validate the requirements. The ultimate result should be a formal, agreed upon requirement specification document that will drive design, development, deployment, and sustainment of the system [17].

> **Question to think about**
>
> *What is the net effect of allowing formal requirements to age and become out of sync with the changes on the program?*

6.1.3 Use of Informal Requirements: Agile Development

Part of the Agile Manifesto pattern is producing working software at the end of each sprint [18]. Working software at the end of the sprint is considered preferable over formal, comprehensive requirements documentation. Since change is a "built-in" part of the agile development environment, requirements for a system developed using the agile method may be informally expressed in non-technical "story boards," which portray what the system should do and how it should interact with both its operational environment and the users who must operate the system. The agile development model uses development iterations (or sprints) lasting 2–4 weeks [19] using the informally developed user stories as requirements for the development sprint. Once the sprint is completed and tested to the customer's satisfaction, requirements documentation commences for the current sprint, based on the "as-built" software. One issue in using informal methods for requirements during sprint development is ensuring that requirements are not conflicting between sprint development teams and ensuring that the low-level requirements formalized at the end of the sprints link up to higher-level requirements and objectives provided by the customer [20]. The agile development process and its subsequent requirements documentation requires substantial customer interaction throughout the entire development process to ensure the product meets customer expectations. During these development sprints, the MDRE engineer(s) must provide multiple support areas to the development team, including [21]:

Identifying emerging requirements coming out of sprint testing.

Ensuring the developed software does not violate standards or policies.

Constantly keep track of possible conflicting requirements and/or features coming out of each sprint across multiple sprint teams [22].

Calculating metrics during each sprint test period and noting emerging performance trends as the software capabilities build across successive sprints.

Creation of test cases that build on each other across sprint teams and successive sprint periods [23].

> **Question to think about**
>
> *What problems may ensue with 4 sprint teams working in parallel without clearly defined functional requirements to guide them?*

6.2 Primary/Contractual Requirements

A primary/contractual requirement typically defines a capability need within the system to meet the objectives the customer is trying to achieve. Primary requirements typically define who, what, when, where, how and constraints as needed to meet an objective [24].

Examples:

- The sales system shall provide automated sales summaries by close of business daily by employee, to maintain an office staff of 25 employees.
- The sales system shall provide automated (how) metrics summaries (what) by close of business daily (when) by employee (constraint), to maintain an office staff of 25 employees (objective).
- The sales system shall provide the capabilities for an office manager (who) to view the daily sales summaries by employee (what).

Included in primary (contractual) requirements provided to the system engineers, the stakeholders typically include learned lessons from other developments or procurements. These requirements may or may not, and probably are not, defined at the top tier of the system or objectives. This requires the systems engineers to properly tier the requirements in the decomposition of the system and provide traceability to the primary requirement set [25]. This may require the systems engineer to derive system tier requirements, if the lower-level tier requirement is not part of a higher-level objective or primary requirement.

6.3 Statement of Work (SOW) and Statement of Objectives (SoO)

6.3.1 The Statement of Work (SOW)

The **Statement of Work (SOW)** defines the overall scope of the development effort. The SOW captures the activities, system deliverables, and a basic time frame in which the system must be delivered. The SOW may provide detailed requirements, but there may be a separate requirements document. The SOW captures all elements of a project including tasks (activities), deliverable items, and a notional schedule for delivery of the system. It is generally a detailed layout of the project that is used to create an overall project plan. There are generally three types of SOWs utilized across industries [26]:

1. **Design-Centric SOW**: The purpose of a design-centric SOW is to convey to the contractor or developer how you want the system developed. Here, the requirements can be at various levels, from high-level architecture, factors like quality and reliability, to detailed measurements of materials. This is risky for the customer to write such a detailed design SOW, since it directs the developers how they should develop the system. If the system does not perform per expectations, the developer can point back to the customer and blame them for providing incorrect requirements on how to develop the system. In general, requirements should say what the system should do, but not how to achieve the requirement [27].
2. **Level-of-Effort-Centric SOW**: In some cases, it is not clear what is required or what is desired by the customer. In these cases, creating a SOW that defines a general set of development guidelines without specific requirements and lays out a "time and materials" description of hours to be put forth at the customer's direction. The notion of Level-of-Effort denotes the contractor is basically an hourly worker performing tasks defined over time by the customer [28].
3. **Performance-Centric SOW**: This is the general type of SOW provided from a customer to a contractor. The Performance-Centric SOW focuses on the overall purpose of the system to be developed, the notional schedule that is expected, the milestone reviews expected, and the expected quality of the final project (which may include performance metrics). What is not in the Performance-Centric SOW is "how the system should be developed." This allows the contractor to design the system, possibly in a variety of ways. If the system is out for competitive bid, each contractor may have entirely different designs, but all perform the tasks and meet the operational needs of the system [29].

6.3.1.1 Writing a Statement of Work

A SOW comprises many sections, each covering different aspects of a project. There are many templates available online and utilizing a template is advised to help get started. The main sections that are normally required for an SOW are outlined below:

1. **Introduction**: A high-level description of the work to be performed and who will be doing the development.
2. **Focus of the Work to be Performed**: First and foremost, this section should highlight why the development project is being initiated, who is funding the development, and what is the overall purpose of the system. It should include a high-level view of the objectives, deliverables, and operational needs (who will operate the system).
3. **Work Scope**: What are the tasks to be performed on the project. This may include the hardware and software needs (at a functional level, not the specific hardware and software to be designed/written). This should include expectations of performance, schedule, and a notional timeline for deliverables [30].
4. **Task Descriptions**: The high-level task descriptions provided in the Scope section should be broken down into manageable tasks and more detail provided in what is expected to be developed and tested during each task. This should include a notion of major milestones that are expected along the development timeline, even though a detailed milestone discussion should be included [31].
5. **Milestones and Deliverables**: This section should describe (in moderate detail) what is due when along the defined schedule. Performance characteristics like preliminary design completed, detailed design completed, prototype completed and tested, should be described and highlighted, with customer reviews (e.g., System Requirements Review) described and laid into the development timeline [32].
6. **Detailed Schedule**: This includes what deliverables are due when in more detail, including how the program is kicked-off (started), what the specific period of performance is expected to be, which major elements shown like design, implementation, system test, delivery, etc.
7. **Standards and Policies**: If there are industry standards and legal policies that must be adhered to, they should be described here. An example would be an electrical system that must adhere to the Underwriters Laboratories (UL) commercial electrical standards. Policies that must be followed may differ at the federal level and be different by state and city. These should be discovered and reviewed during the requirements analysis process, unless specifically directed in requirements by the customer [33].
8. **Requirements**: If there are any high-level requirements that affect the SOW, they should be annotated to ensure the SOW is accepted with these requirements.
9. **Acceptance Criteria**: Here the overall features and attributes the system "must have" to be accepted by the customer should be spelled out in detail. This should include spelling out the deliverables required for acceptance throughout the schedule timeline.

6.3.2 The Statement of Objectives (SOO)

A Statement of Objectives (SOO) is an alternative to a SOW and provides the overall customer objectives and the required support that must be provided by the contractor to achieve the objectives. A modern example of an objective would be "the staff must analyze and report the company's social media metrics" [34]. While not as detailed as the SOW, the SOO should still be compatible with the customer's need statement(s) and the Task list and descriptions. Often, an SOO is prepared and used as the basis for a formal SOW. It allows preliminary discussions to ensure between the customer and potential contractors/developers to establish a feasibility study to understand the technical, cost, and schedule challenges that may hinder development of the system or capabilities set out in the SOO. It is easy to say, "my objective is faster than light travel," but it is quite a different matter to understanding its feasibility.

6.4 Interface Control Documents

All systems must interface to something. Systems are designed to provide capabilities and perform actions for something or someone. Even the Curiosity Rover on Mars must interface with its environment, in this case, the Martian surface. The purpose of an interface control document is to provide a record of detailed interface information about all known interfaces the system must contend with or connect with. This includes requirements, drawings, diagrams, metrics, etc. that define how the system interacts with a given interface (e.g., human operator). An interface control document (ICD) includes all the inputs and outputs for the system being developed. How it gets its input data and how it connects to provide its outputs. It might include information about interfaces between subsystems, between major systems (e.g., a spacecrafts interface with the international space station), or specific interface requirements like connector configurations of voltage limits, or timing relationships between specific signals. The overarching purpose of an ICD is to control and maintain a record of every input and output interface required for a given system, which may change over time. There is much criticism from the agile development community that ICDs are a cumbersome set of documentation that slows development efforts, however, they are useful, and it is not necessary that they be a textual document [35] but may something as simple as an electronic database table of inputs and outputs with basic control logic tied to a set of electronic diagrams.

6.5 Compliance Documents and Policy Compliance

Given that most systems are designed to operate in the world around us (NASA's Curiosity and Perseverance not withstanding), all systems must comply with some regulations. Whether it is zoning regulations or anti-fly-zone compliance, every system show compliance with one or more city, state, or federal policies, regulations, or general guidance documents [36]. A simple example of a compliance document is a "proof of insurance" document that some states require of automobile drivers. Another compliance document required for automobiles is a driver's license. When stopped by the police, you may be asked to produce the compliance documents to prove you follow state laws concerning the operation of a motor vehicle. In general, compliance documentation refers to specific documents or information pertaining to records, reports, observations, or even verbal responses to verify compliance with certain policies or standards related to the system or activity. For a given system, the very design might be governed by certain compliance regulations [37].

Question to think about

Can you think of systems that would require a lot of compliance regulations (e.g., *Nuclear Power Plants)?*

6.6 Concept of Operations (CONOPS)

A Concept of Operations or CONOPS is a document that describes the overall characteristics of a system from the perspective of the operators or users of the system. A CONOPS is useful to communicate quantitative and qualitative characteristics and behaviors to both stakeholders and developers without lower-level requirements or quality attributes of the system (e.g., the color must be blue). A CONOPS communicates the basic set of capabilities the system must exhibit to meet at least the minimum set of objectives the system is being developed to fulfill. Depending on the customer, the CONOPS may be a formal document that is maintained throughout the system's entire life cycle. From the CONOPS, a number of requirements may be derived at the system, subsystem, or component levels. In military terms, CONOPS is a verbal or graphical representation that expresses, in precise terms, the intent of the system, how the system should be developed, and how it should be developed within the available resources. For the Department of Defense (DoD) there are specific requirements on how a CONOPS is to be written

(ISO/IEC/IEEE 15288:2015). At a minimum, a CONOPS includes:

- A statement of goals of objectives for the system to be developed.
- The strategies, policies, and constraints required for the system.
- Statements describing the responsibilities and authorities granted.
- Processes required to deploy or install the system in its operational environment.
- Processes for development, maintenance, and sustainment of the system.

A CONOPS is provided to the developer by the customer. Once analyzed, the development contractor should respond by developing a system Operations Concept (OPSCON) document that describes the will do and will not do characteristics of the to be developed system. In contract to the CONOPS, the OPSCON is a developer-oriented document describing the system characteristics that will be delivered to the customer. The OPSCON basically takes the information from the CONOPS and provides a "I know what your CONOPS said, and here is what it means to us and what we are going to deliver" [38].

Question to think about

Can you think of issues that might arise if an OPSCON is not generated by the developer before building the system?

6.7 Case Study: Requirements Management Disaster

As we have discussed above, often, new systems are extensions to older systems/solutions. Many new systems are developed by modifying existing system designs (both hardware and software) to fast-track development of new capabilities or a new system. Often the current software is adapted to run on new commercial hardware and may include adding commercial software and applications to enhance the capabilities, while trying to integrate a solution at the lowest cost possible as quickly as possible. One of the major issues that can affect new systems is the need to enhance the current/older solution with modernized commercial-off-the-self (COTS) hardware and software, new security hardware and procedures based on new policies and restrictions. A major pitfall with such system is the need to "build more for less cost" [39]. In these instances, solutions are offered to the customer that are refactored versions of current hardware and software solutions that are supposed to provide the new functionality required by the customer. All information related to the actual customer, the contractors, and the application have been removed. But what follows was an actual development effort gone wrong.

6.7.1 A Classic Pitfall

We have discussed the problem of different interpretations in terminology and requirements. What follows is a classic example of differences in expectations and terminology definition differences:

- From the customer side, they are spending $1 Billion on a "new system" and they expect that their new system "***will be new***."
- From the contractor side, their focus is on adding new capabilities to an existing set of solutions "***without major changes***" to their existing product lines to maximize profits and minimize time to market, while providing new capabilities to the customer.

These differences in expectations are likely to result in:

- Massive cost overruns due to rework and meeting the customers' expectations.
- Canceled program.

6.7.2 Background Material

Technical excellence will most always be supported with cost and schedule, but the inverse is not true. Working to cost and schedule, while failing to deliver on technical merit, will always lead to failure.

> If you take care of technical merit, cost and schedule will take care of itself [40].

This case study will focus on a program that was started, went to stop work, restarted, and then finished with a cancelation and no deliverables. This study will expose issues with both the customer and the developer. The customer operates multiple data centers geographically distributed, where for ten (10) plus years, the customer had initiated studies using multiple vendors to find a solution to an integrated system operating across geographical dispersed operational centers. In numerous cases, medium sized development programs were started and stopped due to the mergers in both customer and development organizations. Finally, a formal competition was undertaken to establish a complete strategy to buy, integrate, and field a system of system solutions operating geographically 24x7. The first phase, Concept Definition

Phase (CDP), was initiated with four (4) vendors with extensive experience in the marketplace as well as building and fielding solutions [41]. The competition was executed with the four teams, where the outcomes were eventually shared and scored by the customer and vendors.

The second phase, Conceptual Prototype Phase (CPP), requested the vendors bid and build solution for a set of key capabilities identified by the customer and joint vendor teams. The prototypes were down selected and awarded to three of the four competitors. There was overlap in awards, such that first vendor won five of the prototypes, the second vendor received four, and the third vendor received two of the prototype capabilities. While the prototype phase was underway, and with the third vendors realizing they were in a losing position, the third vendor developed a strategy to eliminate the other two vendors from the competition. In so doing, the third vendor established a "strategic integration center," and developed a system solution using mostly new COTS products from the marketplace, conducted a demonstration, and provided an estimated build cost for the solution (for this review we label it as "The Marketed Price"). The value was so low that it was identified as "trivial cost" by one of the other vendor's Chief Engineers [42].

While the other vendors executed the CPP prototype developments, the contractual offering was awarded to the third competitor as an unsolicited award. To avoid lawsuits and challenges to the award, the customer invited all three CPP vendors to join and build the new system under the leadership of the third vendor. The forced marriage had numerous problems to overcome, namely, all three vendors could build the system, or portions of the system, on their own. This led to an extremely intense in-fight to determine who was responsible for what pieces of technical work share and massive disagreements about what the requirements meant and who was responsible for which requirements. The program was carved up into three major pieces: Program Management, System Development, and System Transition. The system development was subdivided into three main categories, with forced team integration on two of the three integrated development teams.

6.7.3 Program Execution

During contract startup, the integrated development team provided a pre-contract system engineering overview to the customer and the leads from the geographically dispersed operational centers [43]. As part of the post assessment review, two of the four major systems engineering architecture artifacts were found to be unacceptable, the Concept of Operation and the System Requirement Specification. The System Architecture and System Transition overviews were graded by the integrated customer team and found to be acceptable. With outstanding concerns from the systems engineering overview out brief, the program started and was staffed, and development cycles begun. Rather than taking the time to rework the CONOPS and system requirements specification before development began, the systems engineering team made the decision to rework the requirements and CONOPS during the engineering development cycles leading up to the formal Systems Requirements Review (SRR) [44]. To correct the CONOPS, the systems engineering team decided to perform user scenario modeling and requirement deviation derivation, rather than elicitation and requirement analysis which would have led to decomposition, requirement derivation and architecture assessment and refactor.

As requirement engineering was an ever-ongoing process, the program management team developed the final cost of the system, pulling key personnel out of system engineering to help work the bid models, build plan, schedules and like artifacts to manage the cost and schedule of the program. The initial submission to the customer was a 100% increase in cost to "The Marketed Price." This new cost led to multiple cycles of interruption to the requirement engineering efforts, prior to finalizing the "Build to Price" (about 40% higher than "The Marketed Price").

Leading into SRR, the systems and requirements engineering efforts to correct the CONOPS, and system requirements specification failed to be achieved. Also, by SRR. The software and hardware teams were staffed and were already prototyping the system without an established set of hardware or software requirements. During SRR, the main systems engineering artifacts, CONOPS, system requirements, architecture, and transition, were again reviewed with the integrated customer team, and, again, the CONOPS and system requirements specification failed to pass customer approval. This led to multiple meetings, and program realignment of key personnel, to provide more focus on CONOPS and requirement specification corrective actions; however, both the customer and vendors did not want to stop work for 3–6 months to correct the major discrepancies. It was decided to fix the issues during development activities leading to the system Preliminary Design Review (PDR).

6.7.4 The Path to Preliminary Design Review Disaster

Over the next 6 months, multiple cycles on user scenarios and requirements specification were performed in parallel with software and hardware development activities. The requirements associated with software and hardware activities were not fully vetted and approved for development. This led to bottoms-up development, where the software and hardware developers fell back on prior system development efforts and "built what they knew needed to be built." The development schedule was revised to support critical development tasks, and a build plan to satisfy the customer's development objectives. Also, while the program management team and the Office of Chief Engineer (OCE) were building the new critical development schedule and development plan, the development manager formed a parallel team of engineers to decide what they would build in each iterative development cycles independently of the program management team, customer objectives, and requirements changes being constantly added into the system requirements specification.

At the 1-year mark, and 3 months prior to the PDR, the costs and schedule overruns were becoming apparent, as well as the gaps in the system's requirements and user scenarios. The new user scenarios were not helping to solve the CONOPS issues, and the actual development of software and hardware were radically off course relative to the critical capabilities and development plan. The customer requested a corrective action plan, and a program cost, schedule, and technical engineering assessment were performed. With the technical problems well understood by the OCE, the corrective action plan, for the technical merits of the system, was to stop development activities, and pursue a proper top-down requirements decomposition, elicitation, analysis, and unfortunately a scrapping of some of the developed software. The cost and schedule assessment showed a potential grow of cost of about 40%, caused by rework and schedule growth.

The development manager was dismissed for building invaluable software and program management overruled the OCE corrective action for a 3-months period-of-time to allow the development teams a chance at integrating what was built to date. The hardware and software developed bottoms-up failed to integrate, and the technical fixes to the baseline were not achieved due to personnel shortages caused by support to the integration activities.

Fully 1.5 years into the development the program came to a stop. 65 of the staff of 325 were retained to evaluate and correct the technical merit issues, as well as the cost and schedule profile for the program. The OCE was scaled down to four staff members, a prototype staff of 15 members, with the rest of the staff working re-planning of both cost and schedule. During the first 3 months, the CONOPS was redeveloped by knowledgeable engineers and operators from the operational nodes, and formally approved by the customer. Over the next 4 months, the OCE and prototype staff correct the entire requirements set and build of a fully functional prototype for the customer's end system vision, which was also formally approved by the customer. In parallel with the requirements efforts, the prototype software development team working on the prototype removed code developed by the original program staff, corrected software design deficiencies, instituted night builds using automated checks, automated unit test runs, and integrated system startup tests. Their efforts took a nonfunctional hardware and software baseline, removed 30% of the duplicate bottoms-up code, and delivered a prototype fully capable of operating in an operational node. With 2 months of coordination, the program was restarted 9 months after the initial stop work order [45].

6.7.5 The Path to Preliminary Design Review Disaster #2

With approval from the customer for all the big four (CONOPS, Requirements, Architecture, Transition), the program started re-staffing, with a requirement to be ready for a full PDR 6 months after restart. However, as with all developments there is a price to restart; namely, the customer was asked to include some new capabilities by the operational nodes since they were now 2 years behind schedule. The additional requirements needed to add capabilities caused major issues with development. This is due, unfortunately, to forcing vendor work force back together in the forced marriage, which itself came with issues. The prime was no longer in charge of the OCE, two of the sub-contractors were given new work share allowances smaller than the original pre-stop work contracts, but with more responsibilities to development. This led to some interesting behind the scenes negotiations between the program management and the vendors, as well as between program management and the new system engineering team, where both sets of negotiations were carried out without OCE oversight. By the time PDR was to occur, the customer observed the following major issues with the restart effort:

1. The new capabilities and resultant requirements development effort did not match the approved CONOPS.
2. The system readiness was SRR quality, not PDR quality.
3. The staffing and vendors infighting resulted in an inability to control engineering and development efforts.
4. And most importantly, the customer's objectives were at risk again.

Exactly 6 months after restart, and with PDR due, the program went back to stop work for the second time. Now, having started and stopped twice, the customer started accepting technical help for political reasons, and the vendor team was augmented by the addition of a fourth vendor, all with equal work share. Multiple technical meetings were conducted, technical requirements for development were reordered, and a new program plan was developed. This new restart was not going to solve the total objectives of the customer but was going to integrate a new display tier and scheduling solution, and then integrating into the existing operational baselines.

Over the next 6 years, with multiple technical directors from different customer organizations helping, and four vendors with different agenda(s), the program was finally terminated with a cost overrun of 500% of initial "The Marketed Price." The developed software and hardware solutions were scrapped, and the program terminated.

6.7.6 Case Study Assessment

There are multiple failures in this case study. They affect System Acquisition, Teaming, and most importantly multiple drastic failures in Requirements Engineering. While there is a well-founded fact that "The customer is always right, because they are the customer," the truth lies in the ability to convince the customer with facts and data that the path to success should be with sound multidisciplinary requirements and systems engineering.

6.7.6.1 Design Problem

The philosophy of the assignment development for this textbook was to create a set of thought-provoking questions, problems, and design assignments that help drive home the principles, practices, and methodologies of multidisciplinary requirements engineering (MDRE). We endeavored to provide the opportunity for the student to practice the art and science of requirements engineering, helping the student to understand that there is rarely one answer to a requirements engineering problem, and it is only through study and practice that successful requirements engineers are developed. MDRE is a life-long journey, and anyone who tells you they have nothing left to learn does not truly understand the discipline or has never been involved in the development of a complex system. With the recent interest in alternative energies and talk of the "Green New Deal," along with the recent issues with ice storms in Texas taking out the power grid, the assignments for this textbook will follow the theme of effectively dealing with both local alternative energy power grids and the national infrastructure required to facilitate large-scale alternative energy production, sustainment, and management.

6.7.6.2 Design Problem Description: Alternative Energy Power Grid

The purpose of this project is to, over the course of the semester, develop a feasibility study, requirements analysis, requirements elicitation, derivation, and allocation, for a system that will facilitate a "local" alternative energy power grid capable of powering a neighborhood of 500 homes and a large biosphere capable of supporting growing year-round fruits and vegetables for the 500-home neighborhood. The student will have to deal with issues of choosing multiple alternative energies that can provide power throughout all seasonal difficulties, from intense summer heat to freezing winter temperatures and possible ice storms. Control, viability, sustainability, and maintainability will be major factors to understand the entire possibly 30-year life cycle of such a system. How to effectively utilize, store, and manage power distribution, along with when to upconvert "extra" energy to the national power grid, will be issues that drive many of the assignments. Simply relying on the current technologies and energy infrastructures will not move the country and world forward. Hopefully, this will provide insight to the students using this textbook to understand some of the issues that face everyone looking out over the next many decades.

6.7.6.3 DC Power Grid Requirements Discussion

In the initial days of power generation and distribution, DC power was far less practical than the AC distribution system that soon supplanted it. However, even after Tesla's AC system became the predominant system (and rightly so), Edison's DC power did not become obsolete for some time. Eventually power plants eliminated their delivery of DC power, but the overall concept of a DC power grid is not dead. Even our computer systems (large server systems) have uninterruptible power supplies which contain batteries that are charged off the grid and used to upconvert to AC to power devices when the AC power grid goes off-line. So, the notion of a DC power grid is not dead, but generally used as a backup system. The point is, the notion of a DC power grid has remained and now that renewable energies are viable, we should re-examine their use for widespread, local, power distribution.

One of the advantages of a locally generated DC station is energy savings as well as allowing a facility to run more easily off various DC sources, such as Solar Arrays. And for power-hungry installations, the notion of generating the power locally is growing in popularity. Part of the reason is that waste heat from the power generation generators (solar, geothermal, etc.) can be used to warm nearby buildings. By combining the production of heat and power, local power grids can squeeze much more useful energy out of the renewable energy they use, so this approach will certainly become more widespread as time goes on. The purpose of this design problem is to look ahead and provide an architecture and specification for a local DC power generation station that can serve a local community.

Currently many systems exist to control and handle alternative energies. There are systems to control and manage solar power, wind power, geothermal power, and others, but very few are designed to control, manage, store, and distribute energy from a power grid consisting of multiple alternative energy sources. As we saw the issues recently caused by the intense ice storms in Texas, one alternative energy source is not the answer to the growing energy needs/wants in our society. Each alternative energy source has limitations and issues associated with its use. By harnessing multiple alternative energies, and managing the storage of the energies produces, communities can be worry free from power generation. To facilitate utilizing multiple alternative energy sources, a control center must be included in the design that can manage the energy creation, distribution, storage, and return unused energy to the power grid.

6.7.6.4 High-Level Technical Description of the DC Power Grid System

The purpose of the multiple alternative energies DC power grid is to:

- Supply DC power to a customer community of 500 homes and a biosphere for growing food.
- Provide a storage device to store energy in case of a system failure.
- Determine the amount of surplus energy that is above and beyond what is required to power the grid and charge the backup system, and then deliver that surplus energy to the national power grid systems.

Throughout the semester you will be asked to analyze, design, and architect different components of the overall DC power grid system. This will include research into alternative designs, like lithium nanowire storage systems for backup power and/or supercapacitors (also called ultracapacitors). Fig. 6.3 illustrates these technologies connected to the DC Power Grid Regulator Control Center.

6.7.6.5 High-Level Technical Requirements for the DC Grid System

- Design shall utilize at least two alternative energies, with a maximum of 4.
- Total power generation shall be 3.0 megawatts minimum.
- Each of the alternative energy sources shall be capable of generating at least 60% of minimum wattage.
- DC power grid shall generate DC power, with grid voltage being 48, 24, or 12 volts.
- DC power grid shall be capable of servicing at least 500 single family homes and one 4-acre biosphere with hydroponics capable of growing vegetables year-round. (See Fig. 6.4 for a block diagram of a DC-powered home).
- All lights, appliances, and facilities in each home and the biosphere shall be configured for DC operations, if possible. Things like electric ovens may need DC/AC conversion to accommodate their 220v. requirements.
- Only allowed exceptions are the HVAC system and the Washer/Dryer, although DC is preferred.
- There shall be energy storage capability capable of operating the grid of 500 single family dwellings for 7 days if all alternative forms of energy are not producing.
  - Compliance Criteria: Energy storage devices shall be capable of producing 750,000 watts for the entire 7-day period. Nominal energy storage should be batteries or supercapacitors.
- The DC power grid system shall include controlling h/w and s/w to manage the multi-alternative energy strategies and production.
- All energy produced above the required home and energy storage usage shall be upconverted to 220vac for release to the national power grid.
- The DC power grid shall be capable of operating in a temperature range of −40°F to 125°F without degradation of power generation.
- The DC power grid must meet OSHA regulations.

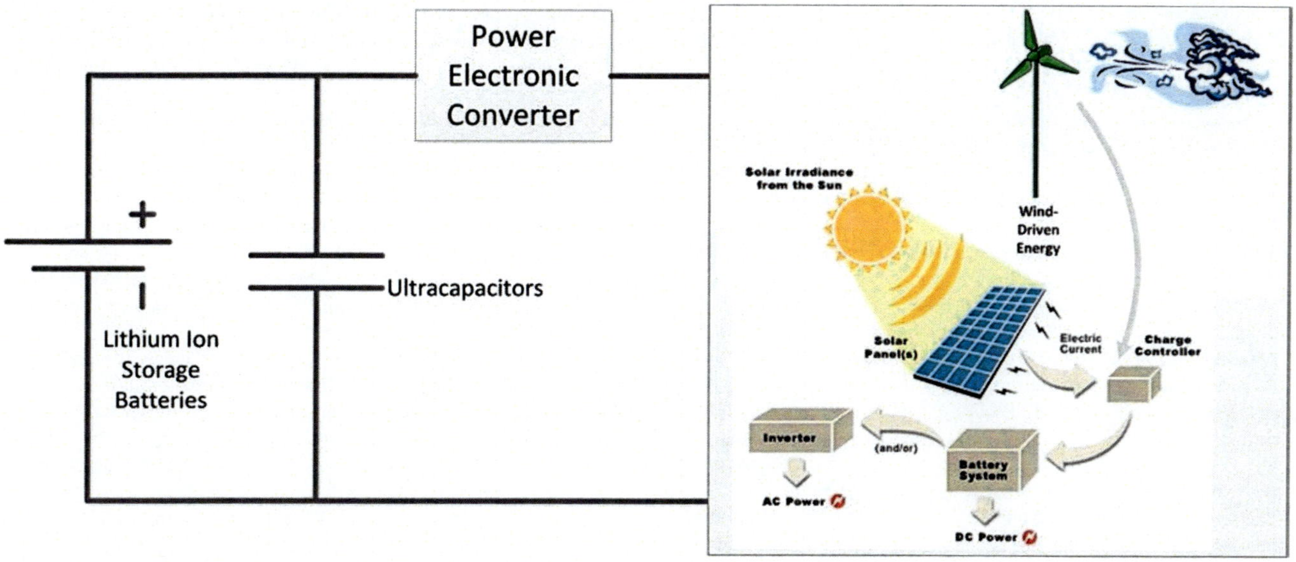

Fig. 6.3 Alternative power management system overview

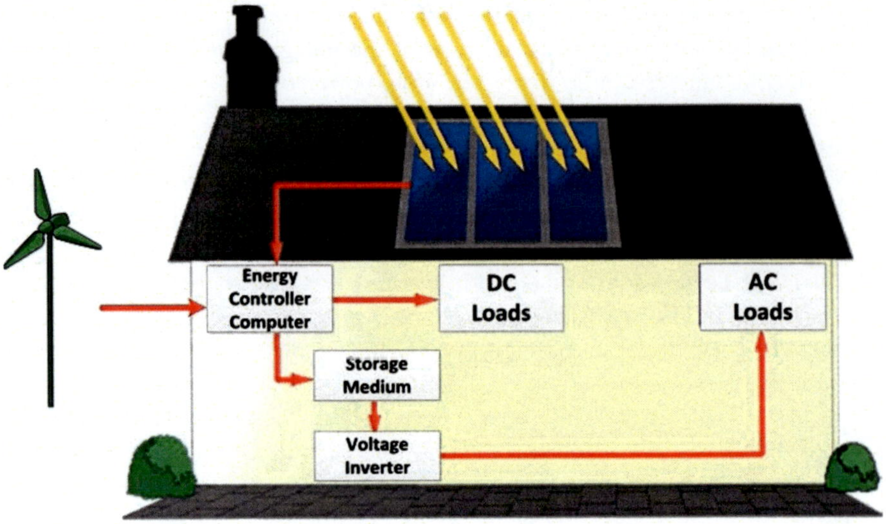

Fig. 6.4 Example of a whole house power grid

6.7.6.6 Assignments

- Using Figs. 6.3 and 6.4 as a graphical CONOPS, develop an OPSCON for your alternative energy system design.
- Using the High-Level Technical Description as a Statement of Objectives, develop a Statement of Work for your alternative energy system.
- Research regulations regarding alternative energy system and describe four compliance regulations your system must meet and how your design will guarantee compliance with these regulations.

References

1. Project Management Institute. (2004). *A guide to the project management body of knowledge (PMBOK®)* (3rd ed.). Project Management Institute.
2. Robertson, S., & Robertson, J. (1999). *Mastering the requirements process*. Addison-Wesley.
3. Standish Group International Inc. (1995). Chaos Reports, Retrieved 03/01/2006 from http://www.standishgroup.com
4. Visitacion, M. (2003, May) A practical guide to requirements management. IBM Rational Seminar.
5. Wiegers, K. (1999). *Software requirements*. Microsoft.
6. Wiegers, K. (2002, June) Software requirements: Eight traps to avoid. Project Management Institute Information Systems Special Interest Group June2002 Webinar.
7. Young, R. (2001). *Effective requirements practices*. Addison-Wesley.
8. Young, R. (2003). *The requirements handbook*. Artech House.
9. Kumar, V. S. (2006). *Effective requirements management. Paper presented at PMI® Global Congress 2006—EMEA, Madrid, Spain*. Project Management Institute.
10. Eberlein, A. and Leite, J. (2002). "Agile requirements definition: A view from requirements engineering," Proceedings of the International Workshop on Time-Constrained Requirements Engineering (TCRE'02), Germany.
11. Paetsch, F., Eberlein, A., Maurer, F. (2003). "Requirements engineering and agile software development," Eighth International Workshop on Enterprise Security, Linz, Austria, 9–11 June.
12. Rompaey, B. and Demeyer, S. (2009). "Establishing traceability links between unit test cases and units under test," 13th European Conference on Software Maintenance and Reengineering, Germany.
13. Palmer, S. R. and Felsing, J. M. (2002). "A practical guide to feature-driven development," The Coad Series.
14. Delgadillo, L. (2007). "Story-Wall: Lightweight Requirements Management for Agile Software Development," 15th IEEE International Requirements Engineering Conference, pp.377–378.
15. J. Hayes, A. Dekhtyar and D. Janzen (2009). "Towards traceable test-driven development," Proceedings of the 2009 ICSE Workshop on Traceability in Emerging Forms of Software Engineering, pp. 26–30.
16. Berenback, B., Paulish, D., Katzmeier, J., & Rudorfer, A. (2009). *Software and systems requirements engineering in practice*. McGraw-Hill Professional. ISBN 978-0-07-160547-2.
17. Hay, D. (2003). *Requirements analysis: From business views to architecture* (1st ed.). Prentice Hall. ISBN 0-13-028228-6.
18. LaPlante, P. (2009). *Requirements engineering for software and systems* (1st ed.). CRC Press. ISBN 978-1-4200-6467-4.
19. McConnell, S. (1996). *Rapid development: Taming wild software schedules* (1st ed.). Microsoft Press. ISBN 1-55615-900-5.
20. Nuseibeh, B. *and* Easterbrook, S. (2000). "Requirements engineering: a roadmap (PDF). ICSE*'00. Proceedings of the conference on the future of Software engineering." pp. 35–46*. CiteSeerX 10.1.1.131.3116. doi:https://doi.org/10.1145/336512.336523. ISBN 1-58113-253-0.
21. Stellman, A., & Greene, J. (2005). *Applied software project management*. O'Reilly Media. ISBN 0-596-00948-8.
22. Wiegers, K., & Beatty, J. (2013). *Software requirements* (3rd ed.). Microsoft Press. ISBN 978-0-7356-7966-5.
23. Abbott, R. J. (1983). Program design by informal English descriptions. *Communications of the ACM, 26*, 882–894.
24. Aguilera, C., & Berry, D. M. (1990). The use of a repeated phrase finder in requirements extraction. *Journal of Systems and Software, 13*, 209–230.
25. Ambler, A. L., & Burnett, M. M. (1990). Influence of visual technology on the evolution of language environments. In E. P. Glinert (Ed.), *Visual programming environments: Paradigms and systems* (pp. 19–32). IEEE Computer Society Press.
26. Balzer, R. (1985). 15 year perspective on automatic programming. *IEEE Transactions on Software Engineering, SE*, 1257–1268.
27. Balzer, R. M., Goldman, N., & Wile, D. (1978). Informality in program specifications. *IEEE Transactions on Software Engineering, SE*, 94–103.

28. Barstow, D. (1984). A perspective on automatic programming. *The AI Magazine, 5*, 5–28.
29. Borgida, A., Greenspan, S., and Mylopoulos, J. (1985). "Knowledge representation as the basis for requirements specifications," in IEEE Computer, pp. 82–90.
30. Bowen, J. P. and Hinchey, M. G. (1994). "Seven more myths of formal methods," Oxford University Computing Laboratory PRG-TR-7-94.
31. Bubenko, J., Rolland, C., Loucopoulos, P., and DeAntonellis, V. (1994). "Facilitating "Fuzzy to Formal" requirements modelling," presented at The First International Conference on Requirements Engineering, Colorado Springs, Colorado.
32. Burns, A., Duffy, A., MacNish, C., McDeremid, J., and Osborne, M. (1995). "An integrated framework for analysing changing requirements," Department of Computer Science, University of York PROTEUS Deliverable 3.2.
33. Chang, S.-K. (1990). Visual languages: A tutorial and survey. In E. P. Glinert (Ed.), *Visual programming environments: Paradigms and systems* (pp. 7–17). IEEE Computer Society Press.
34. Christel, M. G. and Kang, K. C. (1992). "Issues in requirements elicitation," Software Engineering Institute, Carnegie Mellon University CMU/SEI-92-TR-12.Formal and Informal in RE Cybulski 9.
35. Cmu/Sei (1991). "Requirement Engineering and Analysis," Software Engineering Institute, Carnegie Mellon University, Pittsburgh, Pennsylvania CMU/SEI-91-TR30.
36. Cybulski, J. L. "Sharing and reuse in the development of information systems," The University of Melbourne, Department of Information Systems, Melbourne, Research Report 96/4, June 1996.
37. Cybulski, J. L. (1996). "Reusing requirements specifications: review of methods and techniques," presented at 1st Australian Requirements Engineering Workshop, Melbourne, (To appear).
38. Cybulski, J. L., & Reed, K. (1992). A hypertext-based software engineering environment. *IEEE Software, 9*, 62–68.
39. Cybulski, J. L. and Reed, K. (1993). "The use of templates and restricted english in structuring and analysis of informal requirements specifications," Amdahl Australian Intelligent Tools Programme, La Trobe University, Bundoora, Research Report TR024.
40. Dankel, D. D., Schmalz, M. S., and Nielsen, K. S. (1994). "Understanding natural language software specifications," presented at Fourteen International Avignon Conference, AI'94, Paris, France.
41. DeBellis, M. (1990). "The concept demonstration rapid prototype system," presented at Fifth Annual Knowedge-Based Software Assistant Conference, Syracuse, NY.
42. Feather, M. S. (1989). Reuse in the context of a transformation-based methodology. In T. J. Biggerstaff & A. J. Perlis (Eds.), *Software reusability: Concepts and models* (Vol. 1, pp. 337–359). ACM Addison Wesley Publishing Company.
43. Frakes, W., & Isoda, S. (1994). Sucess factors of systematic reuse. *IEEE Software*, 15–19.
44. Fuchs, N. E., Hofmann, H. F., & Schwitter, R. (1994). *Specifying logic programs in controlled natural language* (Vol. 94, p. 17). Department of Computer Science, University of Zurich.
45. Garlan, D. (1990). "The role of formal reusable frameworks," presented at ACM SIGSOFT, International Workshop on Formal Methods in Software Development, Napa, California.

MDRE: Developer Derived/Decomposed Requirements

7

Abbreviations

DOORS	Dynamic Object-Oriented Requirements System
F	Fahrenheit
FAA	Federal Aviation Administration
GPU	Graphics processing unit
HP	Hewlett Packard
HVAC	Heating, ventilation, and air conditioning
MDRE	Multidisciplinary requirements engineering
MDRE	Multidisciplinary requirements engineering
OOA	Object-oriented analysis
OSHA	Occupational Safety and Health Administration
SoO	Statement of objectives
SoO	Statement of objectives
SoW	Statement of work
US	United States
V	Volts
Vac	Volts AC

Definition: Derived requirement—a requirement that decomposes/refines the primary requirement or SoO.

The concept of a derived requirement infers a **requirement that is implied or transformed from a higher-level requirement** [1]. Derived requirements allow the system engineers to elaborate with additional details meet the Statement of Objectives (SoO) or contract. A derived requirement can impose specific solution sets to meet the end needs of the users. A derived requirement can also impose requirements or behaviors of other components at the lower tiers that may impact the performance or functionality of the target requirement. Below is an example of derived requirements:

- The system shall operate in temperatures as low as 0 °F.
- The system shall be fully functional in snow.

Decomposition of the customer's requirements is the first step toward defining the architecture after the systems analysis has been completed [2]. The requirements decomposition process is not set-in stone and is influenced by the type of system being designed, the domain in which the system must operate, the system constraints, and other factors that may be specific to any individual system. The requirements decomposition process radically affects the implementation of the architecture, system design, and system implementation, since different decompositions drive different internal interfaces, different information and data flows, and a different set of Enterprise Services to execute the functionality laid out in the systems analysis and subsequent decomposition [3]. Each of the different decomposition possibilities drives a trio of different time, quality, and cost profile for the development of the system. These must be balanced against the customer technical, performance, and quality requirements, customer-driven schedule, and available cost profile. In the end, the combination of high-level requirements, objectives, goals, constraints, and compliance criteria result in the MDRE team using the elicitation, analysis, derivation/decomposition, and allocation process to:

1. System, hardware, and software requirement specifications.
2. Overall system Development Guidelines, which include things like a Hardware Development and Software Development Guideline document. For example, the software development guidelines might specify the software language or languages that are allowed on the program (e.g., Java, C++) or hardware guidelines like the use blade servers or graphics processing units (GPUs).

Each of these different guidelines has very different effects on the overall implementation of the system design [4].

J. A. Crowder, C. W. Hoff, *Requirements Engineering: Laying a Firm Foundation*,
https://doi.org/10.1007/978-3-030-91077-8_7

Question to ponder

Think of a requirement that would apply to both hardware and software that would need to be decomposed into separate requirements.

7.1 Implied Requirements

As part of the overall requirements elicitation and analysis process, the MDRE team may come across implied requirements, or requirements that are "assumed" by the customer even though they are not listed. It can be challenging to define the entire scope of the project when there are built-in assumptions in the SoW, SoO, or initial high-level customer requirements set that are not provided to the development team. If the MDRE team fails to "discover" these requirements early on, they often show up at deliver of the system to the customer and result in customer's comments like "Well, we thought you would certainly deliver something different than this." This can come from vague requirements where the customer already had a notion of how it would be developed, without relating the information to the development team [5]. It is the responsibility of the MDRE team to define the scope of the system to be developed and, during this process, ensure they have defined all explicit and implicit requirements by communicating clearly what will and will not be included in the system design, implementation, delivery, and sustainment. Interviews with the customer(s) who developed the requirements are crucial to the overall success of the elicitation and requirements analysis phases of the MDRE process.

Question to ponder

Can you think of an example of an implied requirement?

An example of an explicit vs. an implied requirement might be:

Explicit requirement: The system shall have a cast steel valve that can handle 500 lbs. of pressure.

Implied requirement: The valve seals 100%.

It may seem obvious that the valve should seal completely, but without a requirement, a "pretty close to 100%" seal meets the explicit requirement [6].

7.1.1 Case Study "the Front Cover"

This case study involves upgrading an existing Naval communications system for a particular type of Navy vessel. System-level specifications and statement of objectives were very clear on what the system needed to do. The system was designed, hardware and software procured and written, and the system was tested thoroughly. During installation of the system the Navy captain asked the engineers installing the rack of equipment into the ship, "Where's the front cover for the rack?" He was informed that the new system didn't have a front cover, since there were no requirements specifying a front cover. The captain explained that the system must have a front cover to prevent dust and dirt from getting into the electronics and he would not accept delivery of the system without a front cover. His response was, "A front cover should always be an implied requirement for a Navy ship like this." The "fix" was not trivial, since adding a front cover violated the size, weight, and cooling requirements for the system." It required a major redesign and refactoring of the system and equipment to accommodate adding a front cover. A complete understanding of the environment the system must operate in was not done prior to the requirements analysis. Little things can mean big problems when it involves requirements. The MDRE team must be thorough to catch any implied requirements.

7.2 Requirement Derivation/ Decomposition Process

Requirement derivation requires both elicitation and requirement analysis to decompose, derive, and properly allocate requirements into lower and lower levels within the system design architecture. Each step in the requirement derivation process is necessary for the MDRE team to create a complete, linked, and testable system/software requirements specification [7].

Question to ponder

What problems may occur in system development when requirements are decomposed and allocated incorrectly?

7.2.1 Top-Level Requirements Assessment

Here, each of the system-level customer-provided requirements are assessed for overall feasibility, consistency, compliance with system/environmental constraints, and their verifiability. If the MDRE team cannot discern a way to verify a requirement, discussions with the customer must ensue to adjudicate the concerns and agree upon how the requirement is to be verified. Also, if the requirement is valid and verifiable, but violates the cost and schedule constraints, it must be resolved either by schedule and/or budget changes, by getting a waiver on the requirement, eliminating the requirement, or carrying it as a risk to successful development [8].

7.2.2 Requirement Decomposition and Derivation

Object-oriented analysis (OOA) is most often utilized to define a notional top-level functional architecture and drives the derivation, decomposition, and allocation of requirements from the top-level to the lowest level of the system. The OOA process allows the MDRE team and architecture team to decompose the system and is the first step toward defining the complete architecture after the systems/requirements analysis has been completed. In general, if some part of a requirement will be met or instantiated in more than one part of the architecture, the requirement must be decomposed into separate requirements derived for each subsystem/component/etc. and written so they apply directly to those components of the system. Also, if a requirement has information in it from different topics it should be decomposed, and a requirement written for each topic the higher-level requirement includes [9].

Question to ponder

Do you believe a requirement can be allocated to two different subsystems without decomposing it into two different requirements with clarification verbiage?

In general, requirements are derived and decomposed based on the level of architecture decomposition that is necessary to specify the system being developed. This depends on how much detail is required to specify system/subsystem/component level details of the design and development. From these system requirements, hardware and software requirements are derived/ allocated. There purpose is to specify the hardware and software capabilities to be developed specific to the operational environment [10].

The challenge for the MDRE team is ensuring that requirements are properly decomposed and derived for each element within the system, that requirements are not missing or misunderstood, and there exists a clear roadmap that establishes how the element-level requirements and their subsequent decomposition can be used to demonstrate compliance with the customer-provided system-level requirements. Requirement decomposition and derivation is one of the most important aspects of system design, as bad requirements decomposition and derivation is almost impossible to overcome during the development phase without a major amount of rework, including redesign of the various architectures and new development efforts. Bad requirements will have major impacts on the overall cost and schedule of the project. Creating a common requirements framework and management scheme for all requirements decomposition and derivation among each element, therefore, is paramount for program success [11]. Table 7.1 illustrates an example template for all requirements work.

Table 7.1 Requirements decomposition and derivation criteria

Characteristic	Description
Requirement prioritization	Comparison weighting based on overall mission/business needs
Requirement source	Where did the requirement come from: Customer requirements, concept of operations, constraints, laws, agency restrictions (e.g., FAA)?
Requirement usage	Technical functionality alternatives that can be used to execute this requirement
Requirement impact	Performance consequences, potential risks the requirement poses on the system, architectural impacts
Requirement user scenario	How will this requirement be utilized by the operational system and how the people who operate the system will benefit by this requirement

Question to ponder

Many would argue that a spreadsheet is an adequate requirement management tool. What issues can you envision if requirements are only kept in a spreadsheet?

There are many electronic tools to manage requirements; the list below shows a few and will be discussed in detail in Sect. 7.3:

IBM—Rational DOORS Next Generation®.

Intland Software GmbH—codeBeamer Requirements Management®.

Hewlett-Packard—HP Agile Manager®.

Polarion Software—Polarion Requirements (2014 SR2)®.

TechExcel—DevSpec®.

To understand how the requirements in the system should be decomposed, allocated, and utilized across the system architecture, as well as the overall capabilities described by the requirements at each level in the architecture, the MDRE team must understand the following [12]:

Requirement Quality: Here we define quality in terms of the accuracy, compliance, and completeness of requirements and their overall ability to satisfy the system goals, objectives, constraints, etc. This includes topics like data usage, data rate of change within the systems, and synchronous vs. asynchronous data issues [13].

Data Information and Structure: How will the data/information be organized across the system architecture, including the structure to be utilized at each level? Data exchange issues and constraints can radically change the decomposition and derivation of requirements. Data rates and structure can drive specific requirements for how data are formatted [14].

Question to ponder

Is it possible to derive a complete and adequate set of requirements without a thorough understanding of the technology base behind the system to be developed?

One of the major roles for the MDRE team is understanding the technology base required for the system to be developed. For long-term programs, hardware/software technologies pose a continual issue and drive a set of processes and strategies for the MDRE team to manage h/w and s/w requirements during development, test, deployment, and sustainment of the system. Often, for long-term development programs the hardware is obsolete before the system is ever deployed, which may pose serious issues for the develop and the customer. To deal with these issues, the MDRE team will utilize multiple processes to keep track of technology-related issues [15]. These include:

System Technology Roadmap: What technologies are applicable and available when, and how they will be introduced into the system over its life cycle.

System Technology Strategy: The role technology plays in the overall success of the system.

System Life cycle Management Planning: Organizing, controlling, and executing the processes, procedures, protocols, and activities necessary to meet the system requirements, goals, constraints, and the overall performance of the system. For the MDRE team, life cycle management requires constant attention form the early beginnings of the program, to its ultimate retirement when a new system is built and transitioned. This includes ensuring there are requirements derived/allocated for all inputs, functions, and outputs to integrate them to meet all defined system objectives [16].

The handling of requirements is one of the single-most important activities the MDRE team must clearly understand and handle well if program success is at all a priority (of course, it ***always*** is). Managing the requirements is necessary to define the issues that the system designers must deal with in creating the overall architectures and designs required for successful delivery of a system that meets all requirements, goals, and needs [17]. Part of MDRE requirements management is to make sure the requirements at every level in the systems are:

Defined Precisely: Requirements should be clearly worded, sufficiently detailed to eliminate ambiguity, necessary (related to the mission/business needs), and testable in relation to the tier of the system.

Analyzed for Feasibility: Part of requirements planning is to determine the feasibility of the requirements at each level of the system architecture, ensuring that there are no conflicts in the requirements that will make execution of the system problematic. This includes an impact analysis of the major requirements and their impacts on the overall architecture [18].

Question to ponder

What problems can you envision if the requirements are ambiguous, i.e., there are different interpretations as to the requirements meaning?

To accomplish this, the MDRE team must ensure that each requirement:

1. Is written as a single, definitive statement, usually starting with, "The… shall…".
2. Has a unique identifier that is relevant to the requirement type, subsystem, or component to which the requirement applies and allows the requirement to be tracked and traced.
3. Has been allocated to a specific component (called Configuration Items) within the architecture.
4. That each performance requirement has verification requirement(s).
5. Designate the type of testing activity that will be performed to ensure the requirement is testable (e.g., inspection, demonstration).

Once an initial set of requirements has been created, it should be analyzed carefully to make sure the requirements set is [19]:

Complete: Are all higher-level requirements allocated? This may have involved derivation and identification of implied requirements.

Consistent: Is the content, format, and notation for all requirements consistent across the entire requirements set?

Feasible: Can all the requirements be met?

Balanced: What trade-offs were made in creating the requirements, architecture and design that needs to be understood and documented?

Verifiable and Testable: Can each requirement be objectively tested and verified, and does the customer agree to the test/verification method for each requirement?

Human Engineering: Are the requirements and subsequent architecture and design consistent with sound human-machine interface principles? (i.e., can the system be operated by those designated to be operators?).

> **Question to ponder**
>
> Why would consistency of requirements be important?

7.3 Requirement Development Guidelines

As we have discussed above, inadequate requirements and poor requirements management is a leading cause of project/program failure. Often the customer or the contractor's management may have guidelines that must (or should) be followed during the requirements, development, testing, deployment, and sustainment life cycles of a system. This may be as simple as ensuring the requirements do not violate company policies to a complex set of development guidelines like the U.S. Air Force A5R Requirements Development Guidebook from the Directorate of Operational Capability Requirements.[1] And while the A5R guidebook are not directives, they do represent official guidance and procedures to ensure compliance with the Air Force's requirements and acquisition policies. Other guideline documents may include software requirements development guidelines to ensure the software requirements are written to meet certain customer or corporate standards. An example is the Requirements Development Guidebook by Dahlia Biazid, which provides guidance and a roadmap for writing software requirements.[2] We will describe the process for writing requirements in detail in Sect. 7.2 on the Mechanics of Requirements Engineering. The requirements development process and guidelines may be different for designing and developing new systems vs. enhancing or replacing an existing system. Chapter 8 will describe the differences and the processes for requirements development for new vs. existing systems [20].

> **Question to ponder**
>
> How might requirements guidelines be important to a large project, say the new F-35 joint strike fighter?

7.3.1 Design Problem

The philosophy of the assignment development for this textbook was to create a set of thought-provoking questions, problems, and design assignments that help drive home the principles, practices, and methodologies of multidisciplinary requirements engineering (MDRE). We endeavored to provide the opportunity for the student to practice the art and science of requirements engineering, helping the student to understand that there is rarely one answer to a requirements engineering problem, and it is only through study and practice that successful requirements engineers are developed. MDRE is a life-long journey, and anyone who tells you they have nothing left to learn does not truly understand the discipline or has never been involved in the development of a complex system. With the recent interest in alternative energies and talk of the "Green New Deal," along with the recent issues with ice storms in Texas taking out the power grid, the assignments for this textbook will follow the theme of effectively dealing with both local alternative energy power grids and the national infrastructure required to facilitate large-scale alternative energy production, sustainment, and management.

7.3.2 Design Problem Description: Alternative Energy Power Grid

The purpose of this project is to, over the course of the semester, develop a feasibility study, requirements analysis, requirements elicitation, derivation, and allocation, for a system that will facilitate a "local" alternative energy power grid capable of powering a neighborhood of 500 homes and a large biosphere capable of supporting growing year-round fruits and vegetables for the 500-home neighborhood. The student will have to deal with issues of choosing multiple alternative energies that can provide power throughout all seasonal difficulties, from intense summer heat to freezing winter temperatures and possible ice storms. Control, viability, sustainability, and maintainability will be major factors to understand the entire possibly 30-year life cycle of such a system. How to effectively utilize, store, and manage power distribution, along with when to upconvert "extra" energy to the national power grid, will be issues that drive many of the assignments. Simply relying on the current technologies and energy infrastructures will not move the country and world

[1] A5R_Requirements_Development_Guidebook_Vol1.pdf (afacpo.com)

[2] Requirements Development Guidebook by Dahlia Biazid | NOOK Book (eBook) | Barnes & Noble® (barnesandnoble.com)

forward. Hopefully, this will provide insight to the students using this textbook to understand some of the issues that face everyone looking out over the next many decades.

7.3.3 DC Power Grid Requirements Discussion

In the initial days of power generation and distribution, DC power was far less practical than the AC distribution system that soon supplanted it. However, even after Tesla's AC system became the predominant system (and rightly so), Edison's DC power did not become obsolete for some time. Eventually power plants eliminated their delivery of DC power, but the overall concept of a DC power grid is not dead. Even our computer systems (large server systems) have uninterruptible power supplies which contain batteries that are charged off the grid and used to upconvert to AC to power devices when the AC power grid goes off-line. So, the notion of a DC power grid is not dead, but generally used as a backup system. The point is, the notion of a DC power grid has remained and now that renewable energies are viable, we should re-examine their use for widespread, local, power distribution.

One of the advantages of a locally generated DC station is energy savings as well as allowing a facility to run more easily off various DC sources, such as Solar Arrays. And for power-hungry installations, the notion of generating the power locally is growing in popularity. Part of the reason is that waste heat from the power generation generators (solar, geothermal, etc.) can be used to warm nearby buildings. By combining the production of heat and power, local power grids can squeeze much more useful energy out of the renewable energy they use, so this approach will certainly become more widespread as time goes on. The purpose of this design problem is to look ahead and provide an architecture and specification for a local DC power generation station that can serve a local community.

Currently many systems exist to control and handle alternative energies. There are systems to control and manage solar power, wind power, geothermal power, and others, but very few are designed to control, manage, store, and distribute energy from a power grid consisting of multiple alternative energy sources. As we saw the issues recently caused by the intense ice storms in Texas, one alternative energy source is not the answer to the growing energy needs/wants in our society. Each alternative energy source has limitations and issues associated with its use. By harnessing multiple alternative energies, and managing the storage of the energies produces, communities can be worry free from power generation. In order to facilitate utilizing multiple alternative energy sources, a control center must be included in the design that can manage the energy creation, distribution, storage, and return unused energy to the power grid.

7.3.4 High-Level Technical Description of the DC Power Grid System

The purpose of the multiple alternative energies DC power grid is to:

- Supply DC power to a customer community of 500 homes and a biosphere for growing food.
- Provide a storage device to store energy in case of a system failure.
- Determine the amount of surplus energy that is above and beyond what is required to power the grid and charge the backup system, and then deliver that surplus energy to the national power grid systems.

Throughout the semester you will be asked to analyze, design, and architect different components of the overall DC power grid system. This will include research into alternative designs, like lithium nanowire storage systems for backup power and/or supercapacitors (also called ultracapacitors). Fig. 7.1 illustrates these technologies connected to the DC Power Grid Regulator Control Center.

7.3.5 High-Level Technical Requirements for the DC Grid System

- Design shall utilize at least two alternative energies, with a maximum of 4.
- Total power generation shall be 3.0 megawatts minimum.
- Each of the alternative energy sources shall be capable of generating at least 60% of minimum wattage.
- DC power grid shall generate DC power, with grid voltage being 48, 24, or 12 volts.
- DC power grid shall be capable of servicing at least 500 single family homes and one 4-acre biosphere with hydroponics capable of growing vegetables year-round. (See Fig. 7.2 for a block diagram of a DC-powered home).
- All lights, appliances, and facilities in each home and the biosphere shall be configured for DC operations, if possible. Things like electric ovens may need DC/AC conversion to accommodate their 220v. requirements.
- Only allowed exceptions are the HVAC system and the Washer/Dryer, although DC is preferred.
- There shall be energy storage capability capable of operating the grid of 500 single family dwellings for 7 days if all alternative forms of energy are not producing.
  - Compliance Criteria: Energy storage devices shall be capable of producing 750,000 watts for the entire 7-day period. Nominal energy storage should be batteries or supercapacitors.

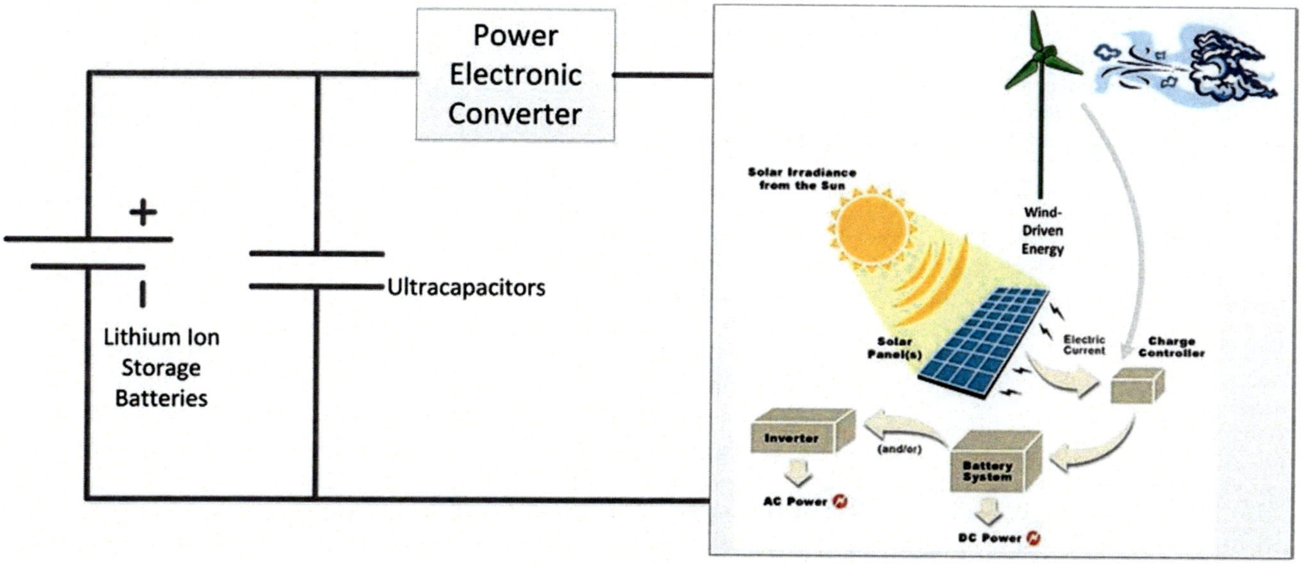

Fig. 7.1 Alternative power management system overview

Fig. 7.2 Example of a whole house power grid

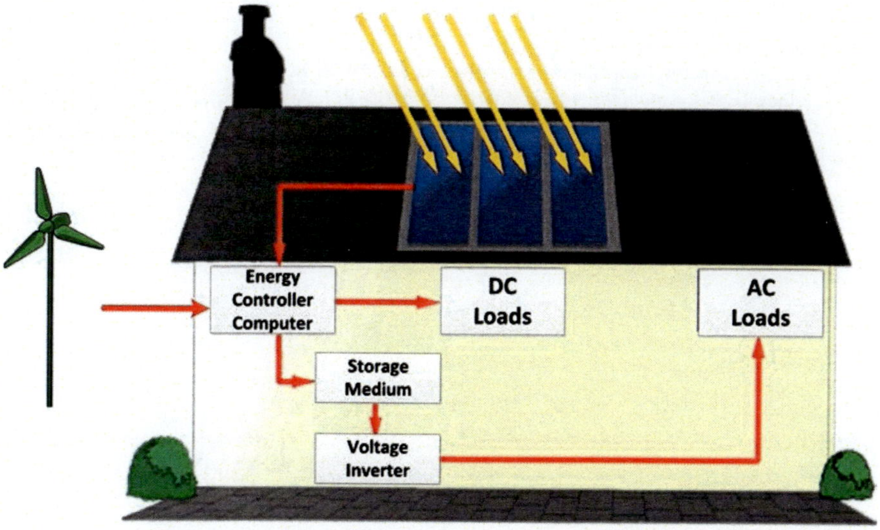

- The DC power grid system shall include controlling h/w and s/w to manage the multi-alternative energy strategies and production.
- All energy produced above the required home and energy storage usage shall be upconverted to 220vac for release to the national power grid.
- The DC power grid shall be capable of operating in a temperature range of −40°F to 125°F. without degradation of power generation.
- The DC power grid must meet OSHA regulations.

7.3.6 Assignments

- Write a 5-page report outlining the possible implicit requirements your system might have, including zoning ordinances and city, state, and federal regulation that might drive implied requirements. If you feel there are no implied requirements, describe how you arrived at that conclusion.
- Discuss the "Front Cover" case study and discuss who you feel should have caught the error and when. Whom do you feel is responsible for paying what it will cast to retrofit the design and deployment with a front cover?
- Using the High-Level Technical Description as a Statement of Objectives develop a functional block diagram for your alternative energy system and decompose and allocate requirements into each subsystem of your architecture.
- Write a 5-page report outlining any requirements development guidelines you feel are appropriate for your system and why. If you feel there are no guidelines, what general guidelines will you follow in developing the requirement specification for your multi-alternative energy system?

References

1. Gotel, O. C. Z., & Finkelstein, A. C. W. (1994). *"An analysis of the requirements traceability problem," presented at The First International Conference on Requirements Engineering*. Colorado Springs.
2. Greenspan, S., Mylopoulos, J., and Borgida, A. "On formal requirements modeling languages: RML revisited," presented at 16th International Conference on Software Engineering, Sorrento, Italy, 1994.Formal and Informal in RE Cybulski 10.
3. Hall, A. (1990). Seven myths of formal methods. *IEEE Software*, 11–19.
4. Hofmann, H. F. (1993). Requirement engineering: A survey of methods and tools. *Institut fŸr Informatik der Universitat ZŸrich, 93*, 05.
5. Hollocker, C. P. (1990). A review process mix. In R. H. Thayer & M. Dorfman (Eds.), *System and software requirements engineering* (pp. 485–491). IEEE Computer Society Press.
6. Hsia, P., Davis, A., & Kung, D. (1993). Status report: Requirements engineering. *IEEE Software*, 75–79.
7. Ibm, "System Application Architecture: AD/Cycle Concepts," International Business Machines Corp. GC26–4531-01, 1991.
8. Johnson, W. L., Benner, K. M., & Harris, D. R. (1993). Developing formal specifications from informal requirements. *IEEE Expert, 8*, 82–90.
9. Johnson, W. L., & Feather, M. S. (1991). Using evolution transformation to construct specifications. In M. R. Lowry & R. D. McCartney (Eds.), *Automatic software design* (pp. 65–91). AAAI Press/The MIT Press.
10. Johnson, W. L., Feather, M. S., and Harris, D. R. (1991). "The KBSA requirements/specification facet: ARIES," presented at 6th Annual KnowledgeBased Software Engineering Conference, Syracuse, New York, USA.
11. Johnson, W. L. and Harris, D. R. (1991). "Sharing and reuse of requirements knowledge," presented at 6th Annual Knowledge-Based Software Engineering Conference, Syracuse, New York, USA.
12. Jones, G. W. (1990). *Software engineering*. John Wiley & Sons.
13. Kado, M., Hirakawa, M., and Ichikawa, T. (1992). "HI-VISUAL for hierarchical development of large programs," presented at IEEE Workshop on Visual Languages, Seattle, Washington.
14. Kaindl, H. (1993). The missing link in requirements engineering. *ACM SIGSOFT Software Engineering Notes, 18*, 30–39.
15. Kang, K. C., Cohen, S., Holibaugh, R., Perry, J., and Peterson, A. S. (1992). "A ReuseBased Software Development Methodology," Software Engineering Institute CMU/SEI-92-SR-4.
16. Kinoshita, T. (1989). A knowledge acquisition model with applications for requirements specification and definition. *SIGART Newsletter*, 166–168.
17. Li, H. (1993) "Reuse-in-the-large: modeling, specification and management," presented at Advances in Software Reuse: Selected Papers from the Second International Workshop on Software Reusability, Lucca, Italy.
18. Litvintchouk, S. D., & Matsumoto, A. S. (1989). Design of ADA systems yielding reusable components: An approach using structured algebraic specification. In T. J. Biggerstaff, A. J. Formal, & Informal in RE Cybulski 11 Perlis (Eds.), *Software reusability: Concepts and models* (Vol. 1, pp. 227–245). ACM Addison Wesley Publishing Company.
19. Lubars, M. D. (1988). Wide-spectrum support for software reusability. In W. Tracz (Ed.), *Software reuse: Emerging technology* (pp. 275–281). Computer Society Press.
20. Madsen, K. H. (1994). A guide to metaphorical design. *Communications of the ACM, 37*, 57–62.

MDRE: New System Vs. Existing System, Extension or Upgrade

8

Abbreviations

AC	Alternating current
C	Centigrade
CONOPS	Concept of Operations
DC	Direct current (or 0 Hz)
F	Fahrenheit
HVAC	Heating, ventilation, and air conditioning
MDRE	Multidisciplinary requirements engineering
OSHA	Occupational Safety and Health Administration
UCAV	Unmanned combat air vehicle
V	Volts
Vac	Volts AC

In Sect. 2.6 we briefly discussed the MDRE process for new vs. existing systems, but we need to take a more detailed dive into the differences between the MDRE process for an entirely new system vs. upgrading or extending an existing, operational system [1]. In the past, the term "legacy system" referred to systems 25 years old or older. In today's world of exponentially increasing technology, a legacy system might be only 2–5 years old but still woefully out of date with today's needs and new technology capabilities. We will explore the differences between MDRE processes and procedures for new systems vs. legacy (existing) systems [2].

8.1 MDRE for Legacy System Upgrades, Extensions, and Technology Refresh

While all the process discussed in the book are applicable to requirements development activities for all projects, there are added processes and care that must be undertaken when a current (legacy) system is in the process of an upgrade, technology refresh, or repurposing/extension. An example of an upgrade/extension of a system is the changes being made to the F-16 fighter to make it an autonomous unmanned combat air vehicle (UCAV) to accompany piloted F-35 fighters. Ethics discussions aside, such an upgrade drives a different or even enhanced set of MDRE procedures than starting from scratch for a new system development. To illustrate the process, we modify the requirements process shown in Fig. 2.2 to accommodate upgrades to legacy systems.

The legacy system upgrade requirement process begins with a feasibility study, but, that feasibility includes the feasibility not only of the finished system, but the feasibility of adapting, enhancing, and/or upgrading the existing system to handle the new or enhanced capabilities required for the final system [3]. Instead of a standard requirements analysis, there must be a requirements applicability analysis to discover which of the original requirements set developed for the legacy system still applies and which requirements must be redefined and added for the new capabilities or enhancements. In Fig. 8.1, there is a dashed line around the two boxes labeled "Definition Changes to Current Requirements" and "Definition of New Requirements." There is a significant effort required for these steps as the MDRE team must determine which requirements to keep, which to re-structure, and which to add. In addition, the elicitation and analysis phase may uncover requirements that are conflicting between the current and future system and must be adjudicated and rewritten. Requirement definitions must be reviewed to ensure that the verbiage still applies and that it has the same meaning as it did when the original requirements for the legacy system were agreed to [4]. All system models developed must be updated, re-run and the results analyzed to under-

> **Question to ponder**
> Do you think it is practical to update an outdated system or should it just be completely replaced?

J. A. Crowder, C. W. Hoff, *Requirements Engineering: Laying a Firm Foundation*,
https://doi.org/10.1007/978-3-030-91077-8_8

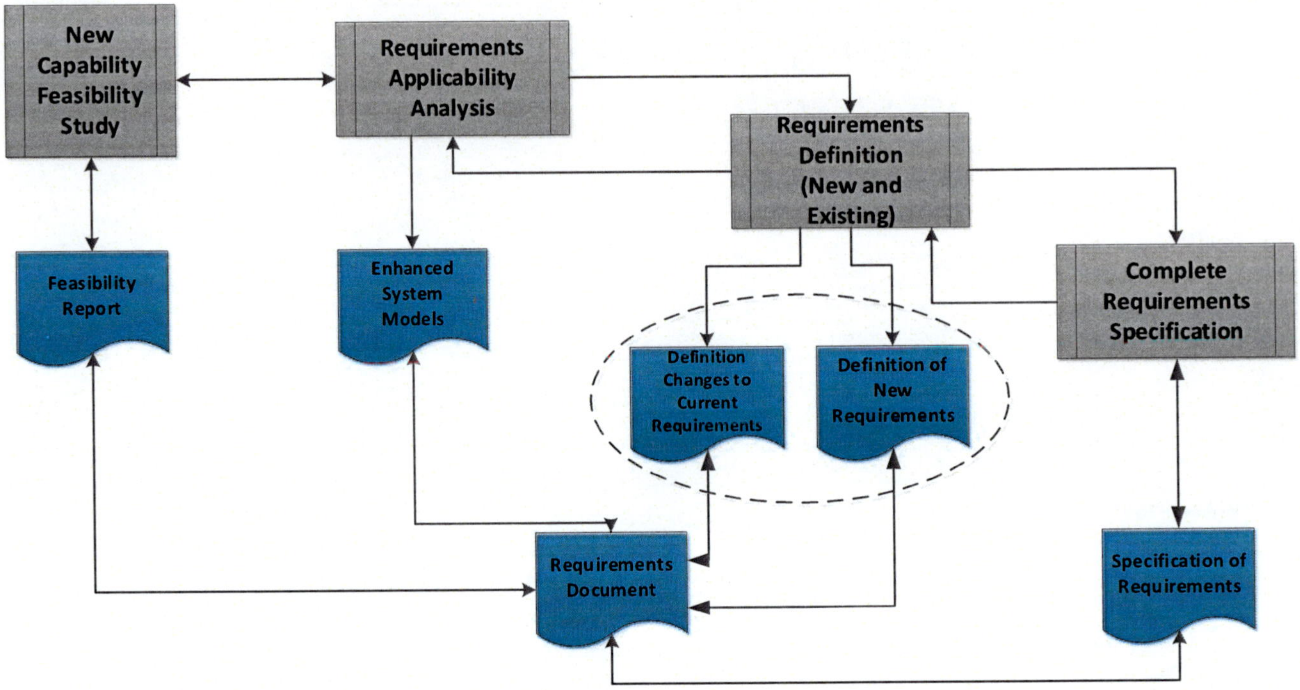

Fig. 8.1 MDRE requirements process for legacy system upgrades

stand how the hybrid (part legacy and part new) system will perform. It is possible for emergent behaviors to arise from the mixture of legacy and new hardware and software that no one could have envisioned.

8.1.1 Requirements Validation for Legacy System Upgrades, Extensions, and Technology Refresh

The requirements validation is critical for any development effort but is particularly crucial for this type of development project. Requirement validation entails analysis to determine whether the current set of requirements is correct, consistent, and complete. Requirement validation is required to make sure that [5]:

- The requirements are appropriate for the system being designed/developed.
- If a system is developed using this requirement set, the system will perform the functions laid out in the SoO and SoW.
- There are no conflicting requirements, i.e., one requirement negates or makes another requirement invalid. For example, one requirement says units should be in °F and another requirement says units should be in °C.
- The requirements are verifiable and testable.

In general, it is not practical to guarantee or prove completeness, instead the validation analysis should look for gaps in completeness, ensuring all functions have valid requirements, all nonfunctional objectives have valid requirements, and all compliance criteria have requirements to ensure adherence to the required compliance documents. In short—requirements validation ensures you are building the right system and the system does what it is supposed to do in the environment in which the system is to be deployed [6]. It is important for the MDRE team to detect and fix requirements issues resulting from:

- Incomplete requirements based on available user scenarios.
- Requirements at low levels that do not trace to higher-level requirements.
- High-level requirements that are not allocated or decomposed.
- Requirements without test cases to verify the requirements.
- Test cases that do not trace to requirements or user scenarios.

Even in the testing phase of development, the test team should be coordinating with the MDRE team to look for requirement validation issues that have not arisen previously. All test cases should tie to functionality and items in the customer CONOPS. Any test cases that test functionality not called out in the CONOPS or requirements should be tagged and reviewed by the MDRE team [7].

8.1.2 Case Study: Requirement Validation Failure—The Tacoma Bridge Project

The Tacoma Bridge failure is one of the most famous bridge failures, and the movie of its failure has been used in universities for decades as an example of design failure [8]. The Tacoma Bridge failure was a classical requirements validation failure. The bridge was designed and built based on scaling up an existing bridge design. The problem was the strait across which the bridge was built had a much stronger wind profile than the environment(s) the previous bridges using that same design were built for. The design engineers for the Tacoma Bridge reused requirements for the existing bridge design(s) and the requirements met standards for bridge design and the bridge met is verification. However, there was never a validation analysis done to ensure the requirements set was appropriate for the environment the bridge was intended for. The result was catastrophic failure of the bridge. It is worth viewing the video to understand how critical requirement validation is to the overall success of a system.

Question to ponder

Look up and view the Tacoma Bridge failure video. Do you think during construction issues should have been noticed before bridge completion? What measures could have been taken prior to construction to ensure the design was sound?

8.2 Design Problem

The philosophy of the assignment development for this textbook was to create a set of thought-provoking questions, problems, and design assignments that help drive home the principles, practices, and methodologies of multidisciplinary requirements engineering (MDRE). We endeavored to provide the opportunity for the student to practice the art and science of requirements engineering, helping the student to understand that there is rarely one answer to a requirements engineering problem, and it is only through study and practice that successful requirements engineers are developed. MDRE is a life-long journey, and anyone who tells you they have nothing left to learn does not truly understand the discipline or has never been involved in the development of a complex system. With the recent interest in alternative energies and talk of the "Green New Deal," along with the recent issues with ice storms in Texas taking out the power grid, the assignments for this textbook will follow the theme of effectively dealing with both local alternative energy power grids and the national infrastructure required to facilitate large-scale alternative energy production, sustainment, and management.

8.2.1 Design Problem Description: Alternative Energy Power Grid

The purpose of this project is to, over the course of the semester, develop a feasibility study, requirements analysis, requirements elicitation, derivation, and allocation, for a system that will facilitate a "local" alternative energy power grid capable of powering a neighborhood of 500 homes and a large biosphere capable of supporting growing year-round fruits and vegetables for the 500-home neighborhood. The student will have to deal with issues of choosing multiple alternative energies that can provide power throughout all seasonal difficulties, from intense summer heat to freezing winter temperatures and possible ice storms. Control, viability, sustainability, and maintainability will be major factors to understand the entire possibly 30-year life cycle of such a system. How to effectively utilize, store, and manage power distribution, along with when to upconvert "extra" energy to the national power grid, will be issues that drive many of the assignments. Simply relying on the current technologies and energy infrastructures will not move the country and world forward. Hopefully, this will provide insight to the students using this textbook to understand some of the issues that face everyone looking out over the next many decades.

8.2.2 DC Power Grid Requirements Discussion

In the initial days of power generation and distribution, DC power was far less practical than the AC distribution system that soon supplanted it. However, even after Tesla's AC system became the predominant system (and rightly so), Edison's DC power did not become obsolete for some time. Eventually power plants eliminated their delivery of DC power, but the overall concept of a DC power grid is not dead. Even our computer systems (large server systems) have uninterruptible power supplies which contain batteries that are charged off the grid and used to upconvert to AC to power devices when the AC power grid goes off-line. So, the notion of a DC power grid is not dead, but generally used as a backup system. The point is, the notion of a DC power grid has remained and now that renewable energies are viable, we should re-examine their use for widespread, local, power distribution.

One of the advantages of a locally generated DC station is energy savings as well as allowing a facility to run more easily off various DC sources, such as Solar Arrays. And for power-hungry installations, the notion of generating the power locally is growing in popularity. Part of the reason is that waste heat from the power generation generators (solar, geothermal, etc.) can be used to warm nearby buildings. By combining the production of heat and power, local power grids can squeeze much more useful energy out of the renewable energy they use, so this approach will certainly become more widespread as time goes on. The purpose of this design problem is to look ahead and provide an architecture and specification for a local DC power generation station that can serve a local community.

Currently many systems exist to control and handle alternative energies. There are systems to control and manage solar power, wind power, geothermal power, and others, but very few are designed to control, manage, store, and distribute energy from a power grid consisting of multiple alternative energy sources. As we saw the issues recently caused by the intense ice storms in Texas, one alternative energy source is not the answer to the growing energy needs/wants in our society. Each alternative energy source has limitations and issues associated with its use. By harnessing multiple alternative energies, and managing the storage of the energies produces, communities can be worry free from power generation. In order to facilitate utilizing multiple alternative energy sources, a control center must be included in the design that can manage the energy creation, distribution, storage, and return unused energy to the power grid.

8.2.3 High-Level Technical Description of the DC Power Grid System

The purpose of the multiple alternative energies DC power grid is to:

- Supply DC power to a customer community of 500 homes and a biosphere for growing food.
- Provide a storage device to store energy in case of a system failure.
- Determine the amount of surplus energy that is above and beyond what is required to power the grid and charge the backup system, and then deliver that surplus energy to the national power grid systems.

Throughout the semester you will be asked to analyze, design, and architect different components of the overall DC power grid system. This will include research into alternative designs, like lithium nanowire storage systems for backup power and/or supercapacitors (also called ultracapacitors). Fig. 8.2 illustrates these technologies connected to the DC Power Grid Regulator Control Center.

8.2.4 High-Level Technical Requirements for the DC Grid System

- Design shall utilize at least two alternative energies, with a maximum of 4.
- Total power generation shall be 3.0 megawatts minimum.
- Each of the alternative energy sources shall be capable of generating at least 60% of minimum wattage.
- DC power grid shall generate DC power, with grid voltage being 48, 24, or 12 volts.
- DC power grid shall be capable of servicing at least 500 single family homes and one 4-acre biosphere with hydroponics capable of growing vegetables year-round. (See Fig. 8.3 for a block diagram of a DC-powered home).
- All lights, appliances, and facilities in each home and the biosphere shall be configured for DC operations, if possible. Things like electric ovens may need DC/AC conversion to accommodate their 220v. requirements.
- Only allowed exceptions are the HVAC system and the Washer/Dryer, although DC is preferred.
- There shall be energy storage capability capable of operating the grid of 500 single family dwellings for 7 days if all alternative forms of energy are not producing.
  - Compliance Criteria: Energy storage devices shall be capable of producing 750,000 watts for the entire 7-day period. Nominal energy storage should be batteries or supercapacitors.
- The DC power grid system shall include controlling h/w and s/w to manage the multi-alternative energy strategies and production.
- All energy produced above the required home and energy storage usage shall be upconverted to 220vac for release to the national power grid.
- The DC power grid shall be capable of operating in a temperature range of −40 degrees F. to 125 degrees F. without degradation of power generation.
- The DC power grid must meet OSHA regulations.

8.2.5 Assignments

- Write a report, 5–10 pages, describing how you intend to design your multi-alternative energy system. If you are expanding on an existing design you find, describe how it is similar and how it is different than your system.
- Write a report 5–10 pages on how you will verify that your system requirements are valid for your multi-alternative energy system. What gaps are you concerned about?

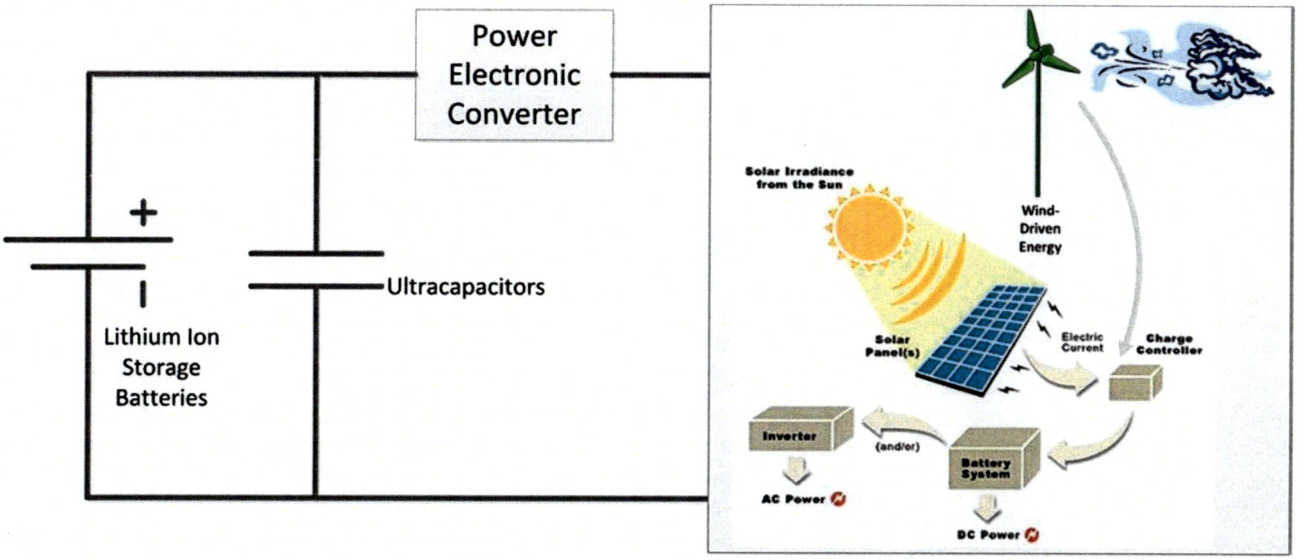

Fig. 8.2 Alternative power management system overview

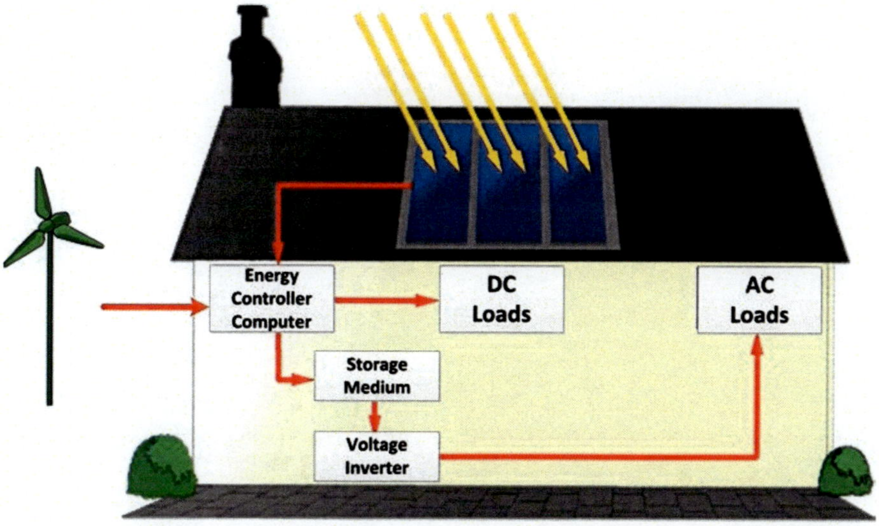

Fig. 8.3 Example of a whole house power grid

- Go out to the internet, find, and view the movies on the Tacoma Bridge. Describe what you see as the major design flaw or requirement gap that ultimately resulted in the catastrophic failure of the bridge.

References

1. Matsumoto, Y. (1989). Some experiences in promoting reusable software: Presentation in higher abstract levels. In T. J. Biggerstaff & A. J. Perlis (Eds.), *Software reusability: Concepts and models* (Vol. 2, pp. 157–185). ACM Addison Wesley Publishing Company.
2. Meyer, B. (1985). On formalism in specifications. *IEEE Software*, 6–26.
3. Morel, J.-M. and Faget, J. (1993), "The REBOOT environment," presented at Advances in Software Reuse: Selected Papers from the Second International Workshop on Software Reusability, Lucca, Italy.
4. Morris, S. J. and Finkelstein, A. C. W. (1994). "Development of multiple media documents," Imperial College, Department of Computing.
5. Myer, B. A. (1990). Visual programming, programming by example, and program visualization: A taxonomy. In E. P. Glinert (Ed.), *Visual programming environments: Paradigms and systems* (pp. 33–40). IEEE Computer Society Press.
6. Poulin, J. (1993). Integrated support for software reuse in computer-aided software engineering (CASE). *ACM SIGSOFT Software Engineering Notes, 18*, 75–82.
7. Prieto-Diaz, R., & Arango, G. (1991). *Domain analysis and software systems modeling*. IEEE Computer Society Press.
8. Reubenstein, H. B. and Waters, R. C. (1989). "The requirements apprentice: An initial scenario," presented at 5th International Workshop on Software Specifications and Design.

9 System Tiers

Abbreviations

AC	Alternating current
API	Application programming interface
DAO	Data Access Objects
DC	Direct current (or 0 Hz)
DoD	Department of Defense
F	Fahrenheit
FAR	Federal Acquisition Regulations
HVAC	Heating, ventilation, and air conditioning
MDRE	Multidisciplinary requirements engineering
OSHA	Occupational Safety and Health Administration
SBS	System breakdown structure
SoO	Statement of Objectives
SoW	Statement of Work
SW	Software
V	Volts
Vac	Volts AC

System tiers can be thought of as segmentation levels each of which represents a capability type. A well-known example of system tiers is represented top-down by: Visualization, Services, Applications and Data. Depending upon the level of fidelity desired, system tiers can be subdivided as needed [1]. Examples are:

- Our "solar system" comprises a sun and 8/9 (give or take a few, depending on your view of Pluto) planets; therefore, you have a sun, and 8/9 planets that comprise the "solar system." Even with cosmology there are disagreements.
- The governance of the United States of America, a very large system, comprises a federal government apparatus, 50 state governments and a few territories, each with their own levels of governance.
- A modern automobile is a system. It comprises multiple sub-systems, components, and units (e.g., System = Automobile, Engine = Sub-system, Water pump = Component, Fan Wheel = unit).

The examples are meant to show, again, that most systems have tiers. MDRE engineers need to be aware that tiers affect the definition and the quality of technical requirements. Hence, the better the definition of the requirements and how tiers affect them, the more potential to effectively validate the requirements [2]. Fig. 9.1 illustrates a high-level view an "n-tier system architecture."

9.1 Explanation of System Tiers

9.1.1 Presentation Tier

This tier responds to service requests from the Service Layer and issues service requests to the Mission/Business Process Tier. The Presentation Layer relieves the Service Layer of concern regarding syntactical differences in data representation within the end-user system [3].

9.1.2 Mission/Business Application Services Tier

This tier responds to service requests from the Presentation Tier and issues service requests to the Mission/Business Processes Tier. The Services Tier provides the mechanism for managing the dialogue between end-user business application processes. It provides and establishes check-pointing, adjournment, termination, and restart procedures.

9.1.3 Mission/Business Processes Tier

This provides services for an application program to ensure that effective communication with another application

J. A. Crowder, C. W. Hoff, *Requirements Engineering: Laying a Firm Foundation*,
https://doi.org/10.1007/978-3-030-91077-8_9

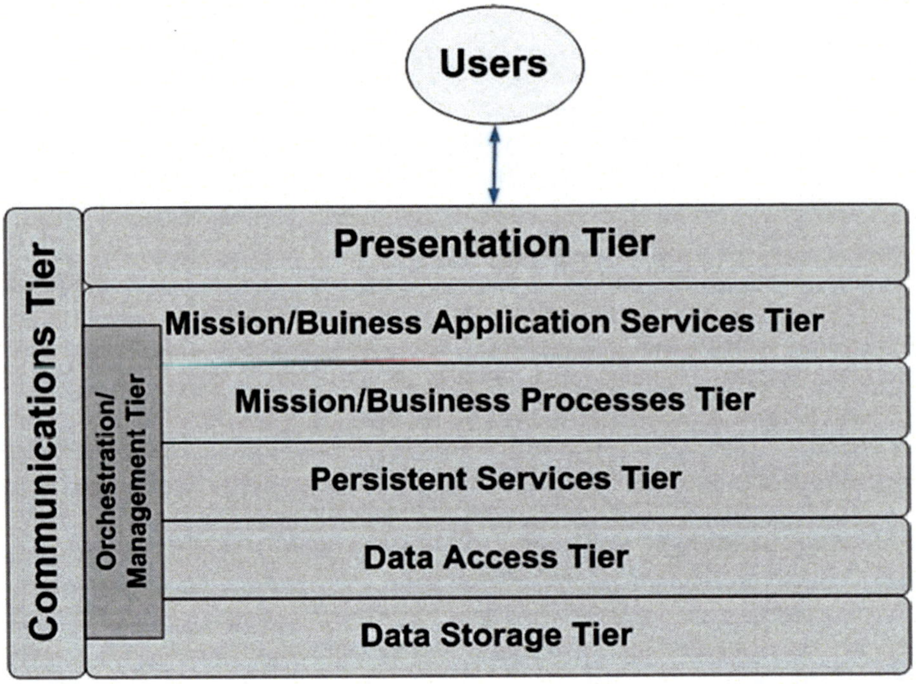

Fig. 9.1 High-level view of a tiered architecture

programs in the enterprise is possible. This processes tier is a service tier that provides the services necessary for the business applications to run and communicate effectively across the enterprise.

9.1.4 Persistent Services Tier

This tier allows application and service developers to work with a standard set of objects that read and save their state to a relational database or other storage formats (e.g., flat files). The Persistent Services Tier can be looked at to access objects within the enterprise storage system. This includes data resource integration that abstracts out the information regarding data object storage such that the Persistent Services Tier contains the concept of Data Access Objects (DAOs). Data Access Objects are used by software classes to handle all interaction with the Data Storage System. An application programming interface (API) is provided so that other classes do not need to know how to create, retrieve, update, or delete a given object. Data Access Objects are defined by their Attributes, Object Key, and Events [4].

9.1.5 Data Access Tier

In the tiered architecture shown in Fig. 7.1, the need to persist data created at runtime is critical. The system's database clusters will normally use a relational database(s). The use of Data Access Objects for data access is indispensable in providing the methods to store database objects and flat files (e.g., log files and legacy data files from legacy systems).

9.1.6 Data Storage Tier

The use of Data Access Objects (DAOs) provides a unified, simple, and transparent persistence interface between the mission/business applications and data stores and can significantly streamline the way the system deals with persistent data. The goal is to abstract away any persistence details and provide a clean, simple, object-oriented API to perform data storage. This drives another concept in Data Management called Data Labelling. Data Labelling is a way of controlling data on a row level. Each data row is given a label used to store information about data sensitivity. A label provides additional sophisticated access control rules in addition to those provided by privilege management [5]. It further mediates access to a data row based on the identity and label of the user and the label of the row. This provides an additional level of privilege management to the overall system. Label-based privilege management is adjusted to conform with the systems privilege management policies dictates by the requirements, constraints, and operational concepts. Together these policies dictate the criteria by which access to an object is either permitted or denied. This category of requirements is crucial to securing a system and access to sensitive data items [6].

9.1.7 Orchestration/Management Tier

This provides automated arrangement, coordination, and/or management of complex services within the Mission/Business application layer and allows more direct access to data and data models for near-real time, computationally intensive Mission/Business applications. This may be done with or without human intervention [7].

9.1.8 Communications Tier

This tier (message handling) is an important part of the overall service-based system, in that it provides various types of messages and notifications for user-to-user, user-to-process, and process-to-process communications. It provides both synchronous operations (generally for query) and asynchronous operations (generally for business transactions) within a single, unified framework [8].

Question to think about

If you are going to try and build a "new system" using the "n-tier architecture" patters, what happens to the development if any of the tiers are removed from the design?

9.2 MDRE Requirements Planning for Tiers

Effective requirements engineering requires a thorough understanding of the categories of requirements that exist within a given system development. An organization may or may not require categorization of requirements by type; however it is good practice to keep the requirements aligned by requirement type. As an example, as one is building a deliverable system, the functional requirements for the system should be aligned separately from the shipping requirements used for delivery [9].

In the case of the Department of Defense (DoD), governed by Federal Acquisition Regulations (FAR), the DoD requires vendors meet numerous specifications and standards to be compliant with the FAR, in addition to the functional requirements assigned to the actual stakeholder needs. As such, requirements management should address all categories of requirements, because failure to comply with all requirements results in a discrepancy within the baseline and comes with legal ramifications [10].

9.3 MDRE Program Structure and Tiers

Certainly not all systems or projects require all the tiers described above. However, each project, whether it is building a system or providing a service, can be broken down into sub-systems or tiers for requirements allocation and decomposition. Whether it is system tiers as described above, or a simpler structure, all constitute a system breakdown structure (SBS) for the project or service [11]. The SBS defines the elements and their related life cycle processes and includes the tiers or elements of the system required to accomplish the requirements, goals, and objectives laid out in the SoO, SoW, and requirements specification. Fig. 9.2 illustrates a high-level tiered SBS for a desktop computer system. While simplistic, the structure shown in Fig. 9.2 provides adequate breakdown to decompose and allocate requirements for building a desktop computer system. The breakdown is generic and in-line with a high-level set of requirements. Notice data storage does not dictate the type of storage which may still be a disk drive or may be a solid-state device [12].

While the number of tiers for a system is different from application to application, most modern systems constitute three basic tiers as illustrated in Fig. 9.3, that is a presentation tier, an application or logic tier, and a data tier. Each layer in an n-tiered architecture is further decomposed to its lowest level of effort, usually down to the component level decomposition. Another example might be a service robot designed to clean a house. This may be broken down into "cleaning tiers" necessary to decompose and allocate different requirements for the robot to understand how to clean each area of the house and for each cleaning function (e.g., vacuuming vs. dusting). Fig. 9.4 illustrates a possible "cleaning tier" architecture, which provides a tiered service architecture rather than a physical system architecture.

Question to think about

Do you believe process/service tiers are as important as functional tiers? What would happen if a process or service was not allocated requirements for a process or service system?

Establishing a tiered architecture defines (at a high-level) categories for requirements decomposition/derivation and allocation down to the lowest level of system/service development. The MDRE team must use the tiers and tier descrip-

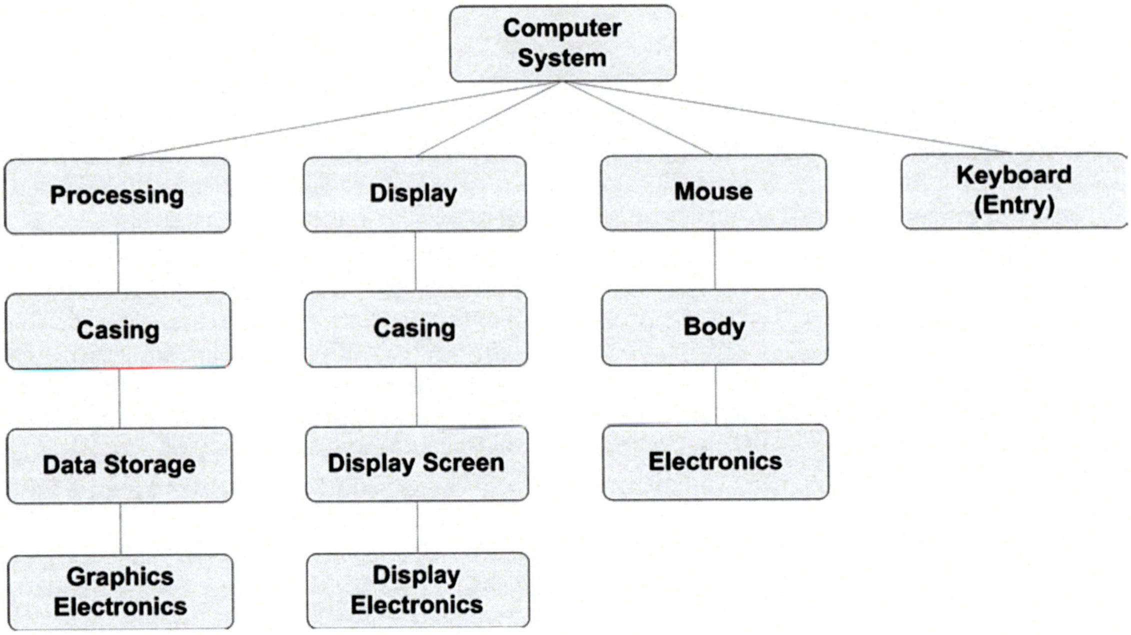

Fig. 9.2 Basic tiered architecture for a desktop computer

Fig. 9.3 Basic three-tiered architecture

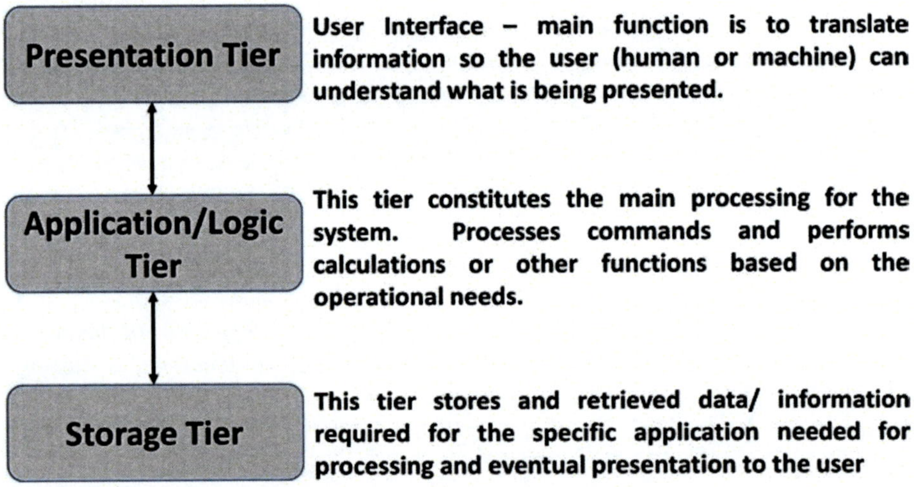

tions to establish a set of requirements up and down the tiers that allows design, implementation, integration, test, deployment, and ultimately sustainment of the system or services. That is why proper MDRE is so crucial to project success. The decomposition of requirements can be created using many methodologies, like [13]:

Deliverable-centric: Where multiple deliverables are needed, each for a different application, the requirements and tiers may reflect the differences between different instantiations of the system. An example would be the Boeing 737 which has different configurations for each airline that buys them. Tiers and subsequent requirements decomposition may reflect the differences in configuration between the 737 models or styles.

Sub-system/component centric: This is the most common tier structure that reflects the separate functional capabilities of the system and allows decomposition and derivation of requirements based on the individual architectural components of the system.

Process centric: In a system that is composed of processes and/or services instead of technical functions (e.g., h/w and s/w), this allows requirements to be decomposed/derived and allocated by individual processes or services as illustrated in Fig. 9.4.

The goal here was not to provide a detailed description of system tiers, but to establish the need for requirements processes/methodologies that reflect the system as it is specified/designed/architected and that MDRE is as impor-

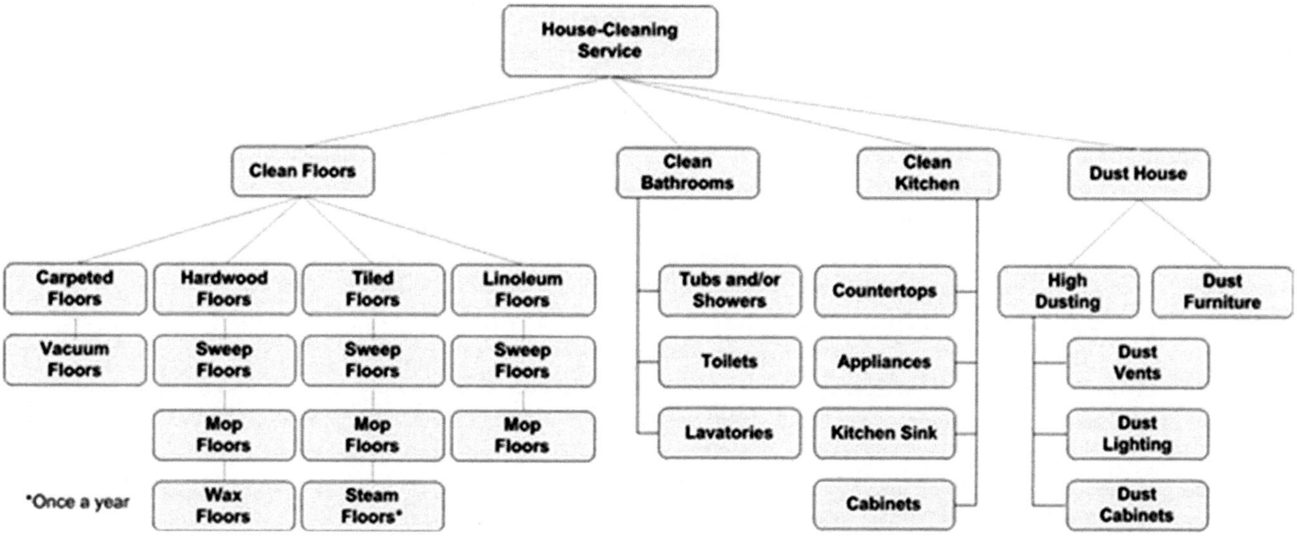

Fig. 9.4 Tiered housecleaning service architecture

tant for housecleaning as it is the F-35 Joint Strike Fighter. Section I has laid the foundation for multidisciplinary requirements engineer and its overall purposes within a system life cycle. Section II will provide the mechanics and methods for creating effective requirements and requirement specifications.

9.3.1 Design Problem

The philosophy of the assignment development for this textbook was to create a set of thought-provoking questions, problems, and design assignments that help drive home the principles, practices, and methodologies of multidisciplinary requirements engineering (MDRE). We endeavored to provide the opportunity for the student to practice the art and science of requirements engineering, helping the student to understand that there is rarely one answer to a requirements engineering problem, and it is only through study and practice that successful requirements engineers are developed. MDRE is a life-long journey, and anyone who tells you they have nothing left to learn does not truly understand the discipline or has never been involved in the development of a complex system. With the recent interest in alternative energies and talk of the "Green New Deal," along with the recent issues with ice storms in Texas taking out the power grid, the assignments for this textbook will follow the theme of effectively dealing with both local alternative energy power grids and the national infrastructure required to facilitate large-scale alternative energy production, sustainment, and management.

9.3.2 Design Problem Description: Alternative Energy Power Grid

The purpose of this project is to, over the course of the semester, develop a feasibility study, requirements analysis, requirements elicitation, derivation, and allocation, for a system that will facilitate a "local" alternative energy power grid capable of powering a neighborhood of 500 homes and a large biosphere capable of supporting growing year-round fruits and vegetables for the 500-home neighborhood. The student will have to deal with issues of choosing multiple alternative energies that can provide power throughout all seasonal difficulties, from intense summer heat to freezing winter temperatures and possible ice storms. Control, viability, sustainability, and maintainability will be major factors to understand the entire possibly 30-year life cycle of such a system. How to effectively utilize, store, and manage power distribution, along with when to upconvert "extra" energy to the national power grid, will be issues that drive many of the assignments. Simply relying on the current technologies and energy infrastructures will not move the country and world forward. Hopefully, this will provide insight to the students using this textbook to understand some of the issues that face everyone looking out over the next many decades.

9.3.3 DC Power Grid Requirements Discussion

In the initial days of power generation and distribution, DC power was far less practical than the AC distribution system that soon supplanted it. However, even after Tesla's AC system became the predominant system (and rightly so), Edison's DC power did not become obsolete for some time. Eventually power plants eliminated their delivery of DC power, but the overall concept of a DC power grid is not dead. Even our computer systems (large server systems) have uninterruptible power supplies which contain batteries that are charged off the grid and used to upconvert to AC to power devices when the AC power grid goes off-line. So, the notion of a DC power grid is not dead, but generally used as a backup system. The point is, the notion of a DC power grid has remained and now that renewable energies are viable, we should re-examine their use for widespread, local, power distribution.

One of the advantages of a locally generated DC station is energy savings as well as allowing a facility to run more easily off various DC sources, such as Solar Arrays. And for power-hungry installations, the notion of generating the power locally is growing in popularity. Part of the reason is that waste heat from the power generation generators (solar, geothermal, etc.) can be used to warm nearby buildings. By combining the production of heat and power, local power grids can squeeze much more useful energy out of the renewable energy they use, so this approach will certainly become more widespread as time goes on. The purpose of this design problem is to look ahead and provide an architecture and specification for a local DC power generation station that can serve a local community.

Currently many systems exist to control and handle alternative energies. There are systems to control and manage solar power, wind power, geothermal power, and others, but very few are designed to control, manage, store, and distribute energy from a power grid consisting of multiple alternative energy sources. As we saw the issues recently caused by the intense ice storms in Texas, one alternative energy source is not the answer to the growing energy needs/wants in our society. Each alternative energy source has limitations and issues associated with its use. By harnessing multiple alternative energies, and managing the storage of the energies produces, communities can be worry free from power generation. In order to facilitate utilizing multiple alternative energy sources, a control center must be included in the design that can manage the energy creation, distribution, storage, and return unused energy to the power grid.

9.3.4 High-Level Technical Description of the DC Power Grid System

The purpose of the multiple alternative energies DC power grid is to:

Supply DC power to a customer community of 500 homes and a biosphere for growing food.

- Provide a storage device to store energy in case of a system failure.
- Determine the amount of surplus energy that is above and beyond what is required to power the grid and charge the backup system, and then deliver that surplus energy to the national power grid systems.

Throughout the semester you will be asked to analyze, design, and architect different components of the overall DC power grid system. This will include research into alternative designs, like lithium nanowire storage systems for backup power and/or supercapacitors (also called ultracapacitors). Fig. 9.5 illustrates these technologies connected to the DC Power Grid Regulator Control Center.

9.3.5 High-Level Technical Requirements for the DC Grid System

- Design shall utilize at least two alternative energies, with a maximum of 4.
- Total power generation shall be 3.0 megawatts minimum.
- Each of the alternative energy sources shall be capable of generating at least 60% of minimum wattage.
- DC power grid shall generate DC power, with grid voltage being 48, 24, or 12 volts.
- DC power grid shall be capable of servicing at least 500 single family homes and one 4-acre biosphere with hydroponics capable of growing vegetables year-round. (See Fig. 9.6 for a block diagram of a DC-powered home).
- All lights, appliances, and facilities in each home and the biosphere shall be configured for DC operations, if possible. Things like electric ovens may need DC/AC conversion to accommodate their 220v. requirements.
- Only allowed exceptions are the HVAC system and the Washer/Dryer, although DC is preferred.
- There shall be energy storage capability capable of operating the grid of 500 single family dwellings for 7 days if all alternative forms of energy are not producing.
  - Compliance Criteria: Energy storage devices shall be capable of producing 750,000 watts for the entire 7-day period. Nominal energy storage should be batteries or supercapacitors.
- The DC power grid system shall include controlling h/w and s/w to manage the multi-alternative energy strategies and production.
- All energy produced above the required home and energy storage usage shall be upconverted to 220vac for release to the national power grid.
- The DC power grid shall be capable of operating in a temperature range of −40 degrees F. to 125 degrees F. without degradation of power generation.
- The DC power grid must meet OSHA regulations.

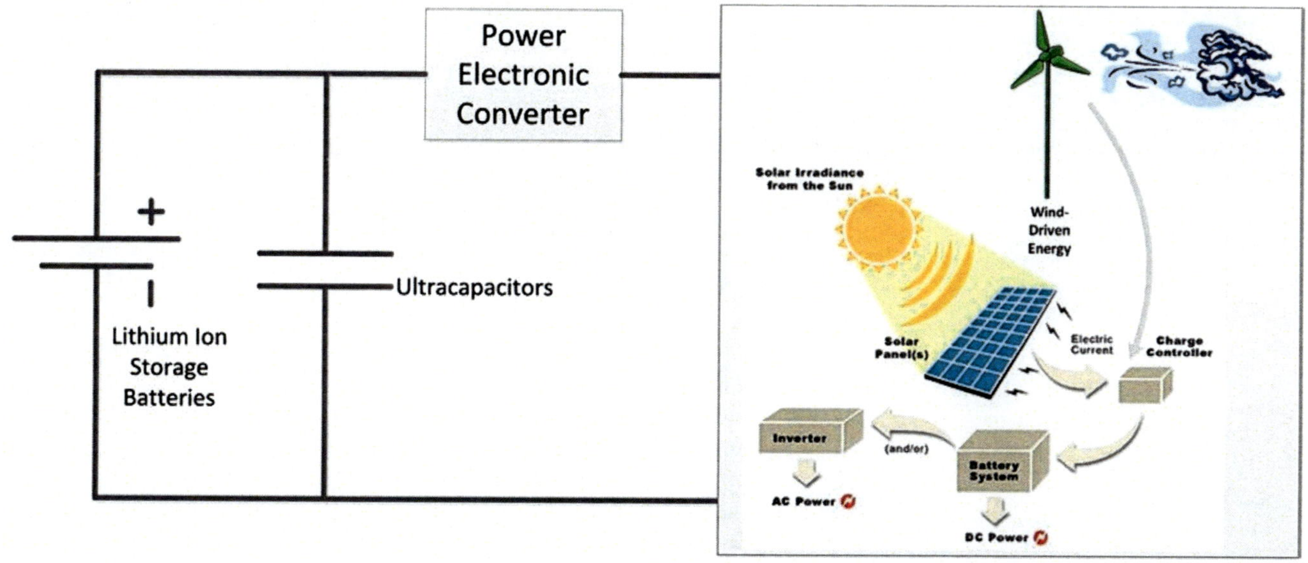

Fig. 9.5 Alternative power management system overview

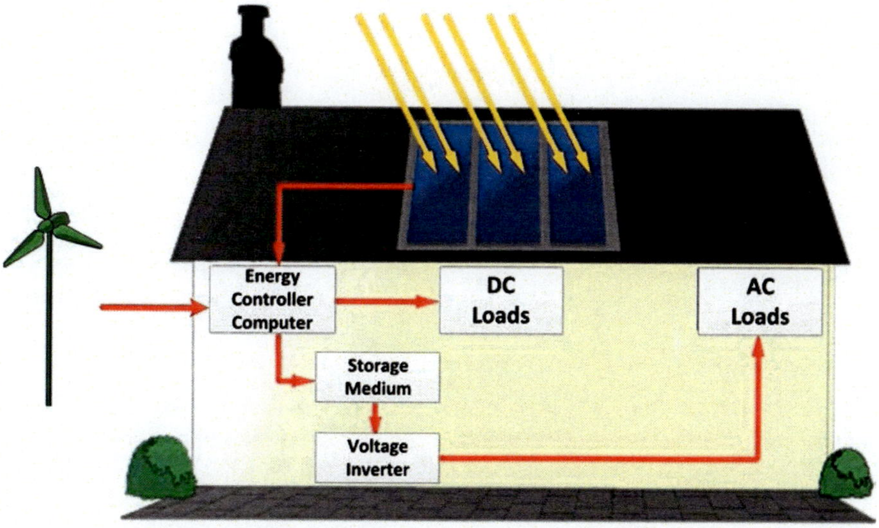

Fig. 9.6 Example of a whole house power grid

9.3.6 Assignments

- Write a report, 5–10 pages, describing the tiers that your system will contain and why each tier is important to your system design.
- For your multi-alternative energy system, answer the following questions:
  - Is a presentation tier necessary for your system and why or why not?
  - What are the application services your system should/will have?
  - What data will you need to acquire, process, store, and retrieve?
  - What type of data storage do you feel is appropriate for your system?
  - What will be your policies on when to and when not to send energy to the local power grid?

References

1. Rock-Evans, R. (1989). *CASE analyst workbenches: A detailed product evaluation*. Ovum Ltd..
2. Saeki, M., Horai, H., and Enomoto, H. "Software development process from natural language specifications," presented at 11th International Conference on Software Egineering, Pittsburgh, Pennsylvania, 1989.Formal and Informal in RE Cybulski 12.
3. Salek, A., Sorenson, P. G., Tremblay, J. P., and Punshon, J. M. (1994). "The REVIEW system: From formal specifications to natural language," presented at The First International Conference on Requirements Engineering, Colorado Springs, Colorado.
4. Srinivas, Y. V. (1991). Algebraic specification for domains. In R. Prieto-Diaz & G. Arango (Eds.), *Domain analysis and software systems Modeling* (pp. 90–124). IEEE Computer Society Press.
5. Vonk, R. (1989). *Prototyping: The effective use of CASE technology*. Prentice-Hall Int.
6. Wood, D. P., Christel, M. G., and Stevens, S. M. (1994). "A multimedia approach to requirements capture and modeling," presented at The First International Conference on Requirements Engineering, Colorado Springs, Colorado.

7. Coram, M., Bohner, S. (2005). The impact of agile methods on software project management. In: 12th IEEE International Conference and Workshops on the Engineering of Computer-Based Systems, ECBS., IEEE 363–370 2.
8. Clark, R. G., & Moreira, A. M. (1999). Formal specifications of user requirements. *Automated Software Engineering, 6*(3), 217–232. 3.
9. Giese, M., & Heldal, R. (2004). From informal to formal specifications in UML. In *Intern. Conf. on the Unified Modeling Language* (pp. 197–211). Springer. 4.
10. Seater, R., Jackson, D., & Gheyi, R. (2007). Requirement progression in problem frames: Deriving specifications from requirements. *Requirements Engineering, 12*(2), 77–102. 5.
11. Mavin, A., Wilkinson, P., Harwood, A., Novak, M. (2009). Easy approach to requirements syntax (EARS). In: Inter. Requirements Engineering Conf., IEEE 317–322 6.
12. Golra, F.R., Beugnard, A., Dagnat, F., Guerin, S., Guychard, C. (2016): Addressing modularity for heterogeneous multi-model systems using model federation. In: Companion Proc. of the Intern. Conf. on Modularity, ACM 206–2118.
13. Boniol, F., Wiels, V.: The landing gear system case study. In: Intern. Conf. on Abstract State Machines, Alloy, B, TLA, VDM, and Z, Springer (2014) 1–18 9.

10 Multidisciplinary Requirements Engineering

Abbreviations

AC	Alternating current
ATM	Automated teller machine
C	Centigrade
DC	Direct current (or 0 Hz)
DDA	Derived/decomposed/allocated
DOORS	Dynamic Object Oriented Requirements System
DREX+	Digital receiver/exciter
F	Fahrenheit
HVAC	Heating, ventilation, and air conditioning
ID	Identification
INCOSE	International Council on Systems Engineering
MDRE	Multidisciplinary requirements engineering
OSHA	Occupational Safety and Health Administration
RMP	Requirements Management Plan
RMT	Requirements Management Tool
SoO	Statement of Objectives
SoW	Statement of Work
UML	Unified Modeling Language
V	Volts
Vac	Volts, AC

As discussed in Chap. 2, successful requirements engineering is a multidisciplinary engineering endeavor that needs to bring together requirements experts that understand all aspects of the system to be developed. This includes architecture, hardware, software, maintenance, sustainment, security, human-factors; those disciplines shown in Fig. 2.1. In short, the MDRE process is the approach that develops realizable solutions for real-world needs [1]. Fig. 10.1 illustrates the entire MDRE process as a set of process/activity tiers, providing more detail as the tiers are decomposed, just like a tiered architecture would show [2].

At the highest level, the MDRE process provides a set of requirements needed to meet the customer's expectations, based on the SoW and SoW given to the development contractor/group for implementation. The highest-level output from the MDRE process must be a set of requirements that are:

- **Quantifiable**: Each requirement's verbiage contains a scale for measurement to determine if the developed system can meet the requirement. Potential solutions to the requirement must be measured against the scale or quantification specified in the requirement. Quantifiable means the requirement must be given the context under which the system will be measured against the requirement.
- **Relevant**: All requirements must pertain to some objective provided in the customer's SoO and/or SoW. The requirement may pertain to "implied" requirements or they may be derived requirements necessary for the proper operation of the system, but all requirements must be there to drive the design for a reason. It may seem obvious that all requirements should be relevant to the system, but often, requirements make their way into system requirement specifications that have no relevance to the design, but were "wants" of the design team, and not asked for by the customer [3].
- **Detailed**: Each requirement should have enough detail to convey what is required in terms of a functionality, attribute, capability, characteristic/behavior (which can be performance), or quality of some part of the system.
- **Unambiguous**: One of the most important aspects of requirements is ensuring that they are unambiguous. The MDRE team must realize that unambiguous is not just about how the requirement is written, but it is also about how the requirement will be read and interpreted. Without communication with the customer and end-users, developers, testers, maintenance, and sustainment personnel, you cannot write unambiguous requirements. Almost anything you write might be misinterpreted by someone. Communication is vital to an unambiguous requirement specification [4]. We will discuss the notion of requirement ambiguity more in Chap. 12 on Requirements Composition.

J. A. Crowder, C. W. Hoff, *Requirements Engineering: Laying a Firm Foundation*,
https://doi.org/10.1007/978-3-030-91077-8_10

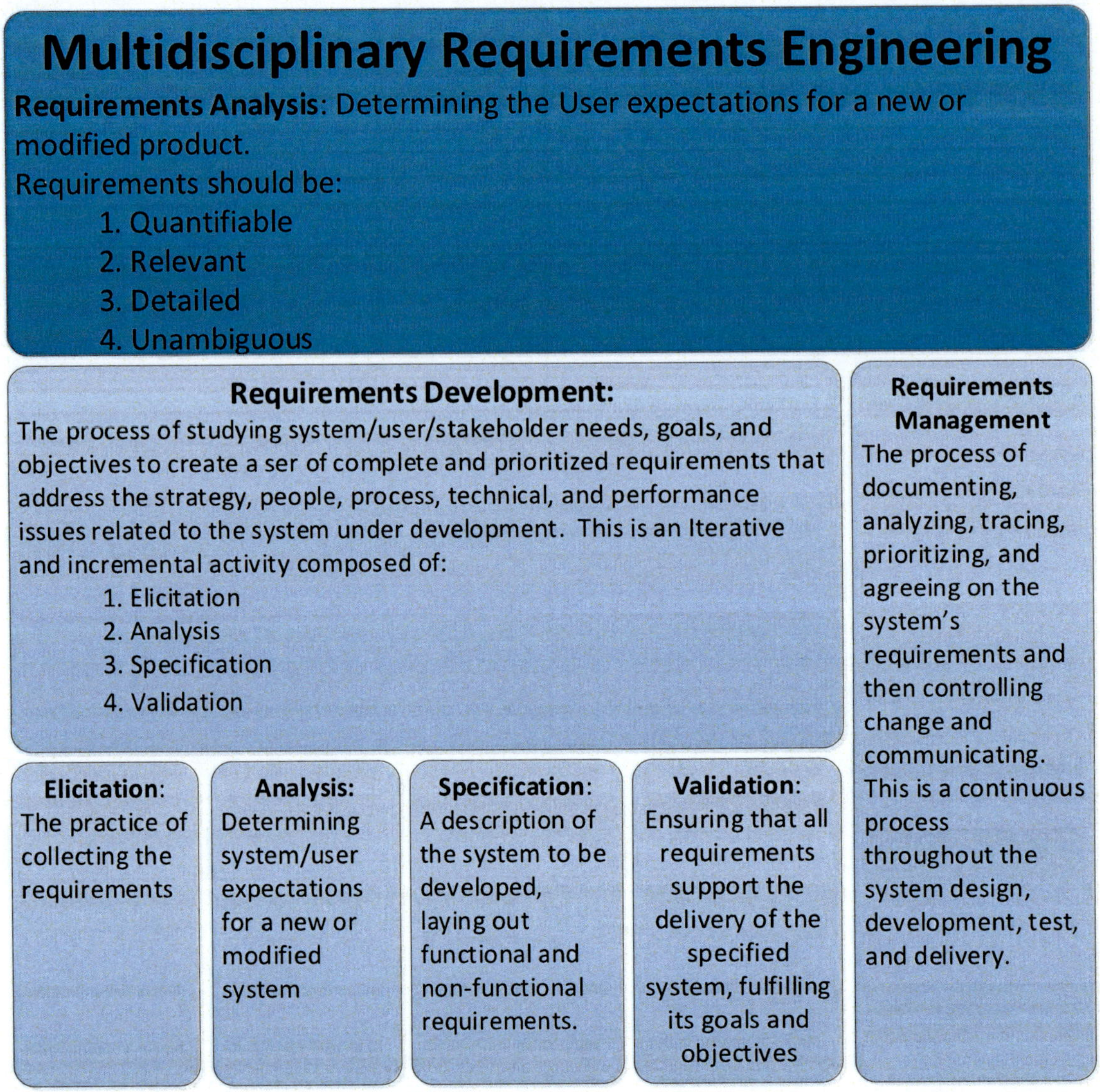

Fig. 10.1 The MDRE process tiers

- **Verifiable**: It is critical that all requirements be verifiable. Without this, how can the contractor demonstrate that the system meets the requirement? It is crucial that the systems engineer consider how the requirement is going to be verified while they are generating the requirement. This is because the language comprising the requirement will likely change to accommodate the verification. The requirement must be written such that it can be verified that the system meets this requirement. That may seem obvious, but consider the requirement, "The system shall be User Friendly." We have all seen requirements like this. This requirement is not verifiable due to not defining specifically what User-Friendly means. Requirements written in the negative, "The system shall never…," are not verifiable because there is no time frame with which to measure the requirement. Unverifiable requirements come from three main sources: (1) The requirement is written poorly, (2) The requirement is incomplete or missing some necessary information to understand how to verify the requirement, and (3) Requirement is not allocated at the level where it can be verified.

Question to think about

Why do you think having requirements that are unambiguous is important? What can happen if performance requirements are ambiguous?

10.1 Requirements Development

One of the most important things the MDRE team must do before beginning the process of developing the system requirements specification is to establish the context of the system/service to be developed. This involves collecting and studying all relevant materials provided not only by the customer, but from compliance documents that are applicable to the system, as well as material to provide a context of the environment and environmental conditions the system is to be deployed within [5]. Naval system deployed on a ship at sea provide entirely different environmental concerns than a vehicle to be utilized in desert terrain. All available information, including user scenarios (if possible, to attain or create) as well as any existing requirements or system specifications available and relevant to the system to be developed should be utilized in the MDRE requirements development process [6]. This is illustrated in Fig. 10.2. We will first discuss the high-level requirements development process (elicitation, analysis, specification, and validation) and then we will look at the inputs and outputs to the MDRE development process shown in Fig. 10.2. We have discussed the notions of SoOs, regulations and standards, and domain information throughout the book and will concentrate on and discuss the importance of both the use of existing system specifications and user scenarios that are required for effective requirements development. We will also examine and discuss the structure and interactions between the outputs shown in Fig. 10.2.

> **Question to think about**
>
> *Do you believe a formal process is needed for creating the requirement specification? What might happen if they were created more* ad hoc*?*

From the discussion in Chap. 2, we established that the MDRE requirements development process involves studying the system objectives, user and stakeholders' needs to create the requirements specification(s) needed to design and implement the system/service being requested. The requirements should provide the information needed to derive the strategies, processes, technical and performance characteristics, and issues needed for system design, development, test, verification, deployment, and sustainment throughout the system's life cycle. As Fig. 10.2 illustrates, creation of the requirements specification, system specification, and system models/simulations [7] is a feedback-driven process that requires inputs from multiple sources. Expounding on the discussion in Chap. 2, the MDRE Requirements Development process involves:

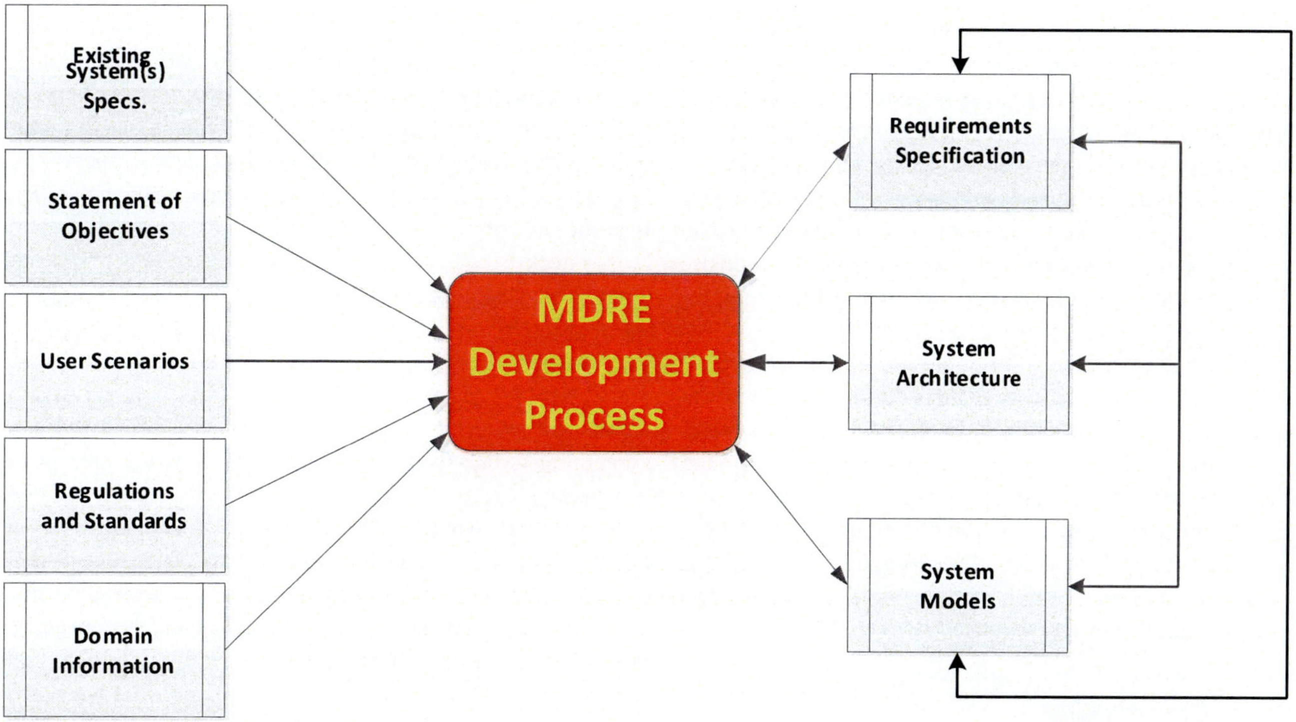

Fig. 10.2 The MDRE development process context diagram

Requirements Elicitation: This involves collecting the available requirements relevant to the system under development. This includes explicit requirements from the customer(s)/stakeholder(s), compliance and legal documents, any existing specifications that are relevant to the system, and any domain-specific requirements. This is not decomposing or deriving requirements, rather, this is the process of collecting existing requirements to determine what areas are missing and what the focus of the requirements analysis process should entail. Requirement elicitation techniques include, but are not limited to, brainstorming, focus groups, interviews, process modeling, prototyping, and requirements workshops [8].

Requirements Analysis: Once the elicitation process is completed, the requirements analysis process is needed to understand the requirement gaps encountered during the elicitation process as well as understanding any implied requirements that exist but are not articulated anywhere. Requirement analysis is necessary to perform two main functions of the overall MDRE process:

Ensure you are designing the correct system.

Ensuring you are designing the system correctly.

Requirements Specification: The end goal of the MDRE process is to create a usable, validated, and testable set of requirements that will drive the architecture, overall design, test, verification, deployment, and ultimately the system sustainment and maintenance throughout the system's life cycle. It is crucial to have an agreed to set of requirements, even at a high level, to establish a system functional concept, system operational concept, and operational/technical boundaries. Even in an agile development, an initial set of requirements is necessary to understand what capabilities to plan into each sprint and across sprint teams. It is understood that as the software is written and tested requirements may change, but those changes must be managed and the overall requirements kept in sync with the agile development or in the end, you will deliver a system where the systems architecture, systems design, software design, and requirements are NOT consistent nor correct, making system test, verification, maintenance, and sustainment all but impossible.

Validation: Validation is an important step in the overall MDRE process, for validation ensures that the requirements support the system. In short, requirements validation says, if we build the system to these requirements, it will provide the capabilities and desires laid out in the customer SoW and SoO. This is very different from verification, as this process verifies that the system meets the requirements. Validation says you are building the right system, verification says you did indeed build what was validated [9].

Question to think about

What could happen if the requirements validation process were skipped?

10.1.1 Existing Specifications

The use of an existing specification can fall into two categories:

- Specifications that are given by the customer, providing high-level requirement for the system to be developed.
- Specifications for existing systems that are "close" to the system to be developed.

In the former case, a customer-provided high-level requirement specification should provide the basis for the MDRE development process and all other requirements should be decomposed, derived, and allocated from and linked to these top-level customer levied requirements. Even implied requirements discovered during the analysis process should be tied to the customer requirement specification. If not, this must be communicated with the customer, and a high-level requirement added to their specification to provide traceability for the newly discovered "implied" requirements that must be formally written and agreed to [10].

Question to think about

What is the danger to a new system if an older requirement specification were used without a comprehensive analysis?

While it may seem obvious that a customer-provided high-level requirement specification should be consistent with the customer-provided SoW and SoO, this is often not the case. It is possible different individuals or different groups of individuals wrote each of the three documents and they may or may not have coordinated them with each other. It is incumbent on the MDRE team to do the analysis and cross-check between the three documents to ensure consistency. And, if any inconsistencies arise, they must be adjudicated before moving on with creating the requirement specification(s) that will be utilized to design, develop, test, verify, deploy, and sustain the system. Failure to do so can result in a system being rejected because it failed to meet something in one of the documents that was not present in the others [11].

In the case where the MDRE team is starting with an existing requirement specification for either a current system being upgraded/extended or an existing requirement specification for a system that is "close" or like the system to be developed, caution must be taken. It is vitally important that part of the analysis considers the differences between the operational environments the two systems are operating within. Also, differences in how the system will be used is important. The analysis must also consider the differences between functional, performance, and non-function requirements. These differences must be understood and agreed to before proceeding with the MDRE requirements development process. Remember the case study above about the Tacoma Bridge disaster. Reuse without an operational environment differences study drove the incorrect design and ultimate failure of the bridge. Just because two systems are "similar" doesn't mean they can be interchanged or reused as-is. There is also a possibility of reusing a discovered implied requirement that is not relevant to the new or upgraded system.

Requirement reusability can be an effective way to jump-start a development effort. However, the MDRE team needs to ensure they do not fall into one of these common issues with reusing requirements [12]:

- Copying and/or linking requirements that are outdated and their dependencies are not relevant to the new/upgraded system.
- Requirements that are tightly coupled to the current product, reusing these requirements can cause confusion and drive design details that will not be compatible with new/updated SoO.
- Requirements generally have data associated with them, including rationale for the requirement, data definitions, compliance documents or constraints. If these are not identical between the two systems, it can drive costly design changes when this is discovered during development or test.

The list below highlights classes of requirements that may be reusable across several projects/programs:

- **Security requirements**: Examples would be password requirements, use of internet, etc.
- **Certain Functional Requirements**: If two systems are similar in operation some of their functional requirements may be similar enough to reuse. Example might be two different ATM machines both requiring the same languages to be displayed.
- **Document Templates**: If the same type of documents is specified for the two different projects, they may have the same requirements for the type of documents and the template(s) that should be used.
- **Data definitions:** It is important to make sure definitions for data items have not changed over time since the original specification was created. Each data item needs to be reviewed for context and meaning within the new system design.
- **Nonfunctional Requirements**: Performance requirements like response time or reliability/availability.

The MDRE team should look for requirements patterns that are similar between systems to help start the requirements development process.

Question to think about

Why are nonfunctional requirements like availability and reliability important to the operational viability of a system?

10.1.2 Operational Assessment: User Scenarios, Use Cases, and Storyboards

Understanding how the system will be used and the types of functions the operators of the system are expecting is important to ensure the final system is usable by those intended operators (whether human or machine). One of the ways to provide this information to the MDRE requirements development process is using user scenarios, use cases, and storyboards. Many engineers believe these terms are interchangeable. However, they are all three useful, but very different constructs for conveying user information [13].

10.1.2.1 User Scenarios

User scenarios are typically a narrative, written in plain language, which describes how a user or operator of a system would interact with the system to perform a specific task that is allowed by the system. An example would be a scenario for a use to access their checking account balance using at ATM. User scenarios, while not relaying technical information to the MDRE team, they do, in general, portray how the system would be used to perform specific tasks, which will then allow requirements to be derived to ensure those functions are possible with the system to be developed. Usability of a system is a major factor in the test, verification, deployment, use, and sustainment of a system. If users cannot use the system to perform the functions required, then the system will not be adopted or used. The user scenario narrative focuses on the needs of the users or operators, not on the technology required to allow users to perform the functions described. The advantage of user scenarios being written in plain language rather than technical language is that, in general, the MDRE team can all understand the scenario regardless of the team member's technical specialty [14]. The lack of "how the system" performs the functions in the background, provides the MDRE team with the flexibility to design the system so these actions can be performed. There can be multiple user scenarios, each written about a separate

Table 10.1 Textual use case example

10. This flow begins when a new user requests access to the system.
10a. An account request form is completed with user information, requested roles and accesses, and supervisor approvals.
10b. The system administrator verifies that the new user has the appropriate need for the system roles requested.
20. The system administrator conducts related system use training to the new user.
20a. General user: Training covers information security rules and regulations, the expected behavior of users, and the consequences of non-compliance.
20b. Specialized user training: Training for individuals who have specific needs for system information (e.g., user credential)
20c. Users are required to sign an acknowledgment of the rules and regulation before access is granted to the system.
30. The system administrator and supervisor approve the account request, including requested roles and privileges.
40. The system administrator creates the new account. Only system administrators are authorized to perform account management functions.
40a. The system administrator assigns a unique user ID and creates an initial password. Upon initial logon, the user will be required to change the password following password policy guidelines.
40b. The system administrator will assign a user certificate for authentication and authorization purposes.
40c. Approved user roles are added to the user accounts based on their authorized duties and roles. Each role will give the user certain permissions and access to perform authorized tasks. Users may be granted more than one role.

Table 10.2 Use case elements

Use case element	Element description
Use case number	ID number used to identify the use case
Use case application	What is the system, element, subsystem, application, etc., for which this use case pertains?
Use case name	Title for the use case. The title should be short and descriptive; should represent the goal the use case is trying to satisfy
Use case description	What system, element, subsystem, etc., functions is this use case executing; done in paragraph form.
Use case primary actor	Who is the main actor for this use case (person, process, or device?
Preconditions	What preconditions must be met before this use case can start?
Trigger	What event(s) trigger this use case
Basic use case flow	The basic flow should describe the events of the use case when everything is perfect (happy path); there are no errors, no exceptions. The exceptions and errors are handled in the "alternate flows."
Alternate flows	Significant alternatives, exceptions, and errors handling.

specific task or action that a user may need to perform with the system. The caution with user scenarios is that it may not be possible to write one for every function a user/operator may interact with the system to perform. User scenarios should be written in such a way that people with non-technical backgrounds can understand and may include cultural information about use as well as circumstances that may result in a user/operator interacting with the system. In general, the information to be gathered before writing a user scenario is:

- Who is the user/operator?
- Why does the user/operator interact with the system?
- What goals does the user/operator have for interacting with the system (e.g., receiving their checking account balance).

10.1.2.2 Use Cases

A use case is a United Modeling Language (UML) construct for also capturing how users (people or other processes) interact and utilize a system. This may involve interacting with the hardware and/or software of the system that implements the functionality laid out in the requirements specification. A use case is generally a more formal construct than the user scenario [15]. There are two types of use case:

A Textual Use Case: This represents a brief step-by-step description of the steps it takes to accomplish a specific function of the system. Use cases are generally written in a more formal and technical structure than a user scenario. Again, generally a user scenario is based on human interaction with a system; however a use case can denote interaction by a person, process, or other device. Table 10.1 below illustrates a use case for laying out the specific steps required to gain access to a computer system [16]. Table 10.2 describes the basic elements for a textual use case.

Use Case Diagram: is a visual behavior diagram which describes the functional requirements exercised by a particular use of the system. The use case diagram illustrates the relationships between actors (elements representing the role of person(s), processes, or devices) and the system. Figure 10.3 illustrates a use case diagram for debug and logging within a system [17].

> **Question to think about**
>
> *What information does a use case diagram contain that is not in a textual use case?*

For use case diagrams a simplified description of the rules for creating a use case diagram are detailed below.

A use case describes a sequence of actions that provide something of measurable value to an actor and is drawn as a horizontal ellipse.

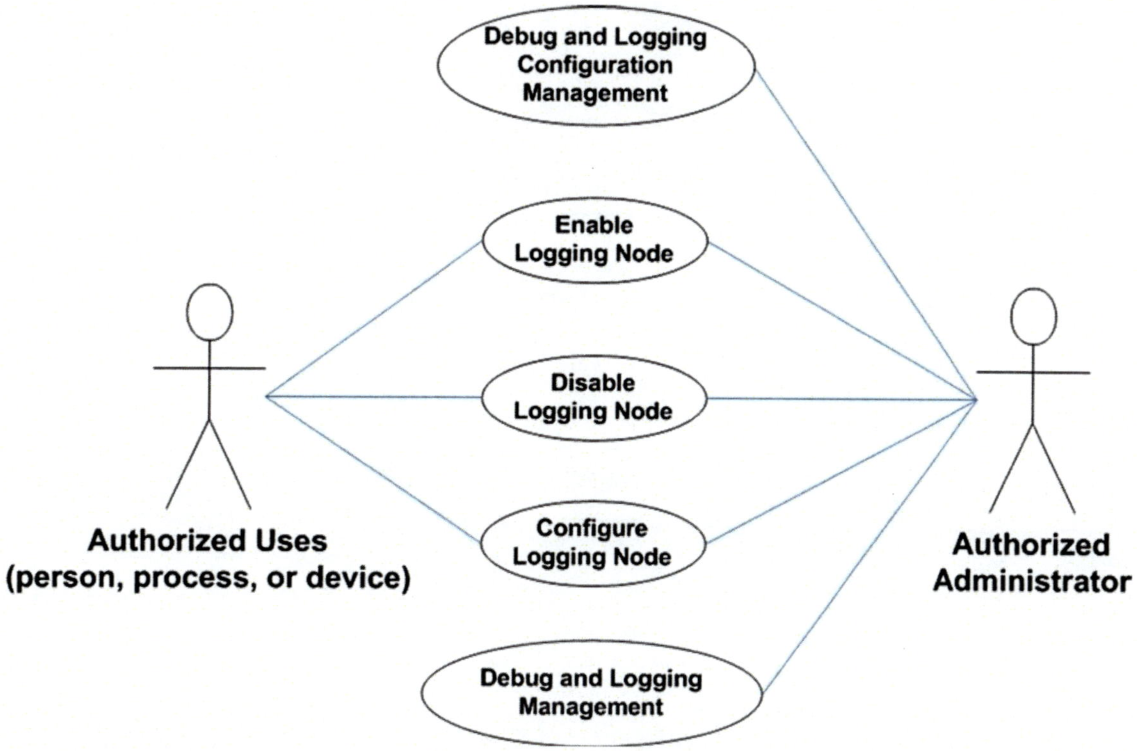

Fig. 10.3 Use case diagram example: debug and logging

An actor is a person, organization, or external system that plays a role in one or more interactions with your system. Actors are drawn as stick figures.

Associations between actors and use cases are indicated in use case diagrams by solid lines. In Fig. 10.3, note the associations between the actors and the actions described in the ellipses.

An association exists whenever an actor is involved with an interaction described by a use case.

Associations are modeled as lines connecting use cases and actors to one another, with an optional arrowhead on one end of the line.

A use case describes how the system should respond to either a human operator or other process within the system. Use cases are written (or diagrammed) with reference to technical steps, including preconditions required before the use case steps are initiated and post-conditions (also called end conditions) describing the expected results of the technical steps laid out in the use case [18]. Use cases describe the system's behavior not the user's behavior indicative of a user scenario. Use cases are important to define the interactions with the system and the goals/objectives required by the system to such interaction to understand the complexities and functionality required to meet these system objectives and outcomes. Use cases help in the analysis processes to determine and ensure there are requirements in the final requirement specification to handle the needs shown in the use cases.

10.1.2.3 Storyboards

Unlike user scenarios and use cases, a storyboard is typically a visual representation of how a user or process interacts with a system through one or more of the system interfaces (e.g., keyboard or touchscreen). A storyboard should articulate in pictures and diagrams the processes a user (again a person, device, or other process) will go through when using the system. Storyboards are useful in determining the requirements for access and utilization of the system and allow brainstorming in ways user scenarios and use cases cannot capture [19]. Storyboards are a good compliment to both user scenarios and use cases in that they provide a sketch of what the user would see when interfacing to the system.

Storyboards focus not on the mechanics (technical steps) to be taken, but rather what the product will do/not do when interfacing with the system. Storyboards can be done with whiteboards, flip charts, post it notes™, or any other medium that can be quickly drawn onto to convey ideas about system usage. Storyboards are a great tool to engage focus groups, illustrating the system and how users will operate the system. This fosters immediate feedback and can be iterated with the focus group to arrive at an interface notion that is useable by most users. Storyboards can also be generated electronically, providing a visual "mock-up" for user interfaces and how the system would be used. Fig. 10.4 illustrates one screen mock-up storyboard for the DREX+ system described throughout

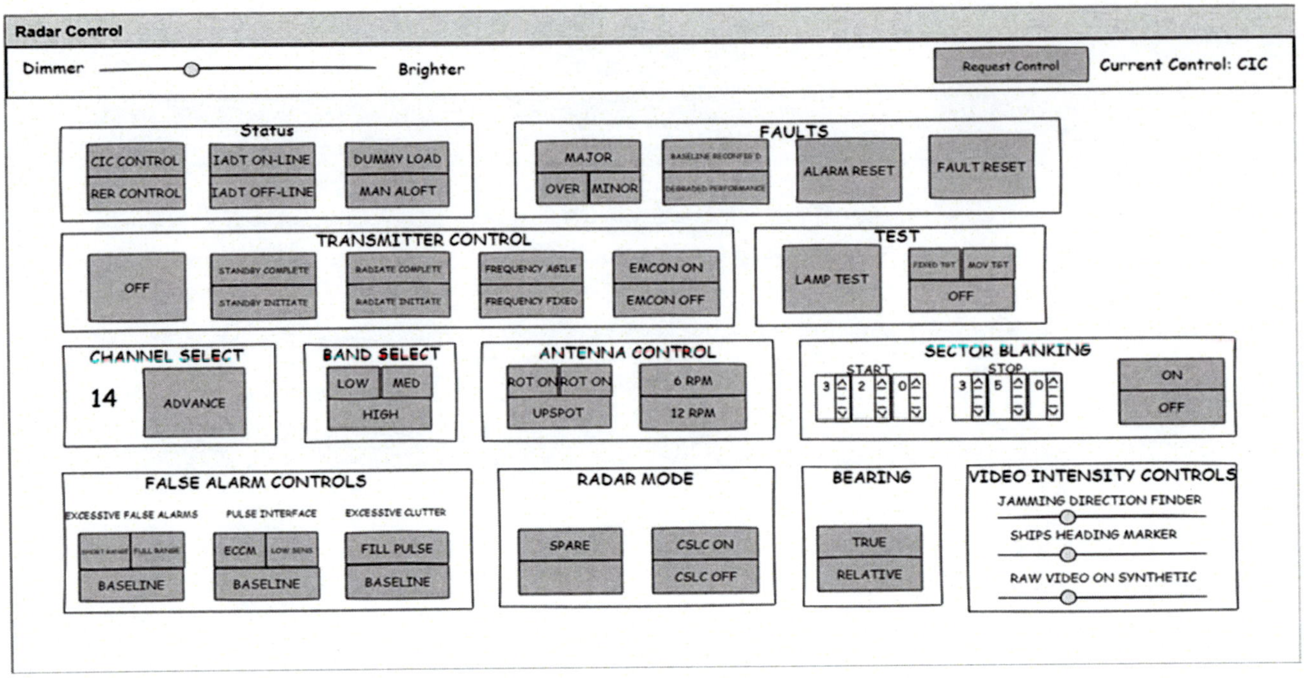

Fig. 10.4 Storyboard mock-up for the DREX+ front screen

the book. This would be used to convey the concepts to users on how the system would be operated, where the controls are envisioned to be placed, and the types of controls that could be provided by the DREX+ system.

> **Question to think about**
>
> *What information does the storyboard in* Fig. 10.4 *relate that a textual description typically would not?*

10.1.3 Requirements Specification

The main output from the MDRE requirements development process is the system requirements specification. This contains all the requirements required to design, develop, test, verify, deploy, and sustain the system. Once the requirements specification has been fully documented (hopefully in a requirements management tool), it becomes the baseline for everything going forward in system design and development [20]. Every requirement in the specification should tie or link to a design element of the system. If there are requirements that do not tie to any part of the system, or parts of the system that have no requirements, these gaps must be taken care of before moving forward with the design. The requirements specification is basically the contract between the development group and the customer on what will be developed and delivered (deployed). The requirements specification should be very detailed and well understood by all customers, stakeholders, and the development team, as this defines what is expected from the system. The development team must understand fully what the customer's problem is, and, using the requirement specification, determine how to provide the capabilities and performance required [21]. From the high-level requirements, either provided by the customer and gathered during the requirements elicitation process, the specification may be used to create a software requirements specification, a hardware requirements specification, and others that are specific to a given technology. At a minimum, whether it is the systems, software, hardware or other requirement specification, the requirements should all answer the same basic questions [22]:

Functional Capability: What does the system do? What is its purpose?

External Interfaces: How is the system accessed, how does the system interact with users (either human or other systems). What assumptions are made about how the system interfaces?

Quality Attributes: What are the nonfunctional requirements pertaining to availability, reliability, maintainability, sustainability, security, portability, manufacturability, and others.

Constraints: What design constraints are or will be imposed on the system. These may be legal, cultural, or political challenges that must be met. An example might be what languages will the system accommodate? What are the operational environment constraints on the system? An example might be the system must operate in an environmental temperature from 30 °C to −70 °C.

Question to think about

Review information on availability and think about why it is such an important nonfunctional requirement to specify?

10.1.4 System Architecture

The system architecture provides a conceptual model of the structure, behaviors, and overall breakdown of the system into subsystems, components, etc. The system architecture organizes the requirement information in a way that allows the design team to reason—or formulate ideas about how to build the system to meet the requirements. Every aspect of the requirement specification needs to be accounted for in the system architecture—including notions of testability and what data are required for testing. You should never design or develop the system if you do not understand how to test the system and what data are required to perform those tests [23]. The process of creating the systems architecture entails generating architecture alternatives and deciding which is the most appropriate architecture to drive the system design and development. The final solution architecture for the system has the features, properties, and characteristics which meet the requirements specification. There may be trade-offs needed to decide between alternatives which may drive the overall life cycle of the system (e.g., maintainability and sustainability) and may determine how the software must be written to support the overall design of the system. The main point here is that there is a symbiotic relationship between the system requirements specification and the system architecture artifacts. These must be consistent and supportive of each. They drive everything else that happens from design to development, test, verification, deployment, and operations.

Question to think about

What would be the problem with not considering alternative solutions and just using the first one available?

10.1.5 System Models

Once the top-level requirements and architecture have been established (at least a draft), system models are used to establish abstract models of the system to present different views needed to ensure all parties (develop, customer, and stakeholders) agree that the system designed meets the objectives and requirements. A system model is not, by itself, an architecture nor a requirement specification. It is simply a tool to represent and express aspect of the system more clearly to all interested parties [24]. Architecture models arise out of the process and information analyses that is part of the MDRE requirements analyses. The architectural model(s) are high-level functional models that are then used to validate the requirements. The models map the architecture against the requirements to ensure the designed system represents the agreed-to set of requirements and that the requirements have been decomposed and allocated correctly across the architecture. The information contained in the high-level models is utilized to create lower-level logical architecture models. The logical models consist of package diagrams, subsystem interdependency models, and information distribution models across the architecture. The logical architecture and logical models define the interaction of operations with the system to ensure all processes are captured and understood. Fig. 10.5 illustrates the relationship between the system architecture, requirements specification, use cases/user scenarios/storyboards, and system models [25]. This same relationship exists at each level of the system decomposition. Lower-level subsystems and components have the same type of relationship to low-level requirements and use cases as does the top, system-level architecture. At each level, models are required to drive the design and ensure functionality, data/information flows, and technical objectives are being met throughout the system.

Overall, system models are useful, in that [26]:

- Modeling diagrams help clarify concepts described in the requirement specification, the architecture, and the use cases/user scenarios.
- Working with system models helps expose inconsistencies in the architecture and/or requirements.
- Communicating with models can relay design concepts less ambiguously than textual descriptions.
- With the right tools, models can be utilized to generate needed artifacts like software code and database schemas.

Figure 10.6 is a simple example of a system architecture model for applications designed for mobile devices (e.g., cell phone and tablet).

There are many types of models that can be generated, based on the architecture, the specific application, and the system level for which the model is being generated. Some examples are provided in Table 10.3. This is not an exhaustive list but provides insight into the types of models that are utilized throughout the different industries.

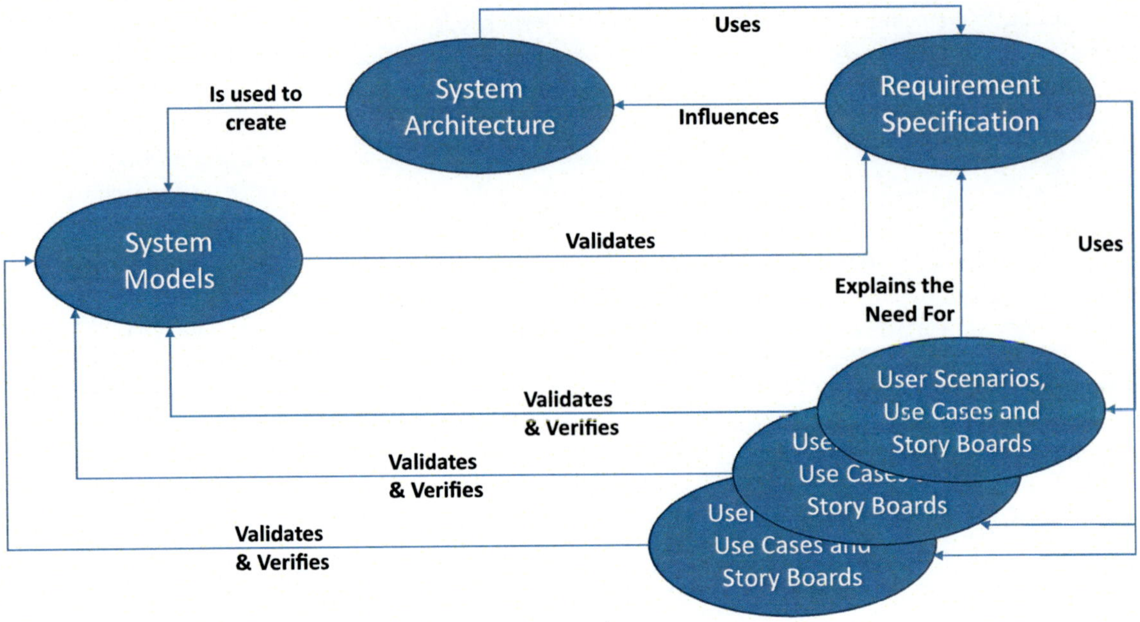

Fig. 10.5 Relationship between system architecture, requirements, use cases/user scenarios, and system models

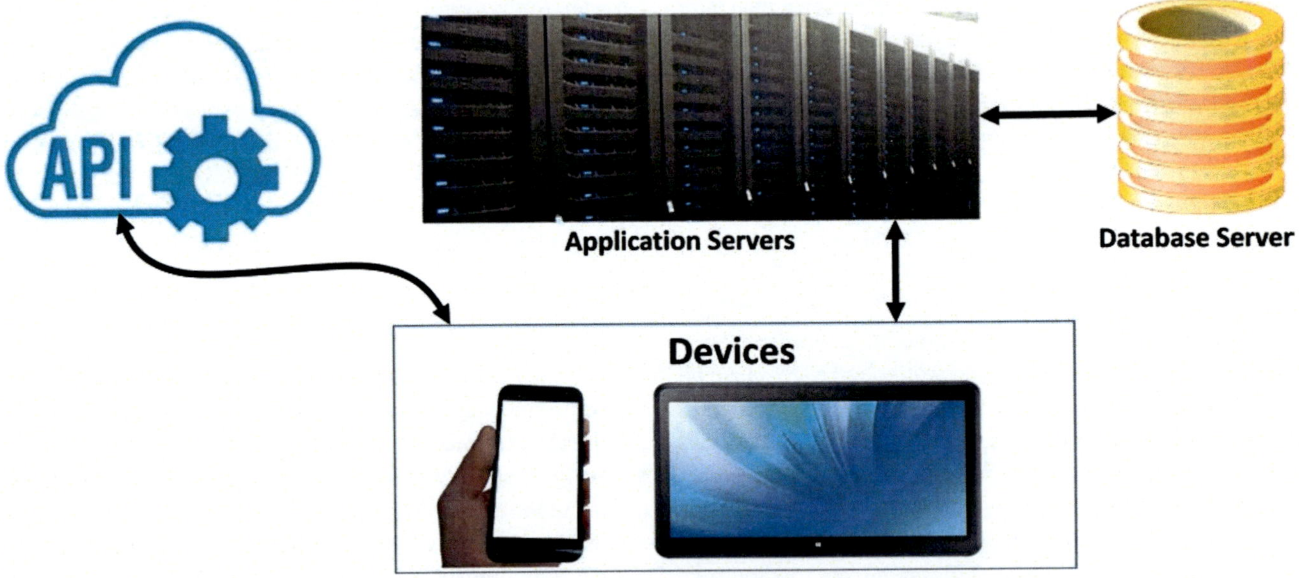

Fig. 10.6 Example of a high-level architecture model diagram

> **Question to think about**
>
> *What could happen if the requirements were not maintained and documented by the MDRE team?*

Once the requirement specification is completed and approved, the requirements must be managed throughout the life cycle of the system. MDRE requirements management is one of the most important functions of the overall MDRE process, and the one that is most often done poorly. Mismanagement of requirements is the single most reported reason for system development failure [27].

10.2 Requirements Management

MDRE requirements management begins before the first requirement exists for the system. MDRE requirements management must be involved in every aspect of the require-

Table 10.3 Examples of design level vs. model types

Design level	Description	Model types
Process	Understanding the context of how the system will be used in operation clarifies what the users/operators of the system need	Class diagrams that describe the processes used during operation
Requirements	Provides the definition of what the system is supposed to do (what are the objectives)	Use cases and use case diagrams, user scenarios and storyboards that clarify what the requirement means in terms of the system and users
Architecture	Provides an overall picture of the system, including the major components and how they interact/interface to each other	Dependency diagrams to describe each interdependent component. System/software designs can be validated against the dependency diagrams to ensure they adhere to the architecture
Software	Diagraming based on the software structures that can be generated from the code itself	Software dependency diagrams show interdependencies between classes. Updated/modified code can be validated against the dependency diagrams. Class diagrams can be used to validate requirement coverage

ments process, which may start writing a Requirements Management Plan (RMP) along with deciding on a template for the requirements and which requirements management software tool will be used throughout the project. In some cases, the requirements management tool may be dictated by the customer to conform with their requirements management systems [28].

The program RMP helps explain how you will receive, analyze, decompose, derive, document, and manage all the requirements within a system throughout the life cycle. The plan usually covers everything from initial information gathering of the high-level project to more detailed product requirements that could be gathered throughout the life cycle of a project. Key items to define in a requirements management plan are the project overview, requirements gathering process, roles and responsibilities, tools, and traceability. The main goal of the RMP is to [29]:

- Describe how the program will identify requirements.
- Prioritize requirements.
- Assign responsibility for requirements management.
- Establish and explain the traceability philosophy and methods.
- Describe the requirements change management process.
  - How change will be managed.
  - How change effectivity (when a change is implemented) will be done.
  - Who has change authority throughout the development cycle?

During the requirements elicitation and analysis phases, it is important to ensure the process and results are captured for reviews. The primary issues that plague MDRE requirements management are incompleteness of the requirements text and inconsistency between requirements. These issues are difficult to find and track automatically and generally require manual viewing. One way to help alleviate this problem is to create a requirements template. Creating or adopting a requirement template can be effective for drafting requirements but will require training for MDRE personnel to understand how the template should be used. It can prevent common mistakes in the requirements formulation, and the template makes it easier to review [30]. It is possible to create machine learning algorithms to review and catch the simplest requirements mistakes; however issues like inconsistency are difficult to handle, even for Natural Language Processing algorithms. Machine learning can be used to analyze the requirement text and suggest improvements based on industry best practices from groups like the International Council on Systems Engineering (INCOSE) [31].

10.2.1 MDRE Requirements Management Approach

The overall requirements management approach depends on the size, scope, and application(s) of the system being developed. In general, requirements can be first divided into:

Program requirements: Requirements identified to meet the overall needs of the system to ensure its readiness and useability. Program requirements are, in general, non-technical requirements like availability, maintainability, etc.

Product requirements: These are requirements required to meet the technical objectives/specifications of the system. These requirements are there to ensure that the performance characteristics are met.

> **Question to think about**
>
> *How might different management styles dictate how requirements are managed?*

•

The MDRE requirement management should manage, document, and review all aspects of requirements engineering shown in Fig. 10.1. MDRE requirements management must be ongoing, throughout the entire life cycle for a given system. This includes ensuring that personnel involved in all aspects of the system design, development, test, deployment, and sustainment report requirement status and raise any issues with requirements across the system. As the development and testing of the system matures, there may be times when it is discovered that requirements need to be adjusted, or requirements added, based on results of design and/or test reviews. Any changes must be controlled by the change management process, discussed below. As discussed, MDRE requirements management must be ongoing throughout the sustainment and maintenance of the system, as conditions may dictate the need for requirements changes and possibly new requirements, which may result in upgrades to the existing system.

10.2.2 MDRE Requirements Documentation and Configuration Management

Requirement documentation must be an integral part of the MDRE process and lack of proper documentation leaves the development program at risk. The top risks associated with lack of proper requirement documentation are [32]:

- **Lack of collaboration and review**: Customer/develop collaboration and review should be an ongoing process. Lack of documentation hinders this process and reduces the visibility into how requirements meet the system objectives.
- **Lack of differentiation between what and how**: Often, miss the essence of requirements, which is to specify the "what" the system should do. Often, requirements include specific design information into the requirement text. This limits the possible solutions for a given requirement and may drive the entire design down a given solution path that may or may not be the correct or optimum one. This can also lead to missed requirements, since directing a solution discourages decomposition of the requirement. System design must be differentiated from requirements, design providing the "how" the system will meet the "what" requirements. It is difficult to manage requirements that look like tasks or design information making it virtually impossible to trace the requirement up and down the requirement hierarchy and throughout the architecture.

Table 10.4 Prioritization definitions

Priority level	Definition
High	Capability critical requirements. These are necessary for overall system success and operations.
Medium	Requirements that support operations/functionality but can be included later or are not essential functionality.
Low	Requirements that are quality or functional enhancements that are not required for basic functionality or operations of the system. Included based on schedule and available resources.

10.2.3 MDRE Requirements Prioritization

Another important part of MDRE requirements management is the prioritization of requirements. Developers need to understand the priorities of the requirements when they are trading off viable design solutions. Also, customers may not understand the scope, time, and/or cost impacts of the requirements they have specified. Collaboration between the design team, customer(s), and stakeholder(s) is needed to decide which requirements are most important, based on all factors. Requirements should be rated to provide context for the design team. A simple high, medium, and low rating is appropriate for at least the first level of prioritization. This rating is based on the importance, impact, and feasibility. Often, these three ratings are designated as "must have," "need to have," and "would like to have" ratings [33]. Table 10.4 provides more formal definitions for these ratings.

Question to think about

What issues might arise from mis-prioritizing requirements?

10.2.4 MDRE Requirements Traceability

One of the important aspects of requirements and requirements management is the ability to trace requirements through the system architecture, system models, and use cases/user scenarios/storyboards. If a requirement cannot be traced to a feature, capability, policy, or attribute of the system, then the requirement should not be there. Also, requirements should be traceable from the top level of the system to the lowest design level of the system. This traceability is essential to verifying compliance with the requirements during test. At the highest level of requirements, each requirement decomposed, derived, and allocated to lower levels must trace back to a top-level requirement. Also, any requirement below the top-level requirements should trace back to one and only one top-level requirement. Top-level requirements are often called "Parent Requirements" and requirements at an architecture level below the top-level (e.g., subsystems) are called "children" requirements. A parent requirement can have multiple child requirements, but a child requirement should have only one parent requirement. Top-level (or primary) requirements typically flow from the customers SoO or from a customer high-level requirement specification. As you decompose the system, the next tiers of the derived/decomposed/allocated (DDA) requirements are children to that tier. Each tier is DDA'd from the prior tier, until you have reached the lowest level of definition needed by the developers [34]. To ultimately satisfy the primary/contractual requirements in the system, the trace down to the lowest development tier needs to be maintained for validation of the primary requirements as you integrate and build up the system from the lowest tier to the deliverable system. The DDA trace becomes part of the requirements information. The basic requirement attributes (which will be elaborated in Chap. 11) are [35]:

Unique ID—to distinguish it from all other requirements.

Requirement Text and/or Object Text—actual language defining technical need.

Allocation—aspect of the system the requirement belongs to (e.g., antenna, engine…).

Validation Criteria—measurable criteria use to accept the capability works as specified.

Decomposition Trace—Parent Requirement ID.

Question to think about

If a child requirement ties to multiple parent requirements, what happens if the child requirement is not satisfied in the design?

Once the key technical attributes are defined, each MDRE organization may have more required and optional attributes. There is no right and wrong number of attributes, but with each additional attribute added to a requirement structure, the more costly the requirement development and maintenance effort will become. Traceability ensures that every function, attribute, objective, or need of the system has requirements that are intended to meet the system needs if the system meets the requirement set. Requirement traceability also ensures that each requirement has a reason to be in the requirements specification. The traceability ensures that each requirement has a home [36]. Missing requirements or unnecessary requirements pose risk to the system. Either functions are built that are not needed through unnecessary requirements in the system or the system is designed and built without needed functionality because the requirements were missing.

We have touched on only a few of the issues with requirements traceability, Chap. 13 will take a deep dive into the subject with many examples and a case study on what happens when traceability is not taken seriously.

10.2.5 MDRE Requirements Management Tools

Utilizing a requirements management tool (RMT) is an important step in the overall MDRE process. Using an RMT greatly increases the potential for efficiency of the MDRE requirements management process by allowing management of [37]:

Requirement changes and versions.

Requirement attributes (e.g., verification method).

Traceability of requirements.

Allowing different views of requirements, based on attribute values.

Controlling access to requirements to prevent unauthorized revisions.

Communication between development groups and customers through common views in the tool.

Monitoring the volatility of requirements through notation of changes, additions, and deletions throughout the system life cycle.

Question to think about

What could happen if versioning were not part of a requirements management strategy?

The use of an RMT increases the efficiency of gathering, documenting, and managing requirements, their attributes (e.g., verification method). Gathering and managing requirements require time and energy and using a RMT allows management of requirements manageable. It is possible to keep track of which requirements are waiting for approval, who has the authority to approve them, and why they have or have not been approved, as well as the effectivity of recently approved requirements.

The RMT allow the MDRE team the ability for accurate and efficient implementation of the Requirements Management Plan (RMP). However, it is important that the RMP drive the RMT and not the other way around. Some RMTs allow creating a textual document of the requirements and some allow web-based views. The purpose of either is to communicate and collaborate between the MDRE team, the design team(s), the implementation team(s), and the customer(s). You want an RMT that is flexible and allows the requirements to be managed at different levels, depending on the complexity of the system and/or design.

Table 10.5 Short list of requirement management tools

Tool	URL
Aha!	www.aha.io/software/produce-roadmap
Codebeamer	www.intland.com/codebeamer
Confluence	www.atlassian.com/software/confluence
DevSpec	www.techexcel.com
Innoslate	www.innoslate.com/requirements-management
Jama connect	www.jamasoftware.com
Jira	www.atlassian.com/software/jira
Modern Requirements4DevOps	www.modernrequirements.com
Monday	www.monday.com
Rational DOORS	www.ibm.com
SpiraTeam	www.inflecta.com/SpiraTeam
Visure requirements	www.visuresolutions.com
Siemens Teamcenter	http://www.plm.automation.siemens.com/en_us/products/teamcenter/
Blueprint requirements Center	http://www.blueprintsys.com/

RMTs can be expensive, but do not have to be. The RMT cost and complexity decision should be determined by the budget and complexity of the system under development. Also, you need to have a thorough understanding of the requirement management needs before you pick an RMT [38]. So, the question becomes, what tools are available and how does the MDRE team pick one? Table 10.5 provides a short, but not exhaustive list of RMTs available for requirements management. Each of these provides the capabilities for:

- Requirements Change Management.
- Requirement Prioritization.
- Data Analytics and Reporting.
- Requirement Status Tracking.
- Task Management.

You may be surprised that Microsoft Excel is not on the list. While Excel can be used on small projects to capture and manage requirements, Excel in NOT a requirements management tool. We are not disparaging Excel in any way because it was not designed to be a requirements management tool. It is possible, using VisualBasic within Excel to add features like tracking, versioning, and access rights; however, using Excel can have major drawbacks for large projects. Using Excel during the initial stages of the MDRE requirements development process can be useful. During elicitation, capturing potential requirements and requirement text in excel offers a quick and easy way to capture the elicitation process. In addition, normally every team member would have Excel on their system. There is a natural tendency to use tools you are familiar with to do everything [39].

However, here are some of the reasons to use an actual RMT for MDRE requirements management:

1. Multiple versions of the requirements document. Having multiple people working with requirements all at the same time in Excel can lead to multiple versions which must be reconciled, and a complete version saved and stored. While it is possible to have a naming convention for versions, it is still difficult to manage in Excel.
2. Tracking changes made to the requirements. Yes, it is possible to write VisualBasic scripts to track changes, but RMTs are specifically designed to do these functions.
3. Traceability of requirements is difficult in Excel. Again, it is possible to write VisualBasic scripts to accomplish this at a rudimentary level, but RMTs are designed to easily link and trace requirements up and down the requirements hierarchy.
4. Visibility into requirements based on multiple factors and attributes. Again, VisualBasic can be used to provide different views, but RMTs handle this much more fluidly, and the views are built into the tool.

One area of MDRE requirements development that needs a deeper dive is understanding the different categories of requirements and how each one is important to the overall system architecture and design. Chap. 11 will explore requirement categories.

10.2.5.1 Design Problem

The philosophy of the assignment development for this textbook was to create a set of thought-provoking questions, problems, and design assignments that help drive home the principles, practices, and methodologies of multidisciplinary requirements engineering (MDRE). We endeavored to provide the opportunity for the student to practice the art and science of requirements engineering, helping the student to understand that there is rarely one answer to a requirements engineering problem, and it is only through study and practice that successful requirements engineers are developed. MDRE is a lifelong journey, and anyone who tells you they have nothing left to learn does not truly understand the discipline or has never been involved in the development of a complex system. With the recent interest in alternative energies and talk of the "Green New Deal," along with the recent issues with ice storms in Texas taking out the power grid, the assignments for this textbook will follow the theme of effectively dealing with both local alternative energy power grids and the national infrastructure required to facilitate large-scale alternative energy production, sustainment, and management.

10.2.5.2 Design Problem Description: Alternative Energy Power Grid

The purpose of this project is to, over the course of the semester, develop a feasibility study, requirements analysis, requirements elicitation, derivation, and allocation, for a system that will facilitate a "local" alternative energy power grid capable of powering a neighborhood of 500 homes and a large biosphere capable of supporting growing year-round fruits and vegetables for the 500-home neighborhood. The student will have to deal with issues of choosing multiple alternative energies that can provide power throughout all seasonal difficulties, from intense summer heat to freezing winter temperatures and possible ice storms. Control, viability, sustainability, and maintainability will be major factors to understand the entire possibly 30-year life cycle of such a system. How to effectively utilize, store, and manage power distribution, along with when to upconvert "extra" energy to the national power grid, will be issues that drive many of the assignments. Simply relying on the current technologies and energy infrastructures will not move the country and world forward. Hopefully, this will provide insight to the students using this textbook to understand some of the issues that face everyone looking out over the next many decades.

10.2.5.3 DC Power Grid Requirements Discussion

In the initial days of power generation and distribution, DC power was far less practical than the AC distribution system that soon supplanted it. However, even after Tesla's AC system became the predominant system (and rightly so), Edison's DC power did not become obsolete for some time. Eventually power plants eliminated their delivery of DC power, but the overall concept of a DC power grid is not dead. Even our computer systems (large server systems) have uninterruptible power supplies which contain batteries that are charged off the grid and used to upconvert to AC to power devices when the AC power grid goes off-line. So, the notion of a DC power grid is not dead, but generally used as a backup system. The point is, the notion of a DC power grid has remained and now that renewable energies are viable, we should re-examine their use for widespread, local, power distribution.

One of the advantages of a locally generated DC station is energy savings as well as allowing a facility to run more easily off various DC sources, such as Solar Arrays. And for power-hungry installations, the notion of generating the power locally is growing in popularity. Part of the reason is that waste heat from the power generation generators (solar, geothermal, etc.) can be used to warm nearby buildings. By combining the production of heat and power, local power grids can squeeze much more useful energy out of the renewable energy they use, so this approach will certainly become more widespread as time goes on. The purpose of this design problem is to look ahead and provide an architecture and specification for a local DC power generation station that can serve a local community.

Currently many systems exist to control and handle alternative energies. There are systems to control and manage solar power, wind power, geothermal power, and others, but very few are designed to control, manage, store, and distribute energy from a power grid consisting of multiple alternative energy sources. As we saw the issues recently caused by the intense ice storms in Texas, one alternative energy source is not the answer to the growing energy needs/wants in our society. Each alternative energy source has limitations and issues associated with its use. By harnessing multiple alternative energies, and managing the storage of the energies produces, communities can be worry free from power generation. To facilitate utilizing multiple alternative energy sources, a control center must be included in the design that can manage the energy creation, distribution, storage, and return unused energy to the power grid.

10.2.5.4 High-Level Technical Description of the DC Power Grid System

The purpose of the multiple alternative energies DC power grid is to:

- Supply DC power to a customer community of 500 homes and a biosphere for growing food.
- Provide a storage device to store energy in case of a system failure.
- Determine the amount of surplus energy that is above and beyond what is required to power the grid and charge the backup system, and then deliver that surplus energy to the national power grid systems.

Throughout the semester you will be asked to analyze, design, and architect different components of the overall DC power grid system. This will include research into alternative designs, like lithium nanowire storage systems for backup power and/or supercapacitors (also called ultracapacitors). Fig. 10.7 illustrates these technologies connected to the DC Power Grid Regulator Control Center.

10.2.5.5 High-Level Technical Requirements for the DC Grid System

- Design shall utilize at least 2 alternative energies, with a maximum of 4.

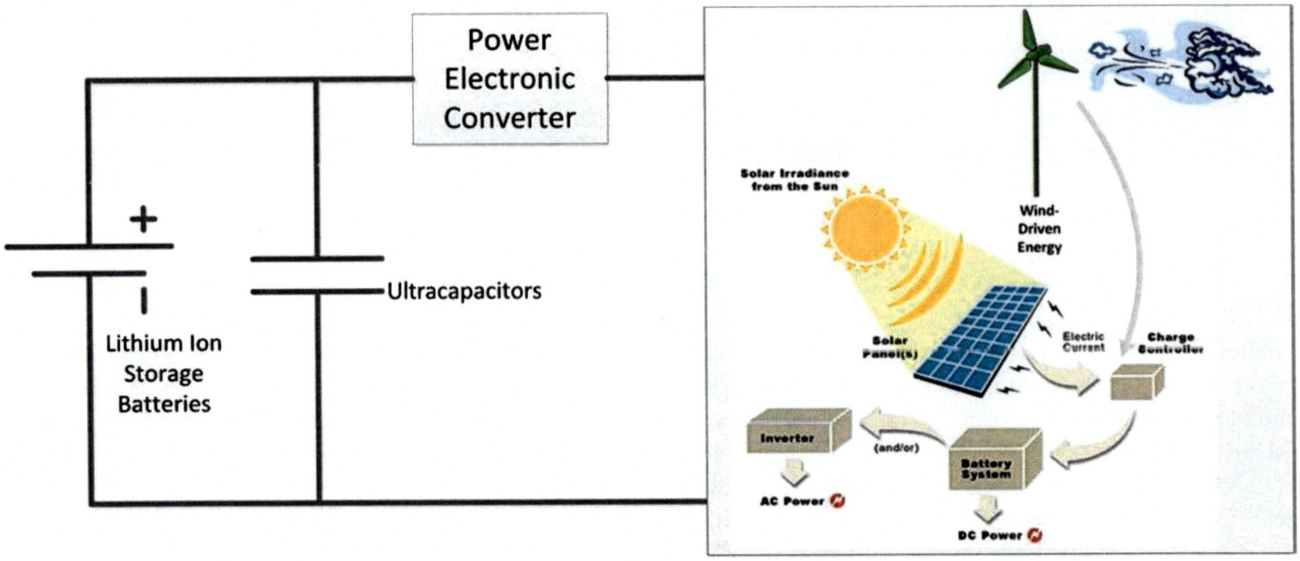

Fig. 10.7 Alternative power management system overview

Fig. 10.8 Example of a whole house power grid

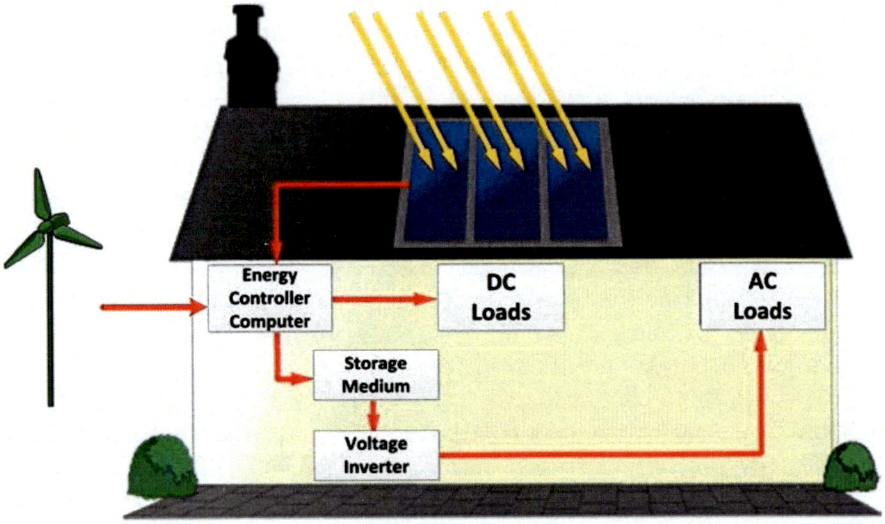

- Total power generation shall be 3.0 megawatts minimum.
- Each of the alternative energy sources shall be capable of generating at least 60% of minimum wattage.
- DC power grid shall generate DC power, with grid voltage being 48, 24, or 12 volts.
- DC power grid shall be capable of servicing at least 500 single family homes and one 4-acre biosphere with hydroponics capable of growing vegetables year-round. (See Fig. 10.8 for a block diagram of a DC-powered home).
- All lights, appliances, and facilities in each home and the biosphere shall be configured for DC operations, if possible. Things like electric ovens may need DC/AC conversion to accommodate their 220v. requirements.
- Only allowed exceptions are the HVAC system and the Washer/Dryer, although DC is preferred.
- There shall be energy storage capability capable of operating the grid of 500 single family dwellings for 7 days if all alternative forms of energy are not producing.
  - Compliance Criteria: Energy storage devices shall be capable of producing 750,000 watts for the entire 7-day period. Nominal energy storage should be batteries or supercapacitors.
- The DC power grid system shall include controlling h/w and s/w to manage the multi-alternative energy strategies and production.
- All energy produced above the required home and energy storage usage shall be upconverted to 220vac for release to the national power grid.
- The DC power grid shall be capable of operating in a temperature range of −40 degrees F. to 125 degrees F. without degradation of power generation.
- The DC power grid must meet OSHA regulations.

10.2.5.6 Assignments

- Write a report, 5–10 pages, describing your requirements management approach, i.e., write a requirements management plan (RMP) for your system development.
- Research Requirements Management Tools (RMTs) and determine which one would be appropriate for your system. Would Excel be a viable choice? Write a report, 5–10 pages explaining your decision on an RMT.
- Write a report, 5–10 pages describing both your elicitation process and your requirements analysis process for your multi-alternative energy system.
- Develop three use cases and two user scenarios for your multi-alternative energy system.
- Develop a storyboard for what your control panel might look like.
- For your multi-alternative energy system, answer the following questions:
  - Can you envision there being a different set of requirements for each type of house? Explain?
  - How many layers of requirements have you identified? How many high-level parent requirements and how many child requirements?

References

1. Rierson, L. (2017). *Developing safety-critical software: A practical guide for aviation software and DO-178C compliance* (p. 10). CRC Press.
2. Abrial, J. R. (2010). *Modeling in event-B: System and software engineering* (p. 11). Cambridge University Press.
3. Abrial, J. R., Butler, M., Hallerstede, S., Hoang, T. S., Mehta, F., & Voisin, L. (2010). Rodin: An open toolset for modelling and reasoning in event-B. *International Journal on Software Tools for Technology Transfer, 12*(6), 447–466. 12.
4. Behutiye, W., Karhapaa, P., Costal, D., Oivo, M., & Franch, X. (2017). Non-functional requirements documentation in agile software development: Challenges and solution proposal. In *Product-focused software process improvement* (pp. 515–522). Springer. 13.
5. Golra, F. R., Beugnard, A., Dagnat, F., Guerin, S., & Guychard, C. (2016). Continuous requirements engineering using model federation. In *24th international requirements engineering conference (RE)* (pp. 347–352) 14.
6. Hebig, R., Giese, H., Stallmann, F., & Seibel, A. (2013). On the complex nature of MDE evolution. In *Model-driven engineering languages and systems. Volume 8107 of LNCS* (pp. 436–453). Springer. 15.
7. Golra, F. R., Beugnard, A., Dagnat, F., Guerin, S., Guychard, C. (2009). Using free modeling as an agile method for developing domain specific modeling languages. In: Proc. of the ACM/IEEE 19th International Conf. on Model Driven Engineering Languages and Systems. MODELS '16, New York, USA, ACM (2016) 24–34 16. ECSS: Space Engineering - Technical Requirements Specification. Standard ECSSE-ST-10-06C, European Cooperation for Space Standardization 17.
8. Arora, C., Sabetzadeh, M., Briand, L., & Zimmer, F. (2015). Automated checking of conformance to requirements templates using natural language processing. *IEEE Transactions on Software Engineering, 41*(10), 944–968. 18.
9. Mashkoor, A. (2016). The hemodialysis machine case study. In: Intern. Conf. on Abstract State Machines, Alloy, B, TLA, VDM, and Z, Springer 329–343 19.
10. Fraser, M. D., Kumar, K., & Vaishnavi, V. K. (1991). Informal and formal requirements specification languages: Bridging the gap. *IEEE Transactions on Software Engineering, 17*(5), 454–466. 20.
11. Van Lamsweerde, A. (2009). *Requirements engineering: From system goals to UML models to software* (Vol. 10, p. 21). John Wiley & Sons.
12. Li, F. L., Horkoff, J., Borgida, A., Guizzardi, G., Liu, L., Mylopoulos, J. (2015). From stakeholder requirements to formal specifications through refinement. In: International Working Conference on Requirements Engineering: Foundation for Software Quality, Springer 164–180 22.
13. Jastram, M., Hallerstede, S., Leuschel, M., & Russo, A. G. (2010). An approach of requirements tracing in formal refinement. In *International conference on verified software: Theories, tools, and experiments* (pp. 97–111). Springer. 23.
14. Jastram, M. (2010). ProR, an Open-Source Platform for Requirements Engineering based RIF. In: Systems Engineering Infrastructure Conference, SEISCONF. 24.
15. Heisel, M., Souqui'eres, J. (1999). A method for requirements elicitation and formal specification. In: Intern. Conf. on Conceptual Modeling, Springer 309–325 25.
16. Wong, P. Y., & Gibbons, J. (2011). Formalizations and applications of BPMN. *Science of Computer Programming, 76*(8), 633–650.
17. Crowder, J., & Friess, S. (2015). *Agile project management: Managing for success*. Springer International. ISBN 978-3-319-09018-4.1.
18. Bahill, A., & Dean, F. (1999). In A. P. Sage & W. B. Rouse (Eds.), *Discovering system requirements. Handbook of systems engineering and management* (Vol. 1999, pp. 175–220). Wiley.
19. ISO/IEC FDIS 25010: 2010 (E)—Systems and software engineering—Systems and software quality requirements and evaluation (SQuaRE) System and Software Quality Models. 2012.
20. The Apache Software Foundation. Apache Wave. http://incubator.apache.org/wave/, 2012.
21. Agustina, F. Liu, S. Xia, H. Shen, and C. Sun. CoMaya: Incorporating advanced collaboration capabilities into 3d digital media design tools. In Proc. CSCW, pages 5–8, 2008.
22. Buschmann, F., Henney, K., & Schmidt, D. C. (2007). *Pattern oriented software architecture volume 5: On patterns and pattern languages*. John Wiley & Sons.
23. Ellis, C. A. and Gibbs, S. J. (1989). Concurrency control in groupware systems. In Proc. SIGMOD, pages 399–407. ACM.
24. Heinrich, M., Lehmann, F., Springer, T., and Gaedke, M.. Exploiting single-user web applications for shared editing—a generic transformation approach. In WWW, pages 1057–1066, 2012.
25. Prakash, A., & Knister, M. J. (1994). A framework for undoing actions in collaborative systems. *ACM Transactions Computing Human Interaction, 1*, 295–330.
26. Rickayzen, A.. Collaborative process modeling. http://scn.sap.com/community/bpm/ business-process-modeling/blog/2012/03/20/, 2012.
27. Sanderson, S.. Knockout : Home. http://knockoutjs.com/, 2012.
28. Shen, H. and Sun, C. (2002). Flexible notification for collaborative systems. In Proc. CSCW, pages 77–86.
29. Sun, C., Xia, S., Sun, D., Chen, D., Shen, H., & Cai, W. (2006). Transparent adaptation of single-user applications for multi-user real-time collaboration. *ACM Transactions Computing Human Interaction, 13*, 531–582.
30. Xia, S., Sun, D., Sun, C., Chen, D., and Shen, H.. Leveraging single-user applications for multi-user collaboration: The CoWord Approach. In CSCW, pages 162–171, 2004.
31. Gotel, O., Cleland-Huang, J., Hayes, J. H., Zisman, A., Egyed, A., Grünbacher, P., Dekhtyar, A., Antoniol, G., & Maletic, J. (2012). In J. Cleland-Huang, O. Gotel, & A. Zisman (Eds.),

Software and systems traceability (pp. 3–22). Springer. https://doi.org/10.1007/978-1-4471-2239-5_1. ISBN 9781447122388.

32. Rempel, P., & Mäder, P. (2015). In S. A. Fricker & K. Schneider (Eds.), *Requirements engineering: Foundation for Software Quality. Lecture notes in computer science* (pp. 81–97). Springer International Publishing. https://doi.org/10.1007/978-3-319-16101-3_6. ISBN 9783319161006.
33. Ralph, P., & Wand, Y. (2009). A proposal for a formal definition of the design concept. In K. Lyytinen, P. Loucopoulos, J. Mylopoulos, & W. Robinson (Eds.), *Design requirements engineering: A ten-year perspective* (pp. 103–136). Springer-Verlag.
34. Chemuturi, M. (2013). *Requirements engineering and management for software development projects*. doi:https://doi.org/10.1007/978-1-4614-5377-2. ISBN 978–1–4614-5376-5. S2CID 19818654.
35. Hood, C., Wiedemann, S., Fichtinger, S., & Pautz, U. (2007). *Requirements management: Interface between requirements development and all other engineering processes*. Springer.
36. Blanchard, B. S., & Fabrycky, W. J. (1990). *Systems engineering and analysis*. Prentice Hall. [Blanchard90].
37. Cohen, L. (1995). *Quality function deployment - how to make QFD work for you*. Addison-Wesley.
38. Dorfman, M., & Thayer, R. H. (Eds.). (1990). *Standards, guidelines, and examples on system and software requirements engineering*. IEEE Computer Society Press.
39. Douglass, B. P., "Designing real-time systems with UML - Part 1," *Embedded Systems Programming*, March 1998.

11 Categories of Requirements

Abbreviations

AC	Alternating current
API	Application programming interface
ATM	Automated teller machine
BIT	Built-in-test
COE	Common Operating Environment
COTS	Commercial off-the-shelf
COVID-19	Coronavirus disease 2019
DC	Direct current (also 0 Hz)
DoF	Definition of Finished
DoR	Definition of Ready
EPA	Environmental Protection Agency
F	Fahrenheit
HSI	Human-systems interface
HVAC	Heating, ventilation, and air conditioning
IDD	Interface Design Document
LRU	Line replaceable unit
MDRE	Multidisciplinary requirements engineering
MPH	Miles per hour
MTBF	Mean time between failure
MTBMA	Mean time between maintenance actions
MTRS	Mean time to restore service
MTTF	Mean time to fail
MTTR	Mean time to repair
NFR	Nonfunctional requirements
OSHA	Occupational Safety and Health Administration
RMP	Requirements Management Plan
RMT	Requirements Management Tool
SEER	Seasonal energy efficiency ratio
SoO	Statement of Objectives
SoW	Statement of Work
ToI	Targets of interest
UI	User interface
V	Volts
Vac	Volts AC

There are three major categories of requirements each possessing an increasing level of detail. Fig. 11.1 illustrates these categories [1]:

- **Mission/business requirements**: These include the high-level statements of goals, objective, and needs for the system or service to be delivered.
- **User or stakeholder requirements**: These describe the needs of the stakeholders, users, and operators of the system or service.
- **System Requirements**: Functional and nonfunctional requirements needed to meet the customer's SoW and SoO.

Figure 11.2 further decomposes the system and stakeholder requirements. One major category of requirements that should be decomposed and derived from the customer mission/business requirements is the transition requirements. Transition requirements define temporary capabilities, conditions, and activities necessary to deploy the system into the operational environment and those conditions under which it can be brought into full operations (operational readiness). Transition requirements are intended to do the following [2]:

- Describe what must happen with people, processes, and technologies required to go from operations today to operations with the new system.
- Describe the training and education necessary for personnel to be able to utilize fully and efficiently the new or upgraded system.
- Define any new data requirements or change to data and/or information required by the new system.

The system requirements decompose into the two main categories most engineers are familiar with, functional and nonfunctional requirements. Figure 11.2 further breaks these down and they will be discussed below.

J. A. Crowder, C. W. Hoff, *Requirements Engineering: Laying a Firm Foundation*,
https://doi.org/10.1007/978-3-030-91077-8_11

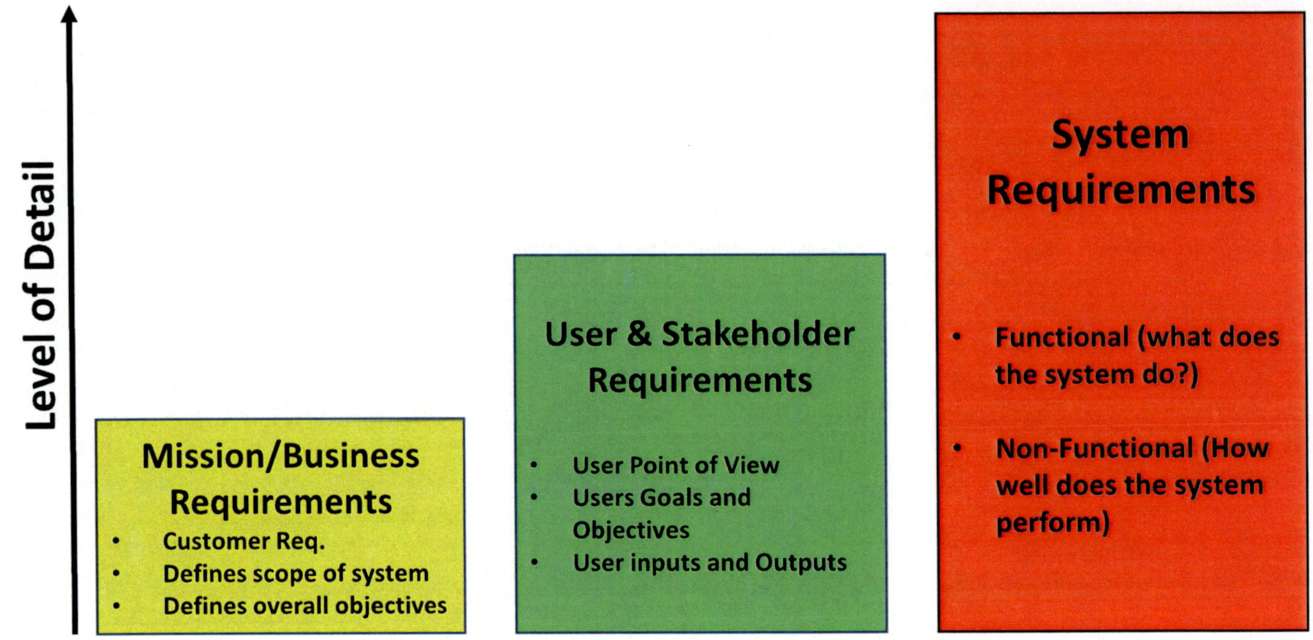

Fig. 11.1 Major requirements categories

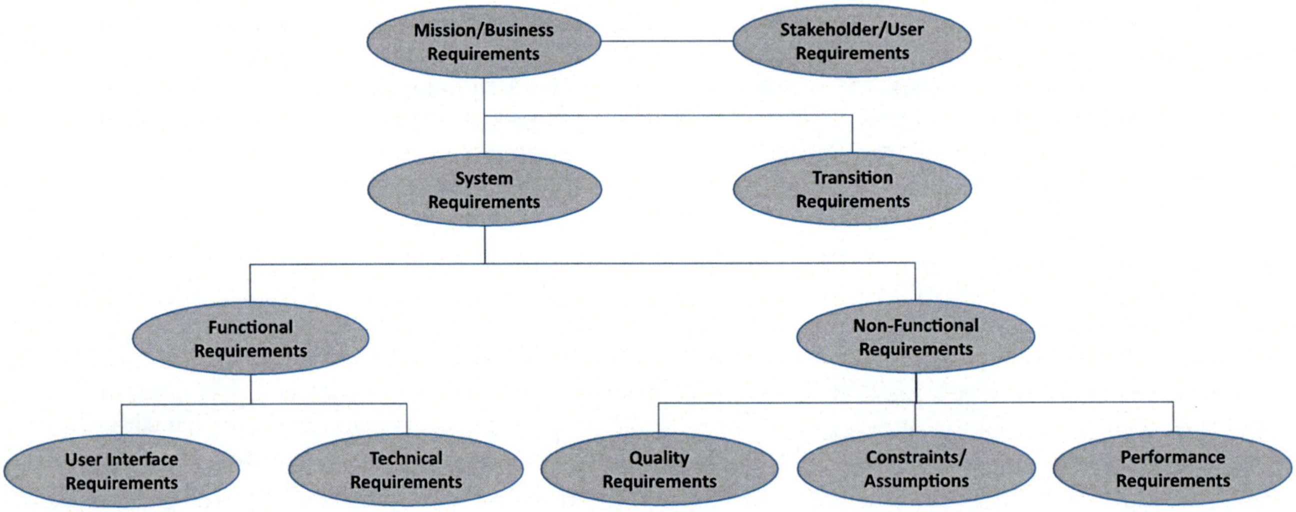

Fig. 11.2 Major requirement categories

11.1 Functional Requirements

Functional requirements are used to describe what the system must do; what behaviors the system must have or what the system should accomplish. Functional requirements decompose into two main categories (see Fig. 11.2), user interface (UI) requirements and technical requirements [3]. Functional requirements, in defining the behavior of a system, may include requirements for data manipulation, calculations (e.g., computing compound interest for a house loan or computing thrust angles for a rocket), and any other functionality specific to the system being developed. Functional requirements specify specific results the system should attain or exhibit. The functional requirements drive the design of the system application architecture while nonfunctional requirements drive the technical architecture, which we will discuss below. The employment of use cases we discussed earlier establish the connection between the requirements and the design, as the use cases validate the system models which, in turn, validate the functional requirements (Fig. 10.5). Each use case, user scenario, and storyboard illustrates the behavioral scenarios described in one or more functional requirements [4].

Typically, a functional requirement comprises a unique identifier number and title, text which defines the requirements, and a rationale for the requirement. These provide a methodology to track the requirement and define why the requirement is needed and the functionality is required by the system. The nonfunctional requirements (Sect. 11.2) impose constraints and quality measures (e.g., performance) on the design of the system. Functional requirements describe what the system must do and drive the creation of technical requirements (Sect. 11.1.2). Implementation of the functional requirements is captured in the system design, both in terms of UI requirements and technical requirements. We will discuss the UI requirements first.

11.1.1 User Interface Requirements

The user interface is an important part of any system and can either make the system easy to use or difficult/complicated. It is important to understand how the system will be accessed, what types of activities people using the system require, and how they are able to handle the tasks they must interface with the system to accomplish. Sometimes called the human-systems interface (HSI), the goal of these requirements is to enable efficient operation and control of the system by humans. The UI requirements should drive toward designs that consider disciplines like ergonomics and psychology to understand how the system will be used and how well users will accept the interface and use of the system [5]. UI requirements need to be detailed enough to understand and design the system for:

- **Operational Details**: Who needs the interface, when and where is it required, and what is expected from the interface (what features)?
- **Interface Element Details**: What kinds of interface fields (e.g., name, address, phone number) are required on the interface, information like sort criteria, etc.
- **Individual Interface Element Details**: What data items are required on the interface, and how these might be displayed; what types of labels might be required? What validation rules are required for the data elements users input through the interface (e.g., phone number must use digits and cannot have letters)?

The UI requirements represent the required characteristics, regardless of the ultimate design, that are necessary for the design and subsequent developed system to be acceptable. The designed, based on the UI requirements, should be captured, and managed with an Interface Design Document (IDD) [6]. The IDD (sometimes called an Interface Specification Document) captures and records the UI design decisions for the external UI for the system design. Management of the UI requirements should ensure that they are consistent with both ends of the interface (the user and the system). Ensuring the data requirements are included is very important and drives the technical design of the UI. The basic information that should be specified by the UI requirements is:

- The sources of required data.
- The data structures required for data exchanges.
- Potential methods for data exchange.
- Timing requirements for the data exchanges.
- Protocols for data interchanges.
- Display of data item(s).

These UI requirements drive the technical requirements for the UI functionality and then drive the technical system design. This leads us into the discussion of technical requirements.

Question to think about

Have you ever interfaced with/used a system where the user interface was difficult to use? Think about the important or proper user interface requirements.

11.1.2 Technical Requirements

We've discussed that the customer high-level SoO and SoW defines the "why" the system must exist, and that the functional requirements define the "what" the system must do to achieve the SoO. The technical requirements derived/decomposed from the functional requirements then define the "how" the SoO will be achieved, as depicted in Fig. 11.2.

The MDRE team decomposes/derives/allocates technical functional requirements to transform the SoO and high-level customer requirements into a set of validated technical requirements in enough detail to allow the system architecture and designers to design and implement a system to meet the SoO [7]. The MDRE technical requirement development process utilizes the customer and stakeholders SoO, SoW and high-level system requirements and produces a complete set of validated technical requirements, expressed as shall statements, that are used to design and implement the system. Creating the technical requirements specification is a feed-back driven, recursive process, developing the product requirements and then lower-level component requirements necessary to satisfy the overall system objectives. This includes identification of all inputs/outputs, constraints, human-system interactions, and interaction with other systems, if applicable. The technical requirements specification communicates to the customer, the stakeholders, and the entire development team the technical details for system implementation. Technical requirements will include requirements for software languages that are and are not allowed, the software framework to be used, which web-browsers the software must be written to support, and the standards it must adhere to. As the technical requirements are decomposed down the architecture, we name the requirements based on the architecture level [8]. Examples are (Fig. 11.3):

The system shall…
The element shall…
The subsystem shall…
The component shall…

Question to think about

Can you think of any technical requirements that would be applicable to your alternative energy system design?

This naming convention approach used in structured developments works well, to bound solutions to specific implementations desired, as well as bounding the solution space. You may wonder how such a naming convention would bound the solution space. The MDRE team is active like a software or hardware architect when making this allocation to specific part of the system architecture. Care must be taken to ensure that the requirement applies to the portion of the system capabilities that the "shall" statements (requirements) define. As the system is decomposed and requirements derived, the derived technical requirements should pertain to one and only one element of the system. While the capabilities may be similar across elements or subsystems, they most certainly will be the same. As the branches of the system decompose from the top tier to lowest level units of software or hardware, those requirements should apply to 1 and only 1 branch of the decomposition. This ensures we have proper decomposition and avoids generic, open-ended requirements in the lowest tiers of the architecture [9].

Customer requirements will often define the highest system tier, but the system engineer must review these requirements carefully, as occasionally, customer requirements contain some requirements that apply directly to lower system tiers. The reasons for this are many but can often be explained simply because the customer has a specific lower-level functionality or behavior that is needed, or that the customer system engineers are subject matter experts in specific lower-level system components. These lower-level requirements should be a small percentage of overall requirements. The reason these are important is these requirements must be handled carefully during the requirement decomposition and derivation processes. In some cases, some of these specific lower-level requirements will require a derived requirement at a higher level to provide proper decomposition [10].

During the MDRE technical requirements phase (which may run through the entire development process if this is an agile or extreme development style) specialty engineering should be involved with the requirements development process to define technical requirements driven by nonfunctional requirements for the system (e.g., availability, reliability). An example would be having the database team develop or at least review data storage and database-oriented require-

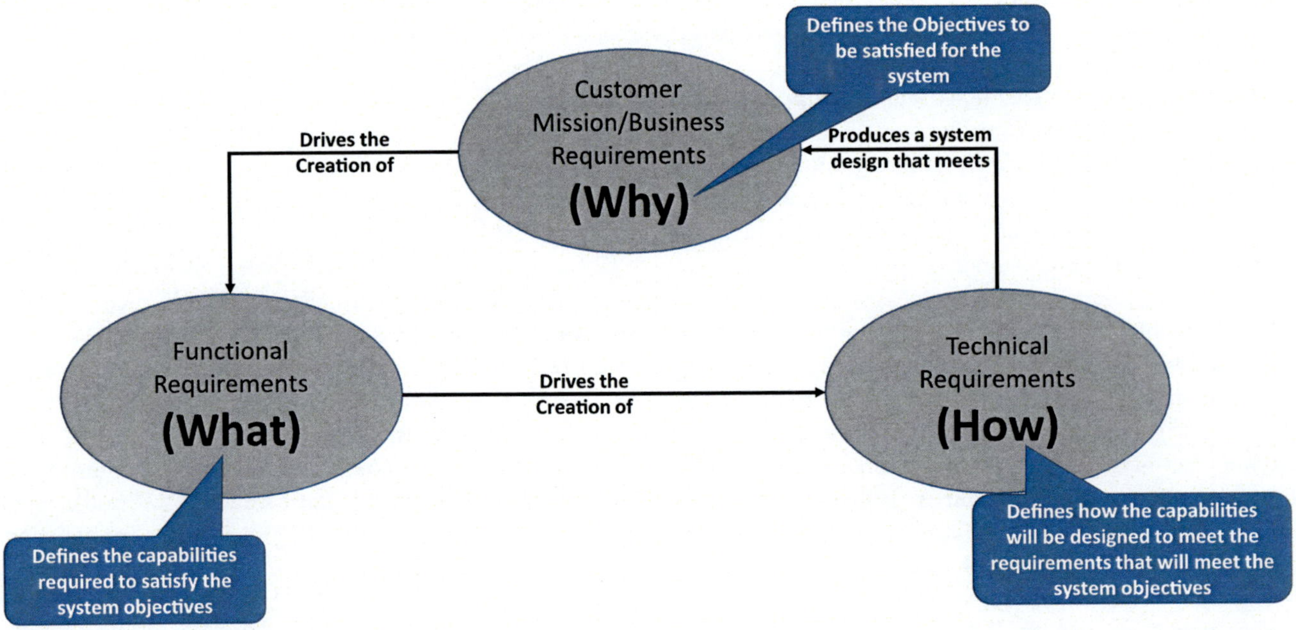

Fig. 11.3 Why, what, and how requirements life cycle

ments. The MDRE team should ensure to define the specialty engineering required functionality required by each element, subsystem, and component of the system such that they can be handed off to the specialty engineering team for design and development.

Question to think about

How useful would an ATM be for users if the availability requirement for it was only 0.25?

With the move to more modern software and hardware approaches, such as Cloud Computing and Service Oriented Architectures, the requirements take on new meanings, but can still utilize the structured approaches to requirements development and management, which still leads to bounding solution spaces offered by new applications and software constructs. Again, care needs to be taken to avoid overly constraining the software and hardware architects and designers by dictating implementation [11].

11.2 Nonfunctional Requirements

Contrasting to functional requirements which to describe what the system must do (see Sect. 11.2) and describe the behaviors the system must have or what the system should accomplish, nonfunctional requirements describe how well, and under what conditions, the system must operate within, in performing the functions elucidated in the functional requirements (see Fig. 11.4). As discussed above, functional requirements drive the overall architecture of the system, while the nonfunctional requirements drive the detailed design of the system.

It is crucial for the MDRE team to ensure that the overall architecture and design consider the complete set of NFRs since they will drive the detailed design of the system. A system that can be allowed to be off-line often will drive a much simpler design than a system that is only allowed to be off-line a few minutes a year. If the design does not account for the NFRs, the hardware and software designs and implementation will not satisfy the NFRs, and it will be very expensive to add them later into the process. Trying to fold in NFRs such as reliability, availability, and maintainability later could cause a complete redesign of the system and cause sever delays in development and deployment of the system. Overall testing and verification of the NFRs drives the Definition of Ready (DoR) for operations and Definition of Finished (DoF) releasing the system for deployment [12].

Question to think about

Research and think about the difference between Definition of Ready and Definition of Finished?

Nonfunctional requirements are generally decomposed into three groups, as shown in Fig. 11.2:
Quality requirements.

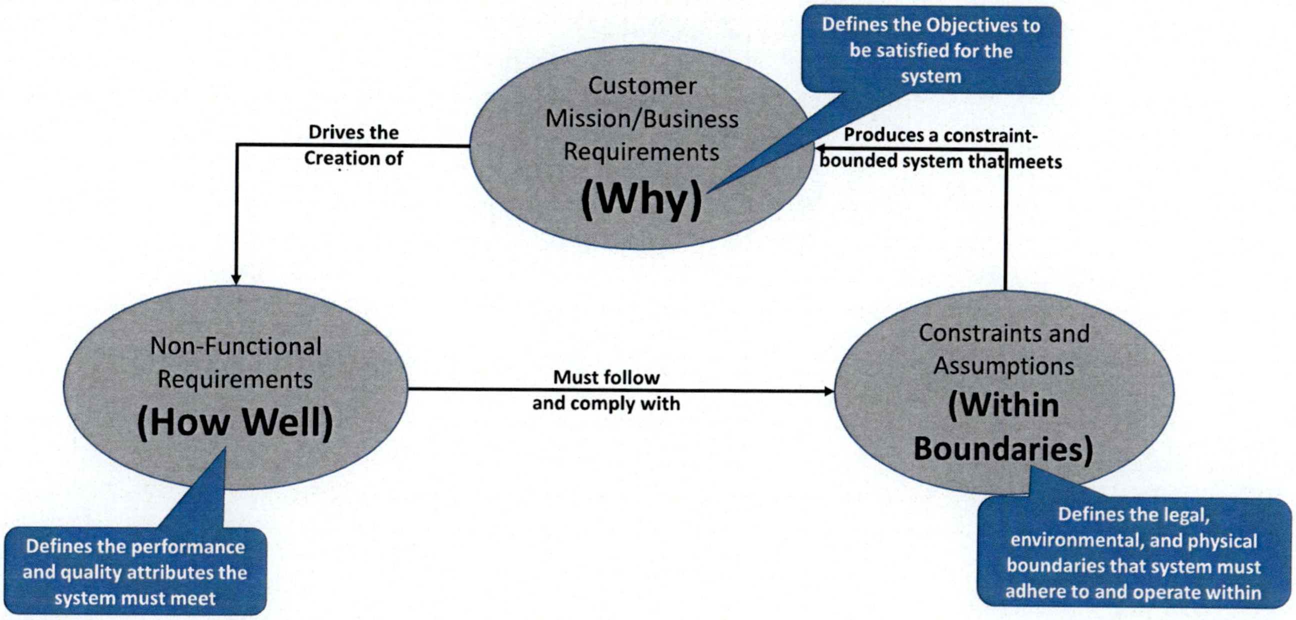

Fig. 11.4 Nonfunctional requirements flow down

Table 11.1 Functional vs. nonfunctional requirements

Parameter	Functional requirements	Nonfunctional requirements
How captured	Use cases	Quality attributes, physical characteristics, power consumption, environmental requirements
End-result	Product capabilities/features/behaviors	Product properties
Capturability	Easy	Easy–Hard
Objective	Verifies functionality of the system, including performance	Verifies physical characteristics, environmental requirements, and quality
Area of focus	Users' requirements	Users' expectations, requirements
Documentation	Describes what the product does, how well it performs its functions, and what behaviors the system has	Describes how well the product functions and how it works, and product physical characteristics
Type of testing	Function testing: System integration, end-to-end testing, API testing, etc.	Performance testing, stress testing, security testing, usability testing, etc.
Test execution	Functional testing precedes nonfunctional testing	Occurs after functional testing
Product info	Product features	Product properties

Performance requirements.
Constraints and assumption.

Each NFR type drives different areas of the overall system design and we will discuss each separately. Table 11.1 illustrates some of the differences between functional and nonfunctional requirements.

11.2.1 Quality Requirements

Quality NFRs, sometimes called quality attributes, are just that, they provide specifications on the quality aspects of the system under development, whether it is a product, a service, or an environment (e.g., Information Technology Network). It is important that the MDRE team capture NFRs, as they provide tangible description of features and how well the system must adhere to these quality attributes. The following is a comprehensive, but not all-inclusive set of quality attributes most developed system must have (either from the customer or through the MDRE elicitation and analysis phases). The MDRE team must decide which of these are applicable and which are not necessary for the given system under design and development:

Availability: System availability is the expected percentage of time the system will be operational and can be utilized to perform the functions it was designed for [13]. Mathematically, availability is the ratio of the expected value of time the system is operational, called uptime, to the aggregate of the expected values of uptime + downtime (the amount of time the system is not available for use):

$$Avail = \frac{E(uptime)}{E(uptime) + E(downtime)} = \frac{E(uptime)}{Total\ Time}$$

Another definition of availability is the ration of the mean time between failures (MTBF), mean time to fail (MTTF), and mean time to repair (MTTR). Fig. 11.5 illustrates the definition of MTTF, MTTR, and MTBF.

Reliability: A measure of enduring and consistent performance in real-world operations. Example of a reliability requirement, "The snare drum shall be designed to maintain its sound for at least 100,000 strikes before requiring tightening the drum-head." Reliability is different from availability in that availability measures the percentage of time the system is operational vs. reliability which measures how long the system will perform its intended functions during operations. Fig. 11.6 illustrates the reliability measure, which considers the total "uptime" calculated from availability [14].

> **Question to think about**
>
> *Would you buy a car where the mean time between failure was 1 week?*

Maintainability: The characteristic of the system design and installation that allows the operational system to operate within normal parameters for a given period if the prescribed maintenance procedures are performed and the required resources are available. Designing for maintainability is important for the overall life cycle operations and sustainability of a system. As the term expresses, maintainability is used to determine how easily the system can be brought to full operational status once a maintenance action is determined to be required, or a common maintenance action is needed (e.g., an oil change in your automobile). There are two types of maintenance actions, preventative maintenance (e.g., oil change is an example), and corrective maintenance defined as maintenance performed on a non-scheduled basis due to an unforeseen error or problem with the system. Corrective maintenance measurements are used in the availability calculation. The basic measurement for maintainability is the mean time to restore service (MTRS) once a maintenance action is required. Maintainability considers many design features that play into how long it takes to restore a system once a problem is found [15]. These are:

Accessibility: How easy/hard is it to reach parts/components/etc. required for maintenance actions?

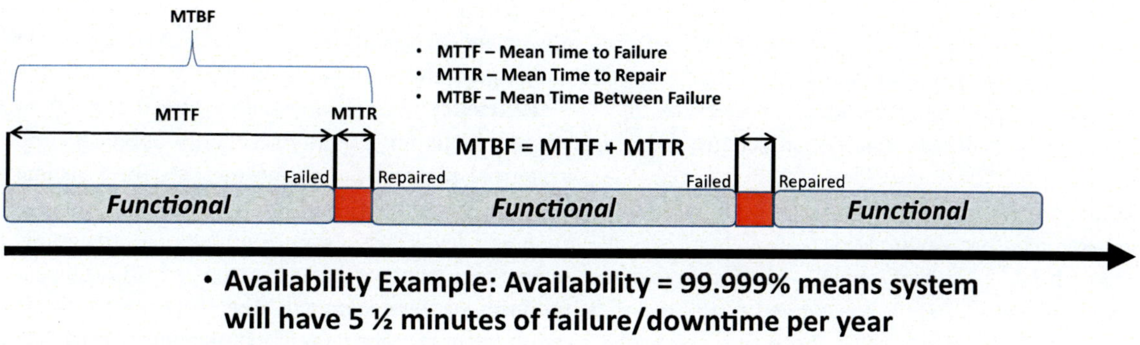

Fig. 11.5 Availability using MTTF, MTTR, and MTBF

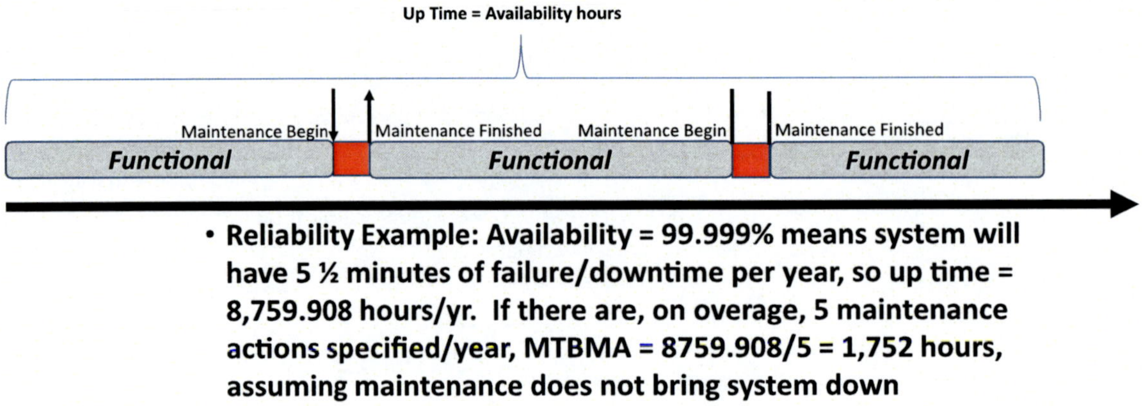

Fig. 11.6 Example of reliability measurement

Visibility: How visible are parts/components that may need to be replaced?

Testability: Can faults within the system be detected and isolated to a given line replaceable unit (LRU)?

Complexity: How many subsystems and components are there in the system that could need maintenance?

Interchangeability: Can parts/components be replaced without extensive calibration?

Sustainability: These requirements come from an analysis of what is required, e.g., resources, for the system to operate for the specified life cycle, in the environment it is intended to operate within. In short, what is required for the operations of the system to be sustained in its environment.

Flexibility: These requirements deal with the system's ability to respond to variable and uncertain events and uncertain demands of loads on the system. Flexibility requirements can also refer to the ability to utilize the system for a variety of operational scenarios, the system's ability to absorb large changes in volume, capacity, and capabilities. How easy is it to add capabilities to the system? An example for flexibility would be the ease in which a sensor processing system can add a new type of sensor to the processing stream [16].

Data Integrity: These are requirements that ensure all data meet a given level of quality and has not been exposed to malicious or even accidental modification or deletion. All input data should be appropriately validated. There should be requirements pertaining to:

Data/information resources must have the capability to determine if they have been modified or deleted from the system, either at rest (stored) or in transit (to/from storage).

Unauthorized modification, alteration, or deletion of a data/information object should result in a security-related event.

All changes to data must be authorized and logged with appropriate backup before modification.

The system should have requirements for data integrity checks.

If data integrity tests (checks) fail, those data should be purged from the system and a security event initiated.

Portability: These requirements deal with the system's ability to be transferred from one environment or computer system to another (e.g., the software shall be portable to any Linux environment).

Manufacturability: This deals with the system's ability to be effectively manufactured given its design, cost, and distribution requirements. For manufacturability, the design team must understand how the system/product will be manufactured and how operator(s) will be trained. The design team must be aware that as the overall complexity of the system increases, it decreases manufacturability, as it will require specialized knowledge for the system to be manufactured. Manufacturability requirements are there to ensure multiple copies of the product/system can be created easily, cost effectively, and at low risk to reliability and availability [17].

Usability: These requirements pertain to expectations designed to ensure that the product or system under development is "easy" or can be operated by the intended users of the systems. Requirements for usability have several different subcategories:

Learnability: How easy is it for users to learn how to operate the system/product/service being developed.

Accessibility: These requirements pertain to how the system is to be accessed and who can access it. As an example, requirements that handle the American with Disabilities Act would come under usability requirements.

Error Tolerance: These requirements exist to ensure that the system performs within a given range and therefore always meets operational needs. An example is, "The ambient temperature shall be 72 °F, ± 1°."

Efficiency: Requirements that drive the design to achieve specific goals for factors like energy usage. Example, "The system shall meet the EPA standards for Seasonal Energy Efficiency Ratio (SEER)" [18].

Accuracy: These are nonfunctional, data, or process related requirements that define the precision which a solution produced by the system will meet. An example of an accuracy requirement is, "The target identification system shall have an accuracy of >98% for all Targets of Interest (ToI)."

Capacity: These requirements deal with the amount of information, services, or objects that can be handled by the system. These requirements drive the design of the system as well as establishing how the system can and will be used. An example capacity requirement is, "The reservation system shall have the capacity to handle 300 simultaneous reservation requests at any time."

Adaptability: These requirements deal with the ability of the system to adapt to a range of capabilities and uses without a major change to the design or implementation. Adaptability measures the system's ability to handle changes in capabilities, user interfaces, performance, sensitivity, and design margins. An example of an adaptability requirement is, "The system shall have the capability to accommodate new sensor types without a redesign of the architecture and without a change to the software."

Survivability: Requirements dealing with the system's ability to continue to function during and after a natural or human-induced disturbance. These requirements should specify the range of conditions over which the system must survive and include the minimum acceptable level of functionality the system must still be capable of after a survivability incident [19]. An example of a survivability requirement is, "The building shall maintain structural integrity for winds up to 180 mph."

Fault Isolation/Health and Status: These requirements pertain to the monitoring, detection, and characterization of faults that can occur within the system, specifying the fault isolation approach for the system. An example of a fault isolation requirement is, "The system shall be capable of monitoring all "Built-in-Test" (BIT) information from all COTS hardware and provide an alert to the operator when the BIT indicates a hardware fault."

Question to think about

How important is fault isolation? What if automobiles had no way to isolate a fault within the computer and/or electrical system?

Scalability: These are requirements dealing with the system's ability to grow in its capacity to meet increasing demands for the products/services that system is designed to deliver or provide. An example of a scalability requirement is, "The system shall be capable of supporting an increase in requested reservations of 25%/year without a change to the architecture or software."

Interoperability: These requirements deal with the system's ability to share information and exchange information with other systems and/or external hardware and software. Interoperability requirements can radically change the overall design of the system to accommodate being interoperable with other systems [20]. An example of an interoperability requirement is, "The system shall be capable of interfacing with both Windows and Linux operating system environments."

Manageability: These requirements deal with the ability of administrators to monitor and affect the system, providing key information needed to debug, analyze, and fix any issues that arise during operations. An example of a manageability requirement is, "The system shall provide an alert to the operator when the system does not complete daily transactional processing within a 2-hour time frame."

11.2.2 Performance Requirements

Almost all systems have some type of performance requirements. A performance requirement specifies how a capability or service must be executed in terms of quantity, accuracy, coverage, timeliness, or readiness. The design and deployment of systems that meet performance requirements does not happen by accident. The only way to ensure the developed and deployed system meets the performance expectation is for the performance targets to be specified in performance requirements. If the performance requirements are not created, and created in an unambiguous language, the system designers will either not consider performance, or add their own interpretation as to what the performance requirements mean. Ambiguous or incorrectly defined performance requirements can result in disputes between the development team and the customer during final testing and verification of the system. It is possible that trade-offs must be made to trade performance vs. budget vs. schedule, but these must be done up-front and agreed to between the customer and the development team. It is up to the MDRE team to ensure they have captured all necessary performance requirements and have adjudicated any questions or disagreements, ensuring the final verbiage of the requirements is understood and agreed to. To help reduce ambiguity in performance requirements, each requirement should have at least the following attributes [21]:

Be outcome focused.

Be measurable.

Have specific criteria.

Produce tangible artifacts to allow measurement.

Question to think about

How different would the design for an automobile be if the performance requirement for the top speed was 25 mph vs. 225 mph?

Overall, performance requirements define how well the system (e.g., hardware and software) accomplishes its goals under specific conditions. Often, performance requirements are based on the system's ability to support user tasks and needs. Performance requirements and their attributes are key elements in the design and testing of a system. The MDRE team must trade off performance attributes with other quality attributes like reliability, robustness, security, availability, reliability, efficiency, and flexibility to ensure inclusion of the performance criteria does not conflict with other quality attributes required for the system design. Performance requirements like high throughput can degrade response time, which can affect useability. The MDRE team must consider the influence of dependencies when establishing performance requirements. For example, in software systems, the system must rely on an operating system and possibly access to an enterprise database system. It would be important in this case to ensure the enterprise database is available on the operating system that is specified. If the throughput requirements are not adequate, it could create a bottleneck accessing data from the enterprise database, greatly reducing usability and response time [22].

During the elicitation and analysis phase of MDRE, gathering, discovering, and deriving performance requirements can be a daunting challenge. This is driven by three basic problems:

- **Scope**: If the scope of the system is at all uncertain or possibly changing over time, it is easy to miss, misinterpret, or have incorrect performance requirements. Establishing a clear scope of the system before gathering performance requirements is essential.
- **Skill**: Performance requirements are only useful is the customer, stakeholders, and developers fully and correctly understand and agree on the scope and verbiage of all performance requirements. The MDRE team must clarify the overall goals and limitations in the performance requirements for the development team to build a suitable system/product/service.
- **Stability**: The MDRE process does not end once an initial set of requirements has been established. Requirement change over time, especially for development efforts that span years. Requirement volatility drives requirements to change over time. Section 11.3 will discuss this in detail. It is essential that the MDRE team has a process for receiving, prioritizing, and assimilating changes into the system requirements specification without disruption to development. Having a requirements management process in place is essential to present overlooked or delayed requirement changes [23].

11.2.2.1 Performance Requirement Analysis Models

Determining performance requirements requires asking questions about how the system will be utilized. Such questions should entail the use of performance analysis models that build scenarios that cover the system activities, their frequency, and importance to the overall user interface and easy of useability of the system [24]. Performance models allow the MDRE team to analyze the system capabilities and interfaces to determine how the system must perform to meet the overall expectations for system use and response. Examples of performance models are [25]:

- **Behavioral Requirement Modeling**: Behavioral requirements are generally captured in process maps or use cases and are used to answer such questions as:
  - How frequently does each activity occur in the system?
  - What is the minimum/maximum/average executions across a given time frame (e.g., hour, day, week)?
  - When must each capability and/or activity be available?
  - What is the minimum acceptable throughput for each activity/capability the system must perform?
  - Is there a projected growth rate for the throughput and/or number of users?
- **Interface Performance Requirement Modeling**: The use of context diagrams and subsequent storyboards are useful in establishing user interface models to understand performance requirement from an interface perspective. Questions to be answered with this type of performance modeling are:
  - What are the triggering events for each interface?
  - How frequently does each triggering event occur?
  - What is the useability needs in terms of ability to complete each task available on the interfaces?
  - How quickly should a new user expect to learn how to use each interface?
  - What are the requirements for things like minimum number of keystrokes, mouse clicks, etc. that are required for each task on each interface?
  - Are there timing requirements for minimum/maximum time allowed for each activity on each interface?
- **Error Handling Requirement Modeling**: These models and subsequent requirements pertain across the system.
  - How should the system respond to error conditions?
  - How frequently are errors expected?
  - How long should it take to recover from an error?
  - How are errors detected?

The information extracted is used to derive the performance requirements for the system. Each performance requirement then must be accurately linked to the functional requirements that provide the capabilities/activities the performance requirement is measured against. The performance requirements should not presuppose a particular design or design implementation or required detailed technical knowledge to understand. They should be written and expressed in simple terms that a subject matter expert can validate [26].

When identifying, deriving, and allocating performance requirements, it is important to identify and provide the reasoning behind each requirement. This should be part of the general MDRE requirements process. It drives capacity planning for the system design. Other information that must be provided, that influences performance requirements, are the constraints and assumptions that the customer, stakeholders, and developers have and make about the system under development.

Question to think about

Is it possible to write requirements without having a notion of how the system will meet the requirement?

11.2.3 Constraints/Assumptions

For any system design, development, and deployment, there are built-in constraints and assumptions that must be dealt with, understood, and folded into the overall design and development of the system. Constraints generally pertain to technical limitations around any solution derived for the system that could result in additional functional requirements to meet the constraint. However, a constraint can also be a range of values of one component of the system which may drive nonfunctional requirements. One example of a constraint would be that the system architecture designed for the system must comply with the Army Common Operating Environment (COE) which would lead to constraints and subsequent requirements that drive the interface design of the system and may cause trade-offs between cost, schedule, and design. Another constraint might occur for a system upgrade that the new system must have a smaller physical footprint than the legacy system. One of the main issues with constraints is that often design teams overlook constraints until an initial solution has been agreed to. Worrying about constraints later in the solution life cycle can lead to major delays due to redesigns required to accommodate the constraints. Every physical system has size, weight, and power restrictions, even if they are implied. Another constraint when dealing with physical system is the need for heating and air-conditioning for the system. For instance, if you are designing and building a new sound system for an automobile, there are definite size, weight, and power restrictions that drive the overall design of the system. The bottom line is that the same amount of effort the MDRE team puts into functional requirements should be put into understanding and considering the constraints on the system up front before design begins [27].

Assumptions are implied requirements that have not been formally created for the system and is un-validated since it has not been captured by the MDRE team or by the customer/stakeholders. When beginning a design, if there are too many assumptions it may indicate that the MDRE team does not really understand what is to be delivered. In that case, the assumptions should become questions back to the customer and stakeholders to understand the assumptions and turn them into requirements. An assumption we have often seen is that users know how to enter data correctly and it is not necessary to error check user inputs. This leads to system crashes when the software does not check the user inputs and ask for the data to be re-entered. There was an incident with a payroll system for a major corporation that crashed because the user was supposed to type in yes or no to a question and typed "Yep" instead. Since the system had no error checking and correction software, the invalid input caused a fatal error condition and shut the system down. The software designers assumed, erroneously, that users would always enter the correct word.

Question to think about

Think about an assumption a user of your multi-alternative energy system might have?

Understanding the difference between functional requirements, nonfunctional quality and performance requirements, and constraints and assumptions is essential for developing a solution and design for any system. It certainly can mean the difference between successful development and deployment of the system and failure of the system to meet its expectations. Constraints can lead to assumptions when the constraints have to do with time-sensitive information. An example is the constraint a building design must meet building codes. The underlying assumption is that the building codes won't change over the course of the time it takes to build and turn over the building to the customer [28]. Another aspect of requirements we will explore in Sect. 11.3 is the relationships between requirements and some of the qualities for various types of requirements.

11.3 Inter-Relationships Between Requirements

Both functional and nonfunctional requirements have certain attributes and qualities that drive the MDRE requirements management process [29]. We felt it was important to highlight the most important and relevant requirement quality categories. These are described below:

Volatile Requirements: Requirements that change either during development, during deployment, or during use of the system. An example would be EPA standards for vehicles that are constantly being updated as new federal policies come into being.

Enduring Requirements: These are requirements that are stable long-term. These are derived from the core activity of an organization for which the system is intended. An example of an enduring requirement is the need for doctors and nurses in a health-care facility. A requirement that would have been thought to be enduring, but is no longer the case, is the requirement that an automobile have a physical human driver, which is no longer the case with a self-driving car.

Question to think about

Were there emergent requirements that happened as a result of Covid-19?

Mutable Requirements: These are requirements that change because of a change in the environment in which the system must operate. An example would be changing the electrical system from carbon-based energy to renewable energy sources.

Emergent Requirements: These are requirements that happen (emerge) as the understanding of the system increases during development efforts. The architecture and design process, with constraints considered, may reveal new requirements that must be created and validated to ensure the system will perform correctly.

Consequential Requirements: These are requirement that result from introduction of the technology required to operate the system, e.g., computer system. Introduction of new technologies and processes may create new possibilities for how the system is developed and executed.

Compatibility Requirements: These are requirements that depend on a specific process or technology. As the processes and technology changes, it may require completely different requirements for the delivered system. Example would be self-driving automobiles must have camera and radar systems that are not required for human-operated automobiles.

Compliance Requirements: Requirements that come from industry or government standards, either local, state, federal, or international. There are requirements the system must comply with for it to be certified. When these standards change, any upgrades to the system must now consider new standards. An example would be new compliance standards for the use of ground-fault interrupter receptacles and arc-fault circuit breakers in new houses and for renovations to existing houses.

Now that we have discussed the MDRE requirements process, including requirements management, requirements development, and requirements categories and descriptions. We will utilize all of that to delve into the actual mechanics and processes of writing valid requirements. Chap. 12 describes the requirement composition process.

11.3.1 Design Problem

The philosophy of the assignment development for this textbook was to create a set of thought-provoking questions, problems, and design assignments that help drive home the principles, practices, and methodologies of multidisciplinary requirements engineering (MDRE). We endeavored to provide the opportunity for the student to practice the art and science of requirements engineering, helping the student to understand that there is rarely one answer to a requirements engineering problem, and it is only through study and practice that successful requirements engineers are developed. MDRE is a life-long journey, and anyone who tells you they have nothing left to learn does not truly understand the discipline or has never been involved in the development of a complex system. With the recent interest in alternative energies and talk of the "Green New Deal," along with the recent issues with ice storms in Texas taking out the power grid, the assignments for this textbook will follow the theme of effectively dealing with both local alternative energy power grids and the national infrastructure required to facilitate large-scale alternative energy production, sustainment, and management.

11.3.2 Design Problem Description: Alternative Energy Power Grid

The purpose of this project is to, over the course of the semester, develop a feasibility study, requirements analysis, requirements elicitation, derivation, and allocation, for a system that will facilitate a "local" alternative energy power grid capable of powering a neighborhood of 500 homes and a large biosphere capable of supporting growing year-round fruits and vegetables for the 500-home neighborhood. The student will have to deal with issues of choosing multiple alternative energies that can provide power throughout all seasonal difficulties, from intense summer heat to freezing winter temperatures and possible ice storms. Control, viability, sustainability, and maintainability will be major factors to understand the entire possibly 30-year life cycle of such a system. How to effectively utilize, store, and manage power distribution, along with when to upconvert "extra" energy to the national power grid, will be issues that drive many of the assignments. Simply relying on the current technologies and energy infrastructures will not move the country and world forward. Hopefully, this will provide insight to the students using this textbook to understand some of the issues that face everyone looking out over the next many decades.

11.3.3 DC Power Grid Requirements Discussion

In the initial days of power generation and distribution, DC power was far less practical than the AC distribution system that soon supplanted it. However, even after Tesla's AC system became the predominant system (and rightly so), Edison's DC power did not become obsolete for some time. Eventually power plants eliminated their delivery of DC power, but the overall concept of a DC power grid is not dead. Even our computer systems (large server systems) have uninterruptible power supplies which contain batteries that are charged off the grid and used to upconvert to AC to power devices when the AC power grid goes off-line. So, the notion of a DC power grid is not dead, but generally used as a backup system. The point is, the notion of a DC power grid has remained and now that renewable energies are viable, we should re-examine their use for widespread, local, power distribution.

One of the advantages of a locally generated DC station is energy savings as well as allowing a facility to run more easily off various DC sources, such as Solar Arrays. And for power-hungry installations, the notion of generating the power locally is growing in popularity. Part of the reason is that waste heat from the power generation generators (solar, geothermal, etc.) can be used to warm nearby buildings. By combining the production of heat and power, local power grids can squeeze much more useful energy out of the renewable energy they use, so this approach will certainly become more widespread as time goes on. The purpose of this design problem is to look ahead and provide an architecture and specification for a local DC power generation station that can serve a local community.

Currently many systems exist to control and handle alternative energies. There are systems to control and manage solar power, wind power, geothermal power, and others, but very few are designed to control, manage, store, and distribute energy from a power grid consisting of multiple alternative energy sources. As we saw the issues recently caused by the intense ice storms in Texas, one alternative energy source is not the answer to the growing energy needs/wants in our society. Each alternative energy source has limitations and issues associated with its use. By harnessing multiple alternative energies, and managing the storage of the energies produces, communities can be worry free from power generation. To facilitate utilizing multiple alternative energy sources, a control center must be included in the design that can manage the energy creation, distribution, storage, and return unused energy to the power grid.

11.3.4 High-Level Technical Description of the DC Power Grid System

The purpose of the multiple alternative energies DC power grid is to:

- Supply DC power to a customer community of 500 homes and a biosphere for growing food.
- Provide a storage device to store energy in case of a system failure.
- Determine the amount of surplus energy that is above and beyond what is required to power the grid and charge the backup system, and then deliver that surplus energy to the national power grid systems.

Throughout the semester you will be asked to analyze, design, and architect different components of the overall DC power grid system. This will include research into alternative designs, like lithium nanowire storage systems for backup power and/or supercapacitors (also called ultracapacitors). Fig. 11.7 illustrates these technologies connected to the DC Power Grid Regulator Control Center.

11.3.5 High-Level Technical Requirements for the DC Grid System

- Design shall utilize at least 2 alternative energies, with a maximum of 4.
- Total power generation shall be 3.0 megawatts minimum.
- Each of the alternative energy sources shall be capable of generating at least 60% of minimum wattage.
- DC power grid shall generate DC power, with grid voltage being 48, 24, or 12 volts.
- DC power grid shall be capable of servicing at least 500 single family homes and one 4-acre biosphere with hydroponics capable of growing vegetables year-round. (See Fig. 11.8 for a block diagram of a DC-powered home).

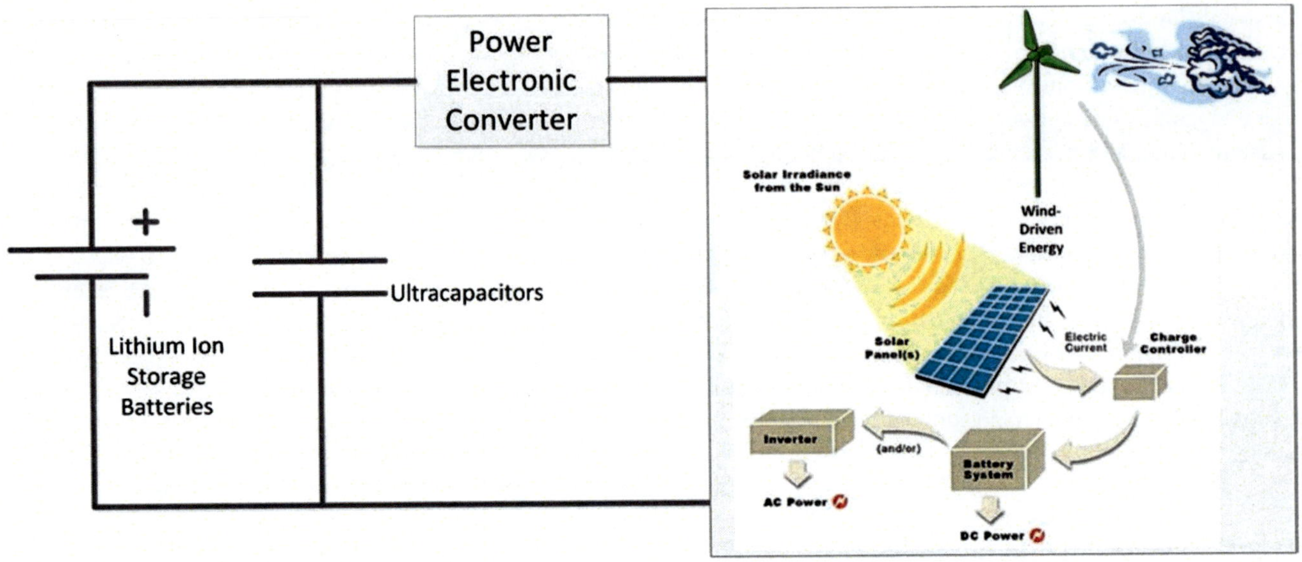

Fig. 11.7 Alternative power management system overview

Fig. 11.8 Example of a whole house power grid

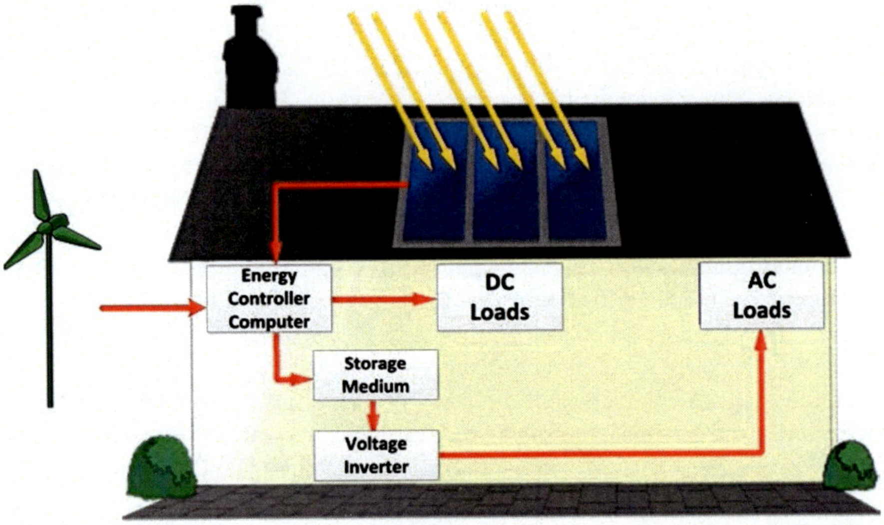

- All lights, appliances, and facilities in each home and the biosphere shall be configured for DC operations, if possible. Things like electric ovens may need DC/AC conversion to accommodate their 220v. requirements.
- Only allowed exceptions are the HVAC system and the Washer/Dryer, although DC is preferred.
- There shall be energy storage capability capable of operating the grid of 500 single family dwellings for 7 days if all alternative forms of energy are not producing.
  - Compliance Criteria: Energy storage devices shall be capable of producing 750,000 watts for the entire 7-day period. Nominal energy storage should be batteries or supercapacitors.
- The DC power grid system shall include controlling h/w and s/w to manage the multi-alternative energy strategies and production.
- All energy produced above the required home and energy storage usage shall be upconverted to 220vac for release to the national power grid.
- The DC power grid shall be capable of operating in a temperature range of −40 °F. to 125 °F. without degradation of power generation.
- The DC power grid must meet OSHA regulations.

11.3.6 Assignments

- Break the requirements you have derived for your multi-alternative energy system and put them into the categories discussed in Chap. 11.

- Write a 5–10-page paper describing how you envision users interface with your system and why you have the user interface requirements you came up with.
- Describe how each of your requirements is necessary and why you needed those requirements.
- Trace your requirements to the use cases you created.
- For your multi-alternative energy system, answer the following questions:
  - Which categories of requirements are relevant to your system and why?
  - What is a reasonable mean time between failure for your multi-alternative energy system? How often would it be ok for the neighborhood and biosphere to be without power?
- Describe three performance requirements that your system should meet. Explain how your design meets these requirements and how the requirements will be verified.
- Describe what constraints/assumptions relate to your system and your specific design.

References

1. Gause, D. C., & Weinberg, G. M. (1989). *Exploring requirements—quality before design*. Dorset House Publishing.
2. Moore, A., & Beckwith, T. (1998). UML's shortfalls in modeling complex real-time systems. *Computer Design*, 53–57.
3. Ramesh, B., Powers, T., Stubbs, C., and Edwards, M. (1995). "Implementing Requirements Traceability: A Case Study," *Proceedings of the Second IEEE International Symposium on Requirements Engineering*, York, United Kingdom, pp. 89–95.
4. Watkins, R., & Neal, M. (1994). Why and how of requirements tracing. *IEEE Software*, 104–106.
5. Lattemann, F. and Lehmann, E., "A methodological Approach to the Requirement Specification of Embedded Systems," *Proceedings of the First IEEE International Conference on Formal Engineering Methods*, Hiroshima, Japan, November 1997, pp.183–191.
6. Patridge, D.. "Where do specifications come from?" *Achievement and Assurance of Safety—Proceedings of the Third Safety-Critical Systems Symposium*, Brighton, United Kingdom, 1995, pp. 302-310.
7. Jaakkola, H., & Thalheim, B. (2011). Architecture-driven modelling methodologies. In A. Heimbürger et al. (Eds.), *Proceedings of the 2011 conference on information modelling and knowledge bases XXII* (p. 98). IOS Press.
8. Paul C. Clements (1996). "A survey of architecture description languages." Proceedings of the 8th international workshop on software specification and design. IEEE Computer Society, 1996.
9. Medvidovic, N and Taylor, R. N. (2000). "A classification and comparison framework for software architecture description languages." Software Engineering, IEEE Transactions on 26.1 (2000): 70-93.
10. Gantt, H. L. (1974), *Work, Wages and Profit*, published by the *Engineering Magazine*, New York, 1910; republished as *Work, Wages and Profits*, Easton, Pennsylvania, Hive Publishing Company, ISBN 0-87960-048-9.
11. Dufresne, T. & Martin, J. (2003). "Process modeling for E-business." Methods for information systems engineering: knowledge management and E-business. Spring 2003.
12. Bubenko Jr J. A. (2007). "From information algebra to enterprise modelling and ontologies—A historical perspective on modelling for information systems". In: *Conceptual modelling in information systems engineering*. John Krogstie et al. eds. pp 1–18.
13. Young, J. W., & Kent, H. K. (1958). Abstract formulation of data processing problems. *Journal of Industrial Engineering, 9*(6), 471–479.
14. Baik, D.-K. eds. (2005). *Systems modeling and simulation: theory and applications: third Asian Simulation Conference, AsiaSim 2004, Jeju Island, Korea, October 4–6, 2004*. Springer, 2005. ISBN 3-540-24477-8.
15. Fulton, R., & Vandermolen, R. (2017). Chapter 4: Requirements—Writing requirements. In *Airborne electronic hardware design assurance: A practitioner's guide to RTCA/DO-254* (pp. 89–93). CRC Press. ISBN 9781351831420.
16. "Supplement 4-A, A Procedure for Requirements Analysis". *Systems Engineering Fundamentals*. United States Government US Army. 2001. ISBN 978-1484120835.
17. Loucopoulos, P. (2005). Chapter 4: Requirements engineering. In J. Clarkson & C. Eckert (Eds.), *Design process improvement: A review of current practice* (pp. 116–139). Springer-Verlag. ISBN 9781846280610.
18. Adams, K. M. (2015). *"3.2 Definitions for functional and non-functional requirements". Non-functional requirements in systems analysis and design* (pp. 45–50). Springer. ISBN 9783319183442.
19. Jönsson, P., & Lindvall, M. (2006). Chapter 6: Impact analysis. In A. Aurum & C. Wohlin (Eds.), *Engineering and managing software requirements* (pp. 117–142). Springer Science & Business Media. ISBN 9783540282440.
20. MITRE Corporate Communications and Public Affairs. "Requirements engineering: Eliciting, collecting, and developing requirements". *The MITRE Systems Engineering Guide*. MITRE Corporation. pp. 304–13. ISBN 978061597442.
21. Stellman, A., Greene, J. (2005). "Chapter 6: Software requirements". *Applied Software Project Management*. O'Reilly Media. pp. 97–130. ISBN 9780596553821.
22. Martins, L. E. G. and de Oliveira, T. (2014) "A case study using a protocol to derive safety functional requirements from fault tree analysis," in Proc. 2014 IEEE 22nd International on Requirements Engineering Conference (RE), pp. 412–419.
23. Sarker, S., Gasson, S., and Haythornthwaite, C. (2005). "Information and communication technologies in support of knowledge management/organizational memory/organizational learning," presented at the System Sciences.
24. Boehm, B. and Basili, V. R. (2005). "Software defect reduction top 10 list," Foundations of empirical software engineering: The legacy of victor R. Basili, p. 426.
25. Assche, F. V., Layzell, P., Loucopoulos, P., & Speltincx, G. (1988). Information systems development: A rule-based approach. *Knowledge-Based Systems*, 227–234.
26. Loucopoulos, P., & Kavakli, E. (1995). Enterprise modelling and the teleological approach to requirements engineering. *International Journal of Cooperative Information Systems, 4*(1), 45–79.
27. Lakshminarayanan, V., Liu, W., Chen, C. L., Easterbrook, S., & Perry, E. (2006). Software architects in practice: Handling requirements. In *Proc. the 2006 Conference of the Center for Advanced Studies on Collaborative Research* (p. 25). IBM Corp.
28. Brinksma, E., Harel, D., Mader, A., Stevens, P., and Wieringa, R., A, (2003). A model-based approach to requirements analysis methods. Addison Wesley, Boston, MA.
29. Davenport, T. H., & Prusak, L. (2000). *Working knowledge: How organizations manage what they know*. Harvard Business Press.

12 Requirements Composition

Abbreviations

AC	Alternating current
ANSI	American National Standards Institute
DC	Direct current (or 0 Hz)
F	Fahrenheit
GM	General Motors
H/W	Hardware
HVAC	Heating, ventilation, and air conditioning
IEEE	Institute of Electrical and Electronics Engineers
MDRE	Multidisciplinary requirements engineering
OSHA	Occupational Safety and Health Administration
S/W	Software
SMART	Specific, Measurable, Attainable, Realistic, and Testable
STD	Standard
TBD	To be determined
TBR	To be resolved
V	Volts
Vac	Volts AC

Throughout the textbook so far, we have dealt with the MDRE process, issues like supportability and manufacturability, how to decompose and allocate requirements, but we have not dealt with how to write requirements. The process of writing good requirements, while there are guidelines we will discuss at length, is a learned skill, generally learned through writing bad requirements and living with the consequences. One of the purposes of this book is to minimize the pain associated with those consequences. As we discussed in Chap. 7, most program failures are due to bad requirements and poor requirements management [1]. In practice, there are many ways to write a good and a bad requirement. No two requirement engineers will write the same verbiage for the same requirement (not withstanding cut and paste). Every different systems engineer has their own style and their own thought process. There are many guidelines and standards that have been written and published over the years by various standards organizations like the American National Standards Institute (ANSI), and, for engineering programs, the Institute of Electrical and Electronics Engineers (IEEE). The ANSI/IEEE STD 830-1984 is one such guide and standard for writing software requirements. Many have acronyms to define the guidelines and attributes of good requirements. The two we will cover in Chap. 12 are SMART (specific, measurable, attainable, realistic, and testable/traceable) and the five Cs of requirements (Correct, Clear, Concise, Cohesive, and Complete). As you can see, while there is overlap between the two requirements guidelines, they are not identical. Even standards organizations don't always agree on what is most important to write requirements.

12.1 Requirements: The WHAT of Systems Design

The main point to understand when writing requirements is that requirements refer to the "what" is being build or developed, not how. Most engineers encounter problems and issues with requirements, either because they were written badly/ambiguously, or because the needed requirements didn't exist. A general rule of thumb when writing requirements is the longer the requirement text, the possibility of the requirement being ambiguous and confusing increases. Requirements should be as short and simple as possible. In addition, short statements are easier to organize, and make for increased readability by all parties that need to use or review the requirements [2]. All parties involved in the system specification, architecture, design, test, operations, and maintenance of a system benefit from a well-written set of requirements.

J. A. Crowder, C. W. Hoff, *Requirements Engineering: Laying a Firm Foundation*,
https://doi.org/10.1007/978-3-030-91077-8_12

Question to think about

Think about the consequences of putting "how" into the requirements set. How does this constrain the design engineers?

Each different engineering discipline will have different views/needs associated with the system requirements specification, but all disciplines need to be able to understand the requirements and it is imperative that all have the same reading on what the requirements mean. Having different groups (e.g., design vs. test) have a different interpretation of the requirements will mean disaster as the system development moves forward. Programs utilizing the agile development method have a different problem. In an agile or extreme development effort, requirements can change rapidly, so taking the time to deliver a full requirement specification is a waste of time. However, requirements management and requirements change management is more important in an agile/extreme development environment than in other development methodologies [3]. In agile/extreme development it is critical to ensure that all the sprint team are in sync with the current requirements baseline and the requirement change process must be followed by all or the teams will develop products that are based on different requirement baselines and will most likely not integrate during integration. For agile/extreme development, the MDRE requirement management team/engineer should collect the requirements incrementally and create user stories or user scenarios to illustrate the current requirements set. Each user story should have an estimated time of completion and a priority. Related user stories should be grouped together and the MDRE team should ensure they reflect a coherent set of requirements [4].

12.2 Writing SMART Requirements

Writing requirements requires skills that are only learned through practice. It requires a clear idea and understanding of the paradigm[1] and objectives involved in the system under development, as well as the environment the system will have to be part of. At its heart, for system engineering and MDRE, a requirement is something that governs what, how well,[2] and under what conditions a system or product will achieve its specified purpose [5]. A good requirement is used to state something that is necessary, verifiable, and attainable. For a requirement to be verifiable, it must make clear how the requirement is to be verified. Typically, the methods of verification are by examination, analysis, test, or demonstration. We will discuss this more later in this chapter.

One of the most important aspects of good requirement writing is to be sure the requirement is necessary to begin with. For functional requirements, this starts with the MDRE team asking the question, "what happens if this requirement does not exist? What is the worst possible outcome?" If you cannot come up with a set of major consequences, re-think writing the requirement. Many MDRE teams follow the "SMART" approach to writing requirements. SMART typically stands for, Specific, Measurable, Attainable, Realistic, and Traceable. There are differences in the definition of SMART among different organizations, for instance using "achievable/actionable/appropriate" rather than attainable, but in context these are basically the same.

12.2.1 SMART Requirements Are Specific

Requirements should be statements that have a specific purpose and should be just that, statements. Many add the terms Short and Simple as part of the "S" in SMART requirements and all three of these are correct and should be part of the thought process for requirements. Long, drawn out paragraphs or multiple sentences lead to ambiguous interpretation as to their meaning. Specific, short, and simple requirements are easier for the engineering teams, customer(s), and stakeholder(s) to agree on. A good requirement should only be able to be understood one way. You should avoid words like "easy," "simple," or "user-friendly"; everyone will have a different view of what each of these words mean as it applies to the system under development. A specific/short/simple requirement should equate to an unambiguous requirement [6]. Each position within the engineering organization has different uses and perspectives on requirements. Each requirement should mean the same to all positions. Table 12.1 below highlights how each role within an engineering development effort utilizes requirements and the requirements specification, from the customer and throughout the development team.

12.2.2 SMART Requirements Are Measurable

All requirements must be tested by some test method and the customer must agree that the system meets each requirement. The test organization must be able to verify that the requirements have been implemented correctly. There must be a way to measure whether the requirement meets or fails

[1]Paradigm: **a framework containing the basic assumptions, ways of thinking, and methodology that are commonly accepted by members of a technical community.**

[2]It's very important to understand the difference between "how" and "how well."

Table 12.1 How requirements are utilized

Role	Description
System customers/ stakeholders	Writes requirements to specify the needs of the system. Customers specify/authorize the changes to requirements
Managers	Uses the requirements and requirement specification to plan and bid the system and to plan the development process
Subcontract managers	Uses the requirements and requirement specification to define what the subcontractor(s) will be developing and the conditions for acceptance
Systems engineer	Uses the requirements and requirement specification to understand what system they are supposed to develop
System test engineers	Uses the requirements and requirements specification to develop validation test for the system under development
System operations and maintenance engineers	Uses the requirements and requirements specification to understand the system and the relationships between components
Hardware and software Developers	Uses requirements to define what they will be developing and to define their respective verification activities
System integration engineers	Uses the requirements' structure and verification to assist in planning and executing the integration of components and subsystem as well as the verification of requirements at each level of the architecture

to meet the requirement. If the requirement language makes it difficult or impossible to measure/verify compliance with the requirement, the system will never pass qualification and be accepted by the customer. Again, ambiguous language makes measuring and verifying requirements difficult if not impossible. Each requirement should come with a verification statement that indicates how the system is to be measured against the requirement and the pass/fail criteria that will be used to measure pass/fail against the requirement [7]. These verification statements should be compiled and provided to the integration and test organization, so they do not have to synthesize how to verify a requirement on their own. It is critical to build the verification statements as the requirements are being generated—as often, the language of the requirement will be affected by how the requirement is to be verified. All maintenance-related requirements must be specified in a measurable, verifiable language. Requirements should all be written not only for measurability, but also to ensure that ripple effects from changes are minimized.

12.2.3 SMART Requirements Are Attainable

Requirements should be written so that they are realistically achievable. This is from not only a technical perspective but also a cost and schedule perspective. The requirement authors must understand not only the application, but also the environment in which the system is to be deployed. There is always a trade-off between cost and benefits for any requirement, especially quality requirements. Writing realistic requirements is crucial to program success. When unattainable requirements are written, the best result is a frustrated MDRE team [8]—not to mention the development teams. When reviewing requirements for attainability, one must ask the simple question, "can this requirement be reasonably done?" The answer to this question comes from the MDRE team understanding how to build, test, and integrate a system with the capabilities specified in the requirements. Deciding that the requirement is achievable is not only from the perspective of realizing the capability, but it also involves understanding if the capability can be measured and maintained.

12.2.3.1 Case Study: Maintenance Nightmare

A great example of poorly written requirements from a maintenance perspective involves the GM line of vehicles, 1975–1980 models of Buick Skylark, Chevy Monza, Pontiac Sunbird, and Oldsmobile Starfire that were equipped with a 3.8-liter V-6 engine. Of course, the requirements specified the need for spark plugs, but what wasn't considered was the fact that the engine was very large, and the engine compartment was very small. The result was, to change the spark plugs on the right side of the engine, it required the engine be disconnected from the motor mounts, a chain had to be attached to the engine, and the engine lifted a few inches to be able to access and change the spark plugs. This was a maintenance nightmare and very expensive just to do a tune-up on these vehicles.

A lack of maintenance perspective can make a system/product unusable in the field due to its issues with maintainability. Remember that "attainable" means in all aspects of the system life cycle.

12.2.4 SMART Requirements Are Realistic

The notion of a realistic requirement goes beyond "is it achievable?". We dealt with that aspect in Sect. 12.2.3. Realistic considers issues like tolerance. Are the tolerances too tight? Are the tolerances defendable and cost-effective? Are the timing constraints realistic for not only the system being built, but also the environment the system operates within? The case study above illustrates one example of this. It was unrealistic to require the engine be disconnected from the motor mounts and lifted with a chain each time the spark plugs needed replaced. What was missing from the requirements for these cars was a sustainment and maintenance requirement for items like spark plugs. Bottom line, a requirement is useless if it is impossible to implement.

Requirement feasibility is related to capability, but also cost and schedule. If the requirement is not technically feasible, then it is not realistic and should not be there. If the requirement will take longer to implement than is possible, the requirement shouldn't be there. If the cost is completely prohibitive, the requirement shouldn't be there. It is acceptable to make a "goal" that stretches the technical feasibility or cost or schedule, but it should be just that, a goal, not a "shall do" requirement.

12.2.5 SMART Requirements Are Testable

Requirements should be written in such a way that the test organization for the program can verify whether the requirement has been implemented correctly or not. Utilization of the verification statements described in Sect. 12.2.2 above is invaluable to the test organization. The subsequent test should either pass or fail. Ambiguously written requirements make it impossible to establish a pass/fail criterion for the requirement. Some words make a requirement untestable. Examples of adjectives that are commonly used in requirements that are troublesome without description of what the term means are: robust, safe, accurate, effective, efficient, expandable flexible, maintainable, reliable, user-friendly, and adequate. Other words, adverbs, that cause issues in requirement testability are, quickly, safely, and in a timely manner. Without descriptions of what is meant by these terms, the test organization has no way to design a test that will show whether the system passes or fails this requirement. An example is the following requirement that we have seen written for a system:

- The system shall be safe.

The question is: safe from what? How safe? What does safe mean in this context?

> **Question to think about**
>
> *Think about how many test procedures for your system you would have to create to test the ambiguous requirement, the system shall be safe?*

12.3 Requirements: The Good and the Bad

The SMART discussion provides some general guidelines for how to construct meaningful and useful requirements, but you might wonder how that plays out in writing actual requirements. What follows are examples of bad requirements and good requirements for five different requirement categories discussed in Chap. 11. Each of these, both good and bad, have made their way into requirement specifications over the last few decades and are examples of real requirements.

12.3.1 User Interface Requirements

Bad/Incomplete Requirement: The system shall provide a workstation for each operator.

Why this is a bad requirement: On its own it does not specify how many operators and does not define the size of the display for each operator. For those operators utilizing a graphics user interface—this requirement may result in displays that are not adequate to perform their respective functions.

Good Requirement: For each user workstation, the configuration of the display fields (what data is displayed and the size of the field) shall be configurable on a user-by-user basis.

Why this is a good requirement: The requirement defines that the display fields to be configurable in size. Different users will display different data depending on their operational tasks.

12.3.2 Technical Requirements

Bad/Incomplete Requirement: The system shall have an instantaneous field of view of 120 degrees.

Why this is a bad requirement: The requirement states a field of view but does not define that field of view relative to how the system is mounted in/on the host vehicle. For instance—a field of view of −45° to +75° would be acceptable under this requirement but would not likely be acceptable to the customer.

Good Requirement: The system shall be designed with an instantaneous field of view of ±60° minimum relative to normal of the antenna face.

Why this is a good requirement: The requirement defines the minimum field of view, but also defines that field of view geometry relative to the sensing antennas.

12.3.3 Quality Requirements

Bad/Incomplete Requirement: The system shall have an availability of 99.99%.

Why this is a bad requirement: The requirement states an availability but does not define how it should be computed. This will likely lead to verification issues between the supplier and the customer if they use differing computational approaches.

Good Requirement: The vehicle shall be designed with an availability of at least 99.99% when calculated using vehicle mean time to failure and mean time to repair.

Why this is a good requirement: The requirement defines the methodology that must be used to compute availability—removing any ambiguity from verification.

12.3.4 Performance Requirements

Bad/Incomplete Requirement: The receiver sensitivity shall be at least −78 dBm.

Why this is a bad requirement: The requirement does not state what mode the receiver sensitivity is measured—and the value could be applied for either acquisition mode or tracking mode. Also, the requirement does not state where the measurement is made from.

Good Requirement: The receiver system shall have a sensitivity, when measured at the receiver input: −78 dBm in acquisition mode (150 MHz Bandwidth) and −85 dBm in tracking mode (20 MHz bandwidth) when measured at the receiver input.

Why this is a good requirement: The requirement defines the sensitivity for both receiver modes and defines where the sensitivity is to be measured.

12.3.5 Constraint/Assumption Requirements

Bad/Incomplete Requirement: The system shall weigh no more than 5 lbs.

Why this is a bad requirement: The requirement is ambiguous in that it implies the system weight of just the LRUs. In a highly weight constrained application, such as an airborne or space-borne system—the difference in system weight of just LRUs vs. system weight of LRUs and interconnect cables could be significant.

Good Requirement: The system shall weigh no more than 5 lbs, including the LRUs and cables connecting the LRUs.

Why this is a good requirement: The requirement defines the conditions under which the system should be weighed and the components that must be included in the total weight. This requirement would be further broken down to provide constraint requirements on each LRU and cables so that the entire assembled unit would be under 5 lbs.

Question to think about

Think about and answer the question of what would be the result of incorrect requirements concerning the total power generation of your multi-alternative energy system?

12.4 Additional Considerations for Writing Requirements

Since there are few courses that involve the mechanics of writing requirements, or technical writing in general, we provide some additional guidelines to the art of writing good requirements. This involves what is known as the Cs of requirements development [9]. These describe how the content of the requirement is written and is additional guidance to the SMART descriptions. The Cs represent guidelines for writing style and language usage, not necessarily requirement content like the SMART guidelines. The Cs are: (1) Correct, (2) Clear, (3) Concise, and (4) Complete.

12.4.1 Correct Requirement Language

No matter the differences between the styles used by different MDRE requirement writers, improper use of words and grammatical mistakes can lead to confusion on the informational content of a requirement. If there are numbers used in a requirement, ensure that they are correct. Correctness refers to grammatical, punctuational, and technical accuracy. Remember, word choice is important. Imprecise terms can lead to ambiguously defined requirements and/or verification criteria [10].

12.4.2 Clear Requirement Language

Even if you believe you and the customer are on the same page when it comes to the requirements objectives for a program/project, the requirements must be written in clear language. The customers, the development team (both H/W and S/W), the operations and sustainment team, and all other teams involved in the program must understand the message for each requirement.

12.4.3 Concise Requirement Language

Word choice is important, not all synonyms have the same interpretation by everyone involved. Imprecise terms can cause confusion and misunderstanding between groups. Also, make sure there aren't redundancies within the text. Remove unnecessary words and make everything in the requirement pertains to the topic of the requirement. Writing concise requirements is more than just minimizing the number of words but involves presenting the information in the easiest format for the developers and customers. Concise requirements are those that identify the problems that needs to be solved by the system under development. Writing concise requirements improves the ability to communicate what needs to be developed, and in doing so maximizes the probability of system success [11].

12.4.4 Complete Requirement Language

For a requirement to be considered complete, the requirement must unambiguously mean one thing and can only be interpreted one way. A complete requirement cannot be broken down into multiple requirements. Another criterion for a complete requirement is that it can be shown to have been met or not met by some form of testing. The last criterion for a complete requirement is that a requirement cannot be considered complete until the customer/stakeholder(s) have agreed to it [12].

12.5 Final Thoughts on Writing Good Requirements

As we have discussed, writing good requirements requires practice. The MDRE team should try to put themselves in both the customer's shoes and the development team's shoes to understand how the requirements must be formed to meet the objectives of both. Ask yourself the question, "As the customer, does this requirement provide the necessary functionality or quality they intended?" Also, ask yourself the question, "As a developer, am I confident what this requirement is asking?" If you can't answer yes to both questions, the requirement needs work. Finally, here is a checklist each requirement should pass before it is finalized [13]:

1. Use of Correct Terms.
   a. Shall = requirements.
   b. Will = facts or declaration of purpose.
   c. Should = goal, not requirement.
2. Use of Correct Format.
   a. Should be in the form "system xxx shall…" verb goes here: do, perform, provide, weigh, or another verb, followed by what should be done.
   b. The requirements must use consistent terminology throughout the requirements set, including lower-level requirements, to avoid confusion with interpretation.
   c. Should include tolerances for qualitative analysis/performance values, including less than, greater than, or equal to, 3 sigma values, etc.
   d. Should be free of implantation discussion. Requirements are what—not how.
   e. Should be free of operational descriptions. Often you see requirements like "The operator shall…" which are most often operational statements, not requirements.
3. Must be grammatically correct.
4. Free of typos, misspellings, and punctuation errors. Be careful, many autocorrect systems don't play well with technical terms and the proper grammar associated with such terms. Don't assume autocorrect will catch mistakes.
5. If applicable, complies with program/customer template and style rules.
6. Minimal to no "To Be Determined" (TBD) entries in the requirements. If values are not locked down, using an estimate with the statement "To Be Resolved" (TBR) is preferable to TBD, and should include the rationale with which it will be resolved.
7. Assumptions that drove the requirement should be listed along with rationale if possible.
8. Requirement Clarity.
   a. Are the requirements clear and unambiguous?
   b. Are the requirements simple and concise?
   c. Do the requirements express one thought and one thought only? There should be no compound requirements.
   d. Does the requirement have one subject and one predicate?
9. Requirement Completeness.
   a. Are the requirements as complete as possible?
   b. Are there any requirements missing? Have any areas been overlooked:
      i. Functional.
      ii. Performance.
      iii. Interface.
      iv. Environmental.
      v. Facility.
      vi. Transportation.
      vii. Training.
      viii. Personnel.
      ix. Operability.
      x. Safety.
      xi. Security.
      xii. Sustainability.
      xiii. Appearance.
      xiv. Physical Characteristics.
      xv. Maintainability.
   c. Have all assumptions been explicitly stated?
10. Requirement Compliance.
    a. Are the requirements at the proper level within the architecture (e.g., system, subsystem, element, component)?
11. Requirement Consistency.
    a. Are the requirements, as stated, without contradictory language?
    b. Is the terminology consistent with the customer's terminology?

   c. Is the requirement terminology consistent with the terminology common to the technology being used in the system?
12. Traceability.
   a. Are all the requirements necessary? Is each lower-level requirement needed to meet the higher-level requirement it is traced to?
   b. Are the requirements all traceable to only one higher-level requirement?
   c. Is each requirement numbered or assigned such that it is uniquely identifiable and referenced?
13. Requirement Correctness.
   a. Is each requirement correct?
   b. Is each assumption correct?
   c. Is each requirement technically feasible?
14. Functionality.
   a. Are all the described functions necessary and sufficient to meet the mission goals and objectives laid out by the customer?
15. Performance.
   a. Are all performance requirements having performance criteria and margins (e.g., timing, throughput, size, latency, accuracy, precision)?
   b. Are the performance requirements realistic, are they feasible?
   c. Are the tolerances correct?
16. Interfaces.
   a. Are all external interfaces clearly defined?
   b. Are all internal interfaces clearly defined?
17. Maintainability.
   a. Does the system have maintainability requirements?
   b. Are maintainability requirements written so that ripple effects from changes are minimized?
18. Reliability.
   a. Are there clearly defined, measurable, and verifiable reliability requirements?
   b. Are there error detection, reporting, and handling requirements?
19. Verifiability/Testability.
   a. Does each requirement have a verification method shown (e.g., demonstration, inspection, analysis)?
   b. Can the verification criteria be executed?
   c. Do the requirements precisely facilitate system test success criteria?
   d. Are the requirements free of unverifiable terms (e.g., portable, fast, appropriate, maximize, minimize, sufficient, easy)?
   e. Are there verification statements written for each requirement?

Finally, writing correct requirements is important and crucial to the success of system development. It doesn't matter how qualified your development team is, doesn't matter what technology you are bringing to create the system if the requirements are not correct [14]. If the development teams are off building the wrong system, or each team is creating subsystems with the wrong interfaces, the project will not succeed, or will require lengthy and expensive redesigns and development efforts. The right systems and right products are those developed with the right requirements. In short, the requirements must bring value to the development effort. Valuable requirements:

Solve a customer's problem or mission objective.

Support the business strategy of both the customer and the developer.

Provide capabilities and services to the end users.

Don't over-solve the mission/objectives.

Question to think about

Think about how your multi-alternative energy system solves the issue of both renewable energy and utility to consumers?

12.5.1 Design Problem

The philosophy of the assignment development for this textbook was to create a set of thought-provoking questions, problems, and design assignments that help drive home the principles, practices, and methodologies of multidisciplinary requirements engineering (MDRE). We endeavored to provide the opportunity for the student to practice the art and science of requirements engineering, helping the student to understand that there is rarely one answer to a requirements engineering problem, and it is only through study and practice that successful requirements engineers are developed. MDRE is a life-long journey, and anyone who tells you they have nothing left to learn does not truly understand the discipline or has never been involved in the development of a complex system. With the recent interest in alternative energies and talk of the "Green New Deal," along with the recent issues with ice storms in Texas taking out the power grid, the assignments for this textbook will follow the theme of effectively dealing with both local alternative energy power grids and the national infrastructure required to facilitate large-scale alternative energy production, sustainment, and management.

12.5.2 Design Problem Description: Alternative Energy Power Grid

The purpose of this project is to, over the course of the semester, develop a feasibility study, requirements analysis, requirements elicitation, derivation, and allocation, for a system that will facilitate a "local" alternative energy power grid capable of powering a neighborhood of 500 homes and a large biosphere capable of supporting growing year-round fruits and vegetables for the 500-home neighborhood. The student will have to deal with issues of choosing multiple alternative energies that can provide power throughout all seasonal difficulties, from intense summer heat to freezing winter temperatures and possible ice storms. Control, viability, sustainability, and maintainability will be major factors to understand the entire possibly 30-year life cycle of such a system. How to effectively utilize, store, and manage power distribution, along with when to upconvert "extra" energy to the national power grid, will be issues that drive many of the assignments. Simply relying on the current technologies and energy infrastructures will not move the country and world forward. Hopefully, this will provide insight to the students using this textbook to understand some of the issues that face everyone looking out over the next many decades.

12.5.3 DC Power Grid Requirements Discussion

In the initial days of power generation and distribution, DC power was far less practical than the AC distribution system that soon supplanted it. However, even after Tesla's AC system became the predominant system (and rightly so), Edison's DC power did not become obsolete for some time. Eventually power plants eliminated their delivery of DC power, but the overall concept of a DC power grid is not dead. Even our computer systems (large server systems) have uninterruptible power supplies which contain batteries that are charged off the grid and used to upconvert to AC to power devices when the AC power grid goes off-line. So, the notion of a DC power grid is not dead, but generally used as a backup system. The point is, the notion of a DC power grid has remained and now that renewable energies are viable, we should re-examine their use for widespread, local, power distribution.

One of the advantages of a locally generated DC station is energy savings as well as allowing a facility to run more easily off various DC sources, such as Solar Arrays. And for power-hungry installations, the notion of generating the power locally is growing in popularity. Part of the reason is that waste heat from the power generation generators (solar, geothermal, etc.) can be used to warm nearby buildings. By combining the production of heat and power, local power grids can squeeze much more useful energy out of the renewable energy they use, so this approach will certainly become more widespread as time goes on. The purpose of this design problem is to look ahead and provide an architecture and specification for a local DC power generation station that can serve a local community.

Currently many systems exist to control and handle alternative energies. There are systems to control and manage solar power, wind power, geothermal power, and others, but very few are designed to control, manage, store, and distribute energy from a power grid consisting of multiple alternative energy sources. As we saw the issues recently caused by the intense ice storms in Texas, one alternative energy source is not the answer to the growing energy needs/wants in our society. Each alternative energy source has limitations and issues associated with its use. By harnessing multiple alternative energies, and managing the storage of the energies produces, communities can be worry free from power generation. To facilitate utilizing multiple alternative energy sources, a control center must be included in the design that can manage the energy creation, distribution, storage, and return unused energy to the power grid.

12.5.4 High-Level Technical Description of the DC Power Grid System

The purpose of the multiple alternative energies DC power grid is to:

- Supply DC power to a customer community of 500 homes and a biosphere for growing food.
- Provide a storage device to store energy in case of a system failure.
- Determine the amount of surplus energy that is above and beyond what is required to power the grid and charge the backup system, and then deliver that surplus energy to the national power grid systems.

Throughout the semester you will be asked to analyze, design, and architect different components of the overall DC power grid system. This will include research into alternative designs, like lithium nanowire storage systems for backup power and/or supercapacitors (also called ultracapacitors). Fig. 12.1 illustrates these technologies connected to the DC Power Grid Regulator Control Center.

12.5.5 High-Level Technical Requirements for the DC Grid System

- Design shall utilize at least two alternative energies, with a maximum of 4.
- Total power generation shall be 3.0 megawatts minimum.
- Each of the alternative energy sources shall be capable of generating at least 60% of minimum wattage.
- DC power grid shall generate DC power, with grid voltage being 48, 24, or 12 volts.

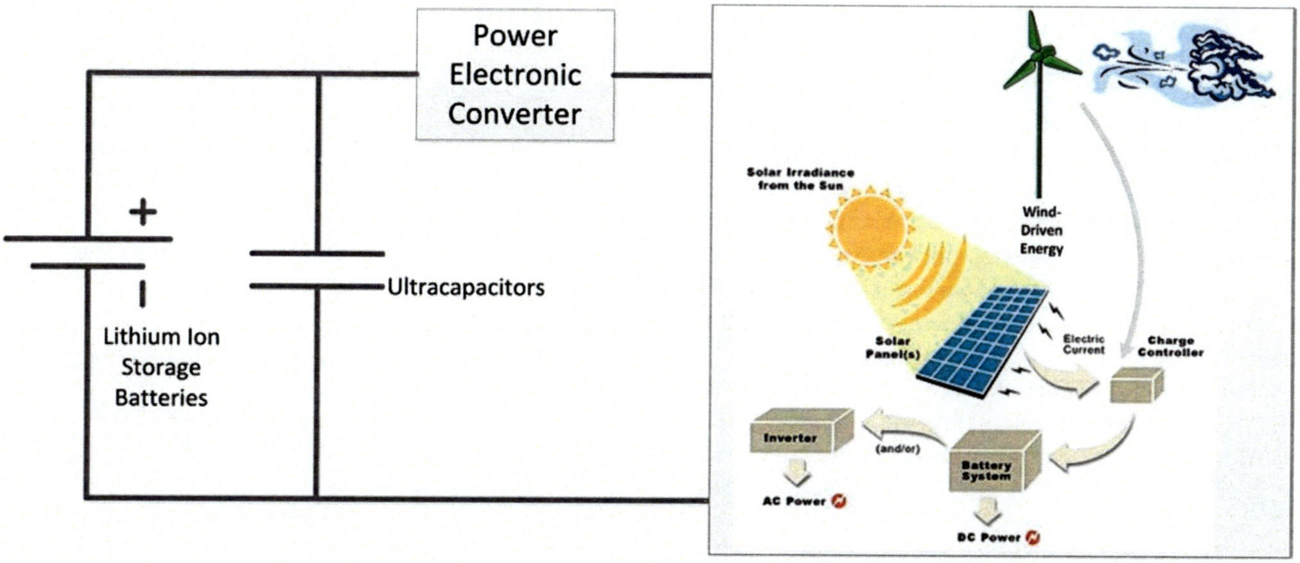

Fig. 12.1 Alternative power management system overview

Fig. 12.2 Example of a whole house power grid

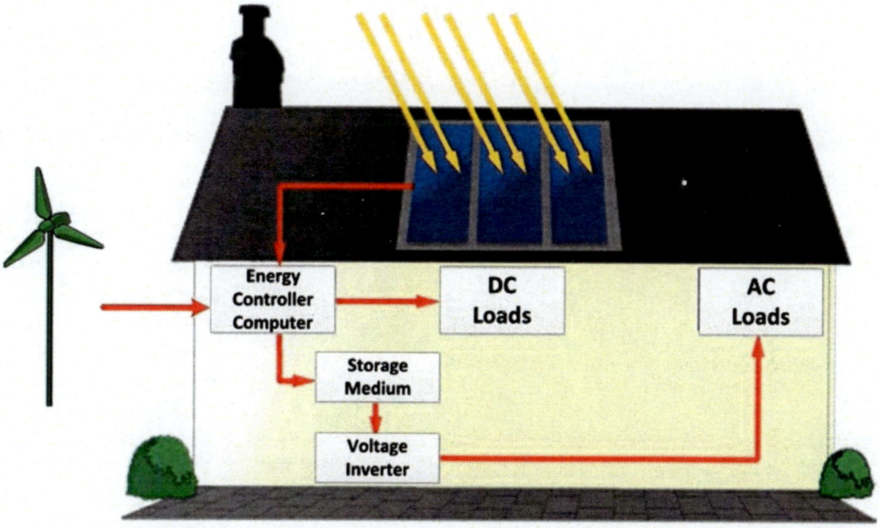

- DC power grid shall be capable of servicing at least 500 single family homes and one 4-acre biosphere with hydroponics capable of growing vegetables year-round. (See Fig. 12.2 for a block diagram of a DC-powered home).
- All lights, appliances, and facilities in each home and the biosphere shall be configured for DC operations, if possible. Things like electric ovens may need DC/AC conversion to accommodate their 220v. requirements.
- Only allowed exceptions are the HVAC system and the Washer/Dryer, although DC is preferred.
- There shall be energy storage capability capable of operating the grid of 500 single family dwellings for 7 days if all alternative forms of energy are not producing.
  - Compliance Criteria: Energy storage devices shall be capable of producing 750,000 watts for the entire 7-day period. Nominal energy storage should be batteries or supercapacitors.
- The DC power grid system shall include controlling h/w and s/w to manage the multi-alternative energy strategies and production.
- All energy produced above the required home and energy storage usage shall be upconverted to 220vac for release to the national power grid.
- The DC power grid shall be capable of operating in a temperature range of −40 degrees F. to 125 degrees F. without degradation of power generation.
- The DC power grid must meet OSHA regulations.

12.5.6 Assignments

- Review the requirements you derived for you multi-alternative energy system and describe how they meet the SMART criteria discussed in Chap. 12.
- Write a 5–10-page report outlining the need for the 5 Cs of good requirement creation.
- For each of the categories outlined in Sect. 12.3, write both a good and a bad requirement for your multi-alternative energy system and describe why each requirement is good or bad.
- Answer the following questions:
  - Why is grammar important for requirements?
  - List the requirements for your system, based on their verification methods:
    - Test.
    - Demonstration.
    - Analysis.
    - Inspection.
- For your performance requirements, what are the tolerances for each requirement?

References

1. Madsen, K. H. (1994). A guide to metaphorical design. *Communications of the ACM, 37*, 57–62.
2. Sedelmaier, Y. and Landes, D. (2014). "Using business process models to foster competencies in requirements engineering," in Proc. 2014 IEEE 27th Conference on Software Engineering Education and Training (CSEE&T), pp. 13–22.
3. Hammani, F. Z. (2014). "Survey of non-functional requirements modeling and verification of software product lines," in Proc. the 2014 IEEE Eighth International Conference on Research Challenges in Information Science, pp. 1–6.
4. Tabassum, M. R., Siddik, M. S., Shoyaib, M., and Khaled, S. M. "Determining interdependency among non-functional requirements to reduce conflict," in Proc. the 2014 International Conference on Informatics, Electronics & Vision, 2014, pp. 1–6.
5. Dabbagh, M., Lee, S. P., and Parizi, R. M. (2014). "Application of hybrid assessment method for priority assessment of functional and non-functional requirements," in Proc. the 2014 International Conference on Information Science and Applications, pp. 1–4.
6. Glinz, M. (2007). "On non-functional requirements," in Proc. the 15th IEEE International Conferences on Requirements Engineering, pp. 21–26.
7. Aceituna, D., Do, H., Walia, G. S., and Lee, S. W. (2011). "Evaluating the use of model-based requirements verification method: A feasibility study," First International Workshop on Empirical Requirements Engineering, pp. 13–20.
8. Pohl, K. (1996). *Process-Centered requirements engineering, Taunton*. Research Studies Press Ltd.
9. Jain, H., Vitharana, P., & Zahedi, F. M. (2003). An assessment model for requirements identification in component-based software development. *ACM SIGMIS Database, 34*(4), 48–63.
10. Creswell, J. W. (2007). *Qualitative inquiry and research design: Choosing among five approaches* (p. 45). SAGE Publications.
11. Loucopoulos, P., & Karakostas, V. (1995). *System requirements engineering*. McGraw-Hill.
12. Mayrhofer, W., Meyer, M., Steyrer, J., Maier, J., Langer, K., and Hermann, A. (2004). "International career habitus–thick descriptions and theoretical reflections," Academy of Management Annual Meeting Symposium on Global Careers and Human Resource Development: Emerging IHRM Perspectives New Orleans, USA.
13. Robertson, S., & Robertson, J. *Mastering the requirements process: Getting requirements right*. Addison-Wesley.
14. Davis, A. M. (1993). *Software requirements: Objects, functions, and states*. Prentice-Hall.

Requirements Traceability

13

Abbreviations

AC	Alternating current
CMMI	Capability Maturity Model Integration
DC	Direct current (also 0 Hz)
F	Fahrenheit
ISO	International Organization for Standardization
MDRE	Multidisciplinary requirements engineering
OSHA	Occupational Safety and Health Administration
RBS	Requirement Breakdown Structure
RTM	Requirements Traceability Matrix
RTT	Requirements Traceability Tool
SI&T	System integration and test
SoO	Statement of Objectives
SoW	Statement of Work
V	Volts
Vac	Volts AC

13.1 The Need for Requirement Traceability

The ability within a project to trace requirements from the top layer (e.g., system level) to the lowest layer (e.g., component level), when done properly, saves time, money, effort for the development team, as well as the SI&T team [1]. It is important for the MDRE team to create an easily traceable set of requirements to provide the following to the development team, management, and customer(s):

- **Allows the final deliverables (deployable system) to be tied directly to the customer's objectives, needs, and goals**:
  - **Traceability from high-level to low-level** traceability (downward tracing) ensures that the high-level objectives for the system are allocated/derived/decomposed to development efforts that will provide the capabilities described in the high-level requirements that are tied to customer objectives. Without complete downward traceability, there is not way to ensure that the developers are creating a system that meets the customer's needs (i.e., are tied to actual deliverables).
  - **Traceability from low-level to high-level** requirements (upward or backward traceability) ensures that every lower-level requirement tie to a higher-level requirement and ultimately to a capability or objective in the Statement of Objectives (SoO) or the Statement of Work (SoW). Any requirement that cannot be traced to a high-level requirement and objective/capability specified by the customer should be closely reviewed as to why it exists in the requirement specification. When customer objectives are traced to requirements, and subsequent data elements and use cases, it can be understood and verified as to how the capability/objective will be realized [2]. Each objective or capability can be reviewed to ensure it will be addressed and the solution can be viewed and explained.
- **Ensures that the development organization doesn't waste time and resources on unnecessary work or rework due to missing or conflicting requirements**:
  - Without requirements traceability, the development teams have the potential to waste valuable time and money backtracking, duplicating efforts, missing capabilities that must be added later, normally at large costs and serious schedule slip [3]. Inadequate traceability has contributed to the failure or delay of many programs over the last few decades [4]. Lack of proper requirements traceability leads to constructing or reconstructing a requirement trace in the middle of an existing specification traceability tree or matrix (discussed below) and is time consuming, prone to errors and mistakes, driving other problems with the development effort.
- **Ensures the system complies with relevant industry standards**:

J. A. Crowder, C. W. Hoff, *Requirements Engineering: Laying a Firm Foundation*,
https://doi.org/10.1007/978-3-030-91077-8_13

- Requirement traceability is a key component in many industry standards like CMMI and ISO 9001:2000 [5]. If the requirements are not traced adequately, industry standards and constraints are not flowed down to lower levels, risking a component or subsystem is developed incorrectly or not in compliance with a constraint or standard, driving possible redesign and much re-work.
- **Allows easier and more accurate impact analysis when requirements change**:
  - Requirement traceability allows impact to changing requirement to be assessed for how a change in one requirement ripples through the system, both up and down the requirements trace. If a customer changes a requirement once the system is in development, or even worse, during system test, traceability enables the MDRE team to quickly understand how many other requirements are affected and assess the full impact of the proposed change. Without the traceability, it will be unclear the complete impact on the system [6].
- Allows for more efficient and effective test program.
  - A complete and comprehensive requirements tracing allows the system integration and test team to design the test program more efficiently to test requirements at the appropriate level—increasing the likelihood that a requirement is verified once and only once. This will eliminate costly re-test at multiple levels. Also, if/when a requirement changes, the retesting will also be more focused and efficient—saving time and cost.

Requirement traceability is one of the foundational processes that drives MDRE. But just knowing traceability is important is not enough, you need to understand how to trace requirements, and the tools available to facilitate good requirements tracing [7].

Question to think about

Think about how your availability requirements might be derived differently for each type of alternative energy. If not traceable to higher-level requirements, how do you determine the overall availability allocation/derivation between energy types?

13.2 The Requirements Traceability Hierarchy

Requirement traceability requires every requirement to be traced to either a lower-level requirement, a higher-level requirement, or both. This bi-directional traceability is the key to effective management of system requirements. It enables the development of an analytical understanding of any system-wide effects of changes to requirements for a given system element, updating requirements documentation with rationale and impacts for approved changes. Requirement traceability allows features and capabilities to be traced back to the high-level requirements and objectives that requirements were derived from or provided by the customer of the system under development. Traceability is useful after the system has been deployed to understand where requirements and features were specified and the rationale for each decisions/requirement in the system. It may allow understanding of the rationale for unused features that the system provides but the operators/users of the system are not using [8]. The requirement traceability hierarchy is just that, a hierarchy of how high-level requirements trace through the system, subsystems, components, hardware, software, etc., down to the lowest level requirements. As requirements are decomposed, derived, and allocated, decisions must be made as to how to allocate constraints such as weight, power, and performance to lower-level parts and components of subsystems. For example, if the weight of a subsystem is specified to be ≤10 lbs. and that subsystem has eight components, requirement must be derived allocating the maximum weight restrictions each of the eight components must comply with. Another example might be a performance requirement that the system must respond within 2 s to a user's command. If the system has four components or subsystems that contribute to that performance requirement, requirements must be derived putting performance constraints on each of the components such that the total response delay is within the higher-level 2-second constraint. The requirement hierarchy is basically a Requirement Breakdown Structure (RBS) that visually displays how the requirements are decomposed, derived, and allocated down to the lowest level needed for system development [9]. An example of an RBS is shown in Fig. 13.1. Figure 13.1 represents part of an overall RBS dealing with user accounts, which can be part of any system that includes software and users who must login to the system to access the software. The top-level requirement breakdown shown is "Manage User Accounts" which would have different designs, depending on the application and environment the system is used within. But, generically, the breakdown or traceability from Req.9.1.1 shown in Fig. 13.1 has 5 lower-level requirements to implement "Manage User Account."

Obviously, these are not the only lower-level requirements that might be needed for a given system, but these represent some common lower-level user requirements. These are:

- **9.1.1.1 Add Users**: For any system that requires users to login to a system, an administrator must have the ability to add users to the list of authorized people/system allowed access.

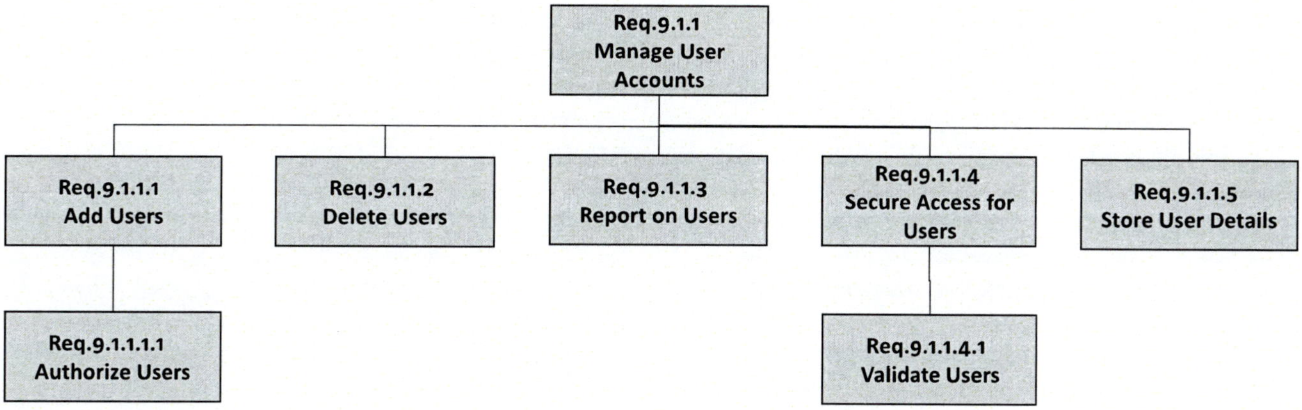

Fig. 13.1 Example of a requirement breakdown structure

- **9.1.1.1.1 Authorize Users**: To add a person or application to the system, authorization credentials like creation of valid username and password must be created, based on rules that would be shown as requirements below the 9.1.1.1.1 level.
- **9.1.1.2 Delete Users**: For users that no longer need access to the system, the administrator must have the ability to remove them from the list of authorized users.
- **9.1.1.3 Report on Users**: It is necessary for the administrator to be able to generate various types of reports for proper administration of a given system, to include: (1) how many incorrect logins have happened, (2) has there been any unauthorized access to the system, and others.
- **9.1.1.4 Secure Access for Users**: The system administrator(s) must secure the system so that only authorized users have access to the system and the information on the system. Requirements would be derived from this, based on the type of system, the level of security needed, and the complexity of who needs access to what type of information available on the system.
  - **9.1.1.4.1 Validate Users**: Once a user has been added to the system, correct login information must be provided and verified before access to the system and information is granted. Again, this requirement would be broken down further with derived requirements for the type of validation, based on the needs and objectives of the system.
- **9.1.1.5 Store User Information**: All user information must be stores (e.g., username and password) for use to authenticate that the person or application trying to access the system has authorization for access.

13.3 Requirements Traceability Best Practices

Writing requirements required discipline for the MDRE team. Here we discuss a few of the best practices the MDRE team must follow to provide good requirements creation, management, and follow-up throughout the life cycle of the system under development.

13.3.1 Unique Requirement Identifiers

For requirements traceability to be effective, every requirement must have a unique, persistent, and unambiguous requirement identifier to track the requirement throughout the design, development, deployment, and operations and sustainment life cycle of the system. Failure to do so brings the potential for capabilities and necessary attributes of the system to be forgotten or lost through inability to view them considering their position and use within the overall system requirement specification [10].

13.3.2 Traceability Ownership

Regardless of the traceability method, whether manual, spreadsheet, or requirements management tool, the MDRE engineer/team must take ownership of the requirements traceability and follow it throughout the development and deployment life cycle. For system deployment, the MDRE team must "hand-off" the maintenance and ownership of the requirements and requirements traceability to the personnel in charge of operations & maintenance of the system. Failure to have knowledgeable MDRE engineers in charge, both from the development side and the operations side, will result in errors and system failures if updates and requirements changes are not handled correctly [11].

13.3.3 Traceability Update Consistency

The MDRE team must assure that requirement updates and changes are handled consistently throughout the development and deployment of the system. Whenever changes or updates occur, an impact analysis must be performed to understand the ripple effects of requirement changes and all requirements at high levels and lower levels the requirement traces to must be impacted and changes reflected in their requirement text. This includes making sure that the party or parties requesting the requirement updates/changes provide all the information required to perform the impact analysis [12].

13.4 The Requirements Traceability Matrix (RTM)

One of the major efforts that goes into any development project is the final testing events and showing how the test verify compliance with the requirements that the system was defined and built against. Within the context of requirements verification and validation, it is fundamental to the program for the MDRE team to ensure full requirement traceability throughout the entire system life cycle. To this end, the Requirements Traceability Matrix (RTM) is most often the right tool. Small projects can get by with "spreadsheet management" of requirements, but medium size (we know, ambiguous term) and larger projects need a formal requirement management tool, which normally includes an RTM to requirements traceability. One of the major capabilities provided by requirements traceability is visualization of the relationship not only between requirements at different levels of the system, but also relationships between artifacts. This includes tying test cases to requirements to ensure that every requirement has a testing event to verify it, and that every test event is tied to at least one requirement [13]. If you have a test event not tied to a requirement, the test event should not exist. The RTM provides one visualization tool that is a table-like correlation matrix that maps requirements (generally depicted in columns) to another artifact like test cases or software modules (code).

The RTM is important because not only the requirements themselves should be traceable, but also the requirements' relationships with all artifacts associated with the requirements. This includes models (we'll talk more about this in Chap. 17 on Model-Based Systems Engineering), analysis results, test cases, trade studies, test procedures, test results, and all other documentation that ties to the requirements. The design artifacts that realize the requirements should be traced to the requirements that drove the development. Any artifact of the system created in the later stages of development, test, and deployment should be traced back by the MDRE team to requirements. There should be nothing in the system that cannot be tied back to a requirement [14]. The MDRE team must choose its design, development, requirements management, and test tools carefully to ensure they work well together. Establishing traceability for requirements dealing with software can be difficult when the requirements are in a RTM, the design artifacts are in tools such as MagicDraw®, Rational Rhapsody®, or others. Also, test verification tools can make it difficult to tie requirements to test cases. The MDRE team should map this out with the development team(s) before launching design and development efforts.

Question to ponder

Research MagicDraw and Rational Rhapsody. Why do you think it would be difficult to provide traceability across both tools?

Utilization of traceability across the system life cycle provides many benefits, several of which are indicated below. Providing traceability beyond just the requirement stack can provide [15]:

Requirement change impact analysis: Traceability links provide dependency information to understand what other requirements are affected by a change to one, along with understanding design artifacts that may be affected by the change.

Coverage analytics: Requirement traceability allows the MDRE team to ensure that requirements have not been missed or overlooked, especially for areas like sustainability, maintainability, and safety.

Project status analytics: Completion criteria can be tracked and traced up and down the traceability stack. Requirements that have not been completed/verified will affect the verification of higher-level requirements. The complete traceability of the requirements allows status analysis to be performed on the system based on completion tags within the RTM.

Test optimization: By linking requirements to source code, test cases, and test results, it is made simple to identify affected parts of the system when test cases fail. It also allows redundant tests to be identified and removed from the system.

We have discussed the reasons for traceability, along with some of the constructs (e.g., RTM) that are useful in system traceability. Now we look at the actual software tools that allow traceability and automation of traceability for effective requirements management.

13.5 The Requirement Traceability Tools

The problem with tools like spreadsheets for requirements traceability is the lack of the ability for collaboration among the MDRE team. Not having the ability for simultaneous work on requirements management tasks is time-consuming and inefficient [16]. Requirements management tools should provide end-to-end solutions that can be utilized through every step in the program, MDRE, design, development, test, deployment, and sustainment and maintenance. A Requirements Traceability Tool (RTT) should come with one or more of the following features:

- **End-to-end traceability**: The RTT should provide traceability between requirements as well as traceability across system artifacts like test cases, use cases, code modules, etc.
- **Security (Permissions)**: Only those authorized to change, modify, add, and/or delete requirements should have permissions within the RTT. Security measures should be fine-grained enough to differentiate who is allowed what within the RTT.
- **Document generation**: The RTT should be able to generate many different documents, including the requirements specification, requirements verification plan, test plans, etc.
- **Collaboration**: Allows multiple authorized MDRE engineers to be working on the requirements simultaneously, keeping track of changes, version numbers, additions, deletions, traceability, etc. across the entire requirements stack.
- **Built-in quality checks**: The RTT should capture and report quality issues like broken or missing links, duplicate links, and traceability issues like no trace to one or more test cases.
- **Reports and diagrams generation**: Once the requirement specification has been captured within the RTT, the tool should allow generation of diagrams and reports like hierarchy diagrams, tree diagrams, etc.
- **Version controls**: Keeping track of the latest versions of all requirements is crucial to effective requirements management. It is a serious issue if different groups within the development effort are using different versions of requirements [17].
- **Automated requirements numbering**: Once the top-level requirements have been loaded into the RTT, the tool should automatically number lower-level requirements based on the numbering scheme from the highest level. This allows complete traceability up and down the requirements stack.
- **Filtering**: The RTT should allow filters to be used to select specific types of requirements for viewing, e.g., performance requirements and sustainability requirements.

No RTT will have every feature, but the MDRE team should understand before acquiring a tool which features are relevant to their program and requirements management needs [18]. The purpose of this discussion is not to endorse or suggest one tool over another. There are many freeware and licensable RTTs out on the market. A little research into them will provide you with a list of tools applicable to current systems being developed. The more expensive RTTs offer automation capabilities for requirements traceability. Such automation techniques catch requirement changes and redundancies that manual searches may miss [19]. While automation tools can reduce the workload on the MDRE team, it is incumbent on the MDRE engineers to diligently monitor and maintain the requirement specification and associated artifacts (e.g., test cases, use cases). The use of automation capabilities and tools within an RTT presumes that requirements management processes and procedures are being practiced, otherwise automation will have little benefit to the program [20]. Many are skeptical of automated traceability tools since few provide all the tools necessary for modern development methods and techniques (e.g., agile development). Regardless of whether a program chooses RTTs that allow automation of traceability, every program will benefit greatly from the use of best practices in their requirements management [16].

Question to ponder

Think about how large a program needs to be in terms of the total number of requirements and requirement levels before a commercial traceability tool is required for requirements management.

13.5.1 Design Problem

MDRE is a life-long journey, and anyone who tells you they have nothing left to learn does not truly understand the discipline or has never been involved in the development of a complex system. With the recent interest in alternative energies and talk of the "Green New Deal," along with the recent issues with ice storms in Texas taking out the power grid, the assignments for this textbook will follow the theme of effectively dealing with both local alternative energy power grids and the national infrastructure required to facilitate large-scale alternative energy production, sustainment, and management.

13.5.2 Design Problem Description: Alternative Energy Power Grid

The purpose of this project is to, over the course of the semester, develop a feasibility study, requirements analysis, requirements elicitation, derivation, and allocation, for a system that will facilitate a "local" alternative energy power grid capable of powering a neighborhood of 500 homes and a large biosphere capable of supporting growing year-round fruits and vegetables for the 500-home neighborhood. The student will have to deal with issues of choosing multiple alternative energies that can provide power throughout all seasonal difficulties, from intense summer heat to freezing winter temperatures and possible ice storms. Control, viability, sustainability, and maintainability will be major factors to understand the entire possibly 30-year life cycle of such a system. How to effectively utilize, store, and manage power distribution, along with when to upconvert "extra" energy to the national power grid will be issues that drive many of the assignments. Simply relying on the current technologies and energy infrastructures will not move the country and world forward. Hopefully, this will provide insight to the students using this textbook to understand some of the issues that face everyone looking out over the next many decades.

13.5.3 DC Power Grid Requirements Discussion

In the initial days of power generation and distribution, DC power was far less practical than the AC distribution system that soon supplanted it. However, even after Tesla's AC system became the predominant system (and rightly so), Edison's DC power did not become obsolete for some time. Eventually power plants eliminated their delivery of DC power, but the overall concept of a DC power grid is not dead. Even our computer systems (large server systems) have uninterruptible power supplies which contain batteries that are charged off the grid and used to upconvert to AC to power devices when the AC power grid goes off-line. So, the notion of a DC power grid is not dead, but generally used as a backup system. The point is, the notion of a DC power grid has remained and now that renewable energies are viable, we should re-examine their use for widespread, local, power distribution.

One of the advantages of a locally generated DC station is energy savings as well as allowing a facility to run more easily off various DC sources, such as Solar Arrays. And for power-hungry installations, the notion of generating the power locally is growing in popularity. Part of the reason is that waste heat from the power generation generators (solar, geothermal, etc.) can be used to warm nearby buildings. By combining the production of heat and power, local power grids can squeeze much more useful energy out of the renewable energy they use, so this approach will certainly become more widespread as time goes on. The purpose of this design problem is to look ahead and provide an architecture and specification for a local DC power generation station that can serve a local community.

Currently many systems exist to control and handle alternative energies. There are systems to control and manage solar power, wind power, geothermal power, and others, but very few are designed to control, manage, store, and distribute energy from a power grid consisting of multiple alternative energy sources. As we saw the issues recently caused by the intense ice storms in Texas, one alternative energy source is not the answer to the growing energy needs/wants in our society. Each alternative energy source has limitations and issues associated with its use. By harnessing multiple alternative energies, and managing the storage of the energies produces, communities can be worry free from power generation. To facilitate utilizing multiple alternative energy sources, a control center must be included in the design that can manage the energy creation, distribution, storage, and return unused energy to the power grid.

13.5.4 High-Level Technical Description of the DC Power Grid System

The purpose of the multiple alternative energies DC power grid is to:

- Supply DC power to a customer community of 500 homes and a biosphere for growing food.
- Provide a storage device to store energy in case of a system failure.
- Determine the amount of surplus energy that is above and beyond what is required to power the grid and charge the backup system, and then deliver that surplus energy to the national power grid systems.

Throughout the semester you will be asked to analyze, design, and architect different components of the overall DC power grid system. This will include research into alternative designs, like lithium nanowire storage systems for backup power and/or supercapacitors (also called ultracapacitors). Fig. 13.2 illustrates these technologies connected to the DC Power Grid Regulator Control Center.

13.5.5 High-Level Technical Requirements for the DC Grid System

- Design shall utilize at least two alternative energies, with a maximum of 4.
- Total power generation shall be 3.0 megawatts minimum.
- Each of the alternative energy sources shall be capable of generating at least 60% of minimum wattage.
- DC power grid shall generate DC power, with grid voltage being 48, 24, or 12 volts.
- DC power grid shall be capable of servicing at least 500 single family homes and one 4-acre biosphere with

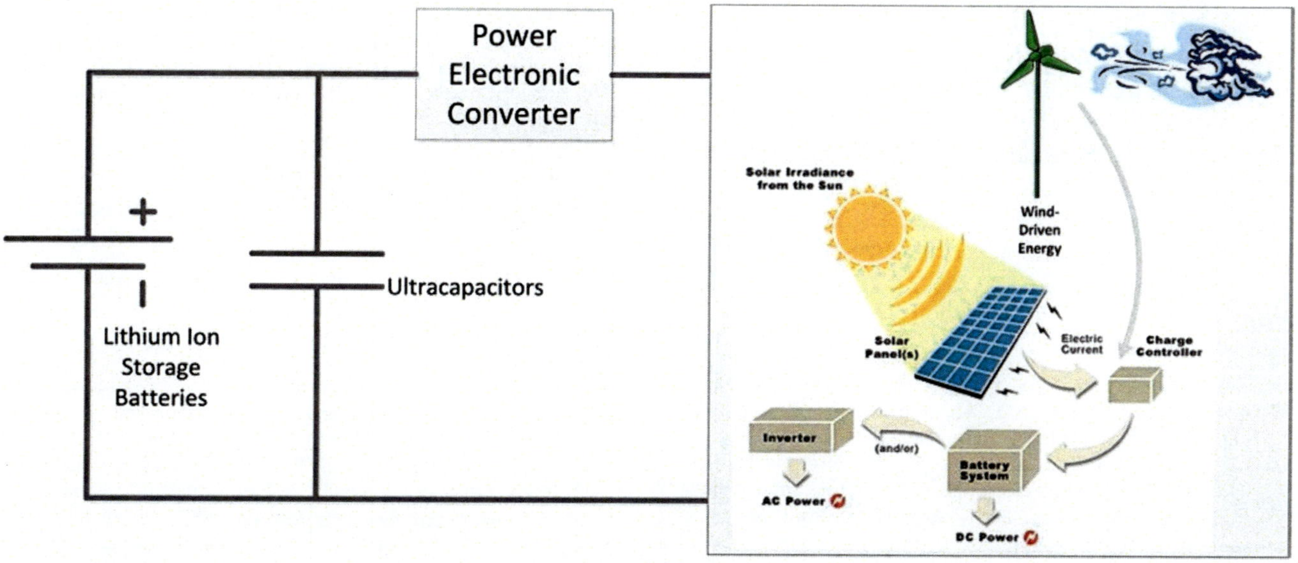

Fig. 13.2 Alternative power management system overview

Fig. 13.3 Example of a whole house power grid

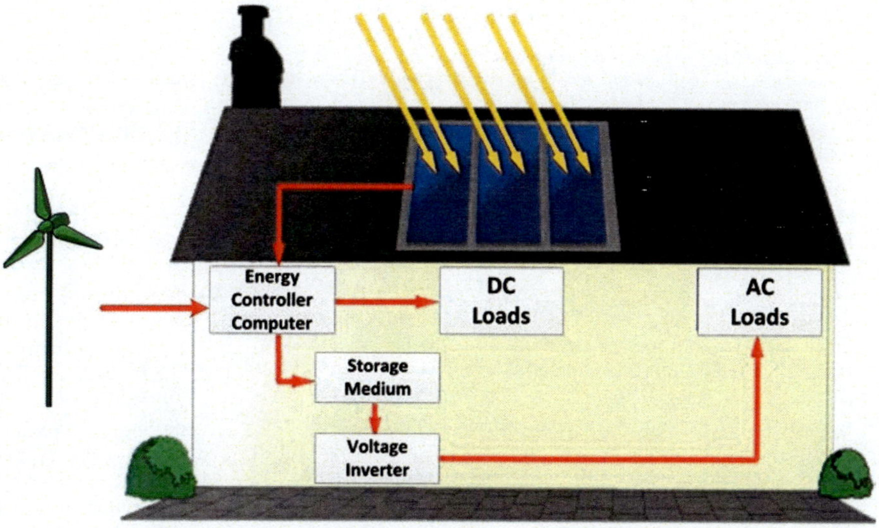

hydroponics capable of growing vegetables year-round. (See Fig. 13.3 for a block diagram of a DC-powered home).

- All lights, appliances, and facilities in each home and the biosphere shall be configured for DC operations, if possible. Things like electric ovens may need DC/AC conversion to accommodate their 220v. requirements.
- Only allowed exceptions are the HVAC system and the Washer/Dryer, although DC is preferred.
- There shall be energy storage capability capable of operating the grid of 500 single family dwellings for 7 days if all alternative forms of energy are not producing.
  - Compliance Criteria: Energy storage devices shall be capable of producing 750,000 watts for the entire 7-day period. Nominal energy storage should be batteries or supercapacitors.
- The DC power grid system shall include controlling h/w and s/w to manage the multi-alternative energy strategies and production.
- All energy produced above the required home and energy storage usage shall be upconverted to 220vac for release to the national power grid.
- The DC power grid shall be capable of operating in a temperature range of −40°F to 125°F. without degradation of power generation.
- The DC power grid must meet OSHA regulations.

13.5.6 Assignments

- Draw a hierarchical Requirement Breakdown Structure diagram for your system requirements.

- Research Requirement Traceability Tools. Write a 5–10-page report describing how you plan on performing requirements traceability and the tool you have decided on and why.
- Write a 5–10-page report describing the test cases for your system.

References

1. Evans, R., Kearney, P., Stark, J., Caire, G. F. J., Gómez-Sanz, J., Pavón, J., Leal, F., Chainho, P. and Massonet P., Project p907, deliverable 3: Methodology for agent-oriented software engineering. Technical Information Final version, European Institute for Research and Strategic Studies in Telecommunications (EURESCOM), September 2001, Available from http://www.eurescom.de/public/projectresults/P900series/907ti1.asp
2. Feather, M., Fickas, S., Finkelstein, A., & van Lamsweerde, A. (1997). Requirements and specification exemplars. *Automated Software Engineering, 4*, 4.
3. Kannenberg, A. and Saiedian, H. "Why software requirements traceability remains a challenge." *CrossTalk: The Journal of Defense Software Engineering*. July/August 2009.
4. Cuddeback, D. "Automated requirements traceability: The study of human analysts." Master's thesis for California Polytechnic Institute. http://digitalcommons.calpoly.edu/theses/317/. Accessed October 20, 2010.
5. Myers, M. D., & Avison, D. (1997). Qualitative research in information systems. *Management Information Systems Quarterly, 21*, 241–242.
6. Jung, H. W., Kim, S. G., & Chung, C. S. (2004). Measuring software product quality: A survey of ISO/IEC 9126. *IEEE Software, 21*(5), 88–92.
7. Zhang, X. L., Ding, C., and Wong R. K., "Non-functional requirement analysis and recommendation for software services," in Proc. 2013 IEEE 20th International Conference on Web Services (ICWS), 2013, pp. 555–562.
8. Bouain, A., El Fazziki, A., and Sadgal, M., "Integration of non-functional requirements in a service-oriented and model-driven approach," in Proc. the 2014 IEEE Eighth International Conference on Research Challenges in Information Science, 2014, pp. 1–8.
9. Rashwan, A., Ormandjieva, O., and Witte, R., "Ontology-based classification of non-functional requirements in software specifications: A new corpus and SVM-based classifier," in Proc. 2013 IEEE 37th Annual Conference on Computer Software and Applications, 2013, pp. 381–386.
10. Hickey, A. M. and Davis, A. M. (2003). "Requirement elicitation and elicitation technique selection: model for two knowledge-intensive software development processes," in Proc. the 36th Annual Hawaii International Conference on System Sciences, p. 10.
11. Silingas, D. and Butleris, R., "UML-intensive framework for modeling software requirements," in Proc. the 14th International Conference on Information and Software Technologies, 2008, pp. 334–342.
12. Santos-Neto, P., Resende, R., and Pádua, C., "Requirements for information systems model-based testing," in Proc. the 2007 ACM symposium on Applied computing, 2007, pp. 1409–1415.
13. Aranda, J., Easterbrook, S., and Wilson, G., "Requirements in the wild: How small companies do it," in Proc. 15th IEEE International Conference on Requirements Engineering, 2007, pp. 39–48.
14. Kwasnik, R. B. H. The role of classification in knowledge representation and discovery. *Library Trends, 48*(1), 22–47.
15. A Guide to the Business Analyst's Body of Knowledge® (BABOK® Guide), Version 2.0, International Institute of Business Analysis, Toronto, Ontario, Canada, ©2005, 2006, 2008, 2009.
16. Weigers, K. E. (2003). *Software requirements: Practical techniques for gathering and managing requirements throughout the product development cycle*. Microsoft Press.
17. Gottesdiener, E. (2005). *The software requirements memory jogger* (p. 293). GOAL/QPC.
18. Gotel, O., Cleland-Huang, J., Hayes, J. H., Zisman, A., Egyed, A., Grünbacher, P., Dekhtyar, A., Antoniol, G., & Maletic, J. (2012). In J. Cleland-Huang, O. Gotel, & A. Zisman (Eds.), *Software and systems traceability* (pp. 3–22). Springer. https://doi.org/10.1007/978-1-4471-2239-5_1. ISBN 978144712238.
19. *Systems and software engineering – Vocabulary. Iso/Iec/IEEE 24765:2010(E). 2010-12-01. pp. 1–418.* doi:https://doi.org/10.1109/IEEESTD.2010.5733835. ISBN 978–0–7381-6205-8.
20. *IEEE Guide for Developing System Requirements Specifications. 1998 Edition IEEE STD 1233. 1998-12-01. pp. 1–36.* doi:https://doi.org/10.1109/IEEESTD.1998.88826. ISBN 978–0–7381-1723-2.
21. Ross, R. G. (2009). *Business rule concepts: Getting to the point of knowledge*. 3rd ed. Business Rule Solutions, 36.

14 Requirements Change Management

Abbreviations

AC	Alternating current
CM	Configuration management
DC	Direct current (or 0 Hz)
EPA	Environmental Protection Agency
F	Fahrenheit
H/W	Hardware
HVAC	Heating, ventilation, and air conditioning
MDRE	Multidisciplinary requirements engineering
OSHA	Occupational Safety and Health Administration
RCM	Requirements change management
RTM	Requirements Traceability Matrix
S/W	Software
V	Volts
Vac	Volts AC

Over the last two decades, it is estimated that 70% of programs involving change management efforts have failed due to poor change management processes [1]. And while there are multiple tools that can be utilized by the MDRE team to manage requirements and requirements change management (RCM), the reason change management fails so often is not the process or the tools, but it is getting a requirements organization to commit to keeping up with change management and owning the processes and procedures throughout the system life cycle. It doesn't matter if the program is following a waterfall development, a spiral development, or an agile/extreme development model, change must be captured and processed, and the ripple effects captured, and the designs and development adjusted accordingly. The purpose of a requirements change is to improve the product either through enhancement because a requirement was missing, or because something changed in the environment of customer needs. The MDRE team must keep a record of the change, what changed, when and why it changed, and who changed it [2]. Both Configuration Management (CM) and Requirements Management are important to capture and process requirements changes in the system. CM and Requirements Management are not the same thing, although many programs and engineers confuse the two. The requirements management discussion we presented in Chap. 10 can help enormously with the change and configuration of requirements, their traceability, and their relationship with other design, development, and test artifacts.

Changes to requirements during software development can be one of the greatest issues for systems/software development programs. Most systems now involve some amount of software (the resurgence of analog systems not withstanding). If the development project involves agile development, changing software requirements is a way of life and expected. It is vitally important that RCM is put in place before the development efforts begin to capture the possible rapid changes to requirements that may happen between agile sprints.

> **Question to ponder**
>
> Consider how requirements changes could affect your multi-alternative energy system design. What EPA requirements changes might have a large effect on your design?

14.1 Change Management Vs. Configuration Management

For all development efforts that are lengthy (more than 6 months) requirements changes are always a possibility. Software and systems design, development, test, deployment, and sustainment are always subject to change. The need (requirements) for the systems and their capabilities grows, shrinks, and adapts, depending on the environment(s) in which they operate [3]. Requirements are the underlying

J. A. Crowder, C. W. Hoff, *Requirements Engineering: Laying a Firm Foundation*,
https://doi.org/10.1007/978-3-030-91077-8_14

basis and driving force behind any development project (or at least should be); identifying, recording, controlling, and tracking the change in requirements is essential. All changes to the system must be tracked and controlled, whether requirements changes or other changes like a change in parts supply vendors, or a change to a design part. Such changes, while not requirement changes, can greatly affect the deployment of a system. All changes affect cost and schedule in different ways, and all should be tracked, either in Configuration Management (CM) or Requirements Change Management (RCM). The difference is that CM tools are not designed to manage requirements and the information at the granularity needed for RCM (e.g., a single attribute change). RCM tools have evolved to provide some level of CM for the data they must track [4], providing the ability for version control and the capability to create requirement baselines. A requirement baseline is a snapshot in time of the current set of approved requirements that have been reviewed, approved, and agreed upon by all parties and have been committed for a system release version [5].

The ability of RCM tools to provide some level of CM can cause a conflict between RCM and CM since both take ownership of some of the same information. Some CM organizations may insist that the requirements be controlled and approved from within the CM organization, despite requirements changes being handled and tracked from within the MDRE team. An example of this conflict is the fact that the RCM tools, using traceability matrices, can track and tie requirements to code. RCM tools support links to the rest of the program, such as requirements to design and requirement to test cases. These relationships require a level of granularity not generally available in CM tools. It is important that the CM group and the MDRE RCM personnel work together to provide an integrated solution for both CM and RCM, based on the size of the project, the number of requirements, and the overall complexity of the design [6].

14.2 RCM and Agile Development

The agile development methodology is studied here in more detail than other development methods due to its intrinsic nature of change between increments. Most aspects of agile development projects are managed incrementally (between development sprints) and that includes requirements. As working software is produced, tested, and demonstrated at the end of each sprint, the results can drive requirements changes for the next set of sprints. Because requirement changes are changing continuously as development is ongoing, change is looked upon as necessary and is encouraged since the changes are driven by a desire for continual improvement in the developed software [7].

The risk to agile development comes from not tracking and managing the requirements as they change and evolve over consecutive sprints and between sprint teams. The MDRE RCM team must ensure all changes are coordinated across sprints and sprint teams, so all development efforts are developing to the same set of requirements. In addition, requirements change applied, but not yet chosen for development in the current set of sprints may introduce the need for design changes that have not been folded into the current development sprints. This could lead to major and costly rework. The requirements change analysis effort is crucial in agile development efforts. It is also possible that requirements are documented in forms other than textual in a sprint development. Care must be taken to choose an RCM tool that can adequately manage non-textual requirements (e.g., use case diagrams, activity diagrams, state diagrams, process flows). The RCM tools must be able to manage graphically represented requirements as well as textual requirements [8].

14.3 Requirements and Change History

One of the most fundamental forms of RCM is recording and tracking the change history of the requirements. It is not enough to track the changes to requirements; it is also necessary to trace the history and the rationale for changing requirements. This type of tracking should be done in the background and happen automatically as the requirements are changed and managed across the life cycle of the system. Understanding the changes to requirements and why they were changed is important to the sustainment and maintenance personnel to understand how the system requirements and ultimately the delivered system changed over the development period. It is important to understand how requirements changes affected not only the design and development of a system, but also how it may affect maintenance and sustainment of the system. Small changes to physical characteristics of the system can have a major effect on the ability to reach parts and may change the size, power, cooling, and air conditioning requirements. Such factors are often overlooked in the change analysis associated with requirements changes management [9].

14.4 The Perils of Deleted Requirements

The problem of deleting information of any kind from a system, requirement or otherwise is that the information doesn't exist anymore. And while many will argue that the entire point of deleting information is to get rid of it because it is no longer needed, in the case of requirements that isn't necessarily true. Deleting a requirement indicates that the capability, attribute, or feature is no longer needed. Once a requirement becomes part of the baseline, it affects the design, the way the system is tested and maintained. Deletion of a requirement will also impact the requirements tracing—so that must be taken into account also. Careful thought must go into a decision to eliminate a requirement and to completely analyze the ripple effects through the system if this requirement no longer exists. Once the system design is changed due to a deleted requirement. If the customer reviews the before and after design, it may not be apparent why the design changed. Part of RCM is not only managing the changes to requirements, but it includes managing the deletion of requirements. This may take the form of not deleting a requirement from the requirements, but simply leaving it in the requirements set and marking it deleted with the rationale behind the deletion. Now the design development reflects the deletion, but the information is still available. If the RCM tool that is chosen does not provide this capability, the MDRE team must decide on a strategy for tracking deletions and rationale [10].

This type of tracing of deleted information applies to other artifacts of the system design, like linked relationships. Links between requirements and other artifacts pose difficult decisions for the MDRE team. When a requirement is deleted, should all links to the requirement be deleted also? The answer may seem obvious, but it isn't. There may be reasons to leave a ghost link in place to understand why certain other artifacts exist. There is no hard and fast rule for this, the discussion here is merely to point out that these decisions must be made consciously by the MDRE team and in concert with the CM team. Some tools may send an alert when a link is broken and some RCM tools will not allow a requirement to be removed until all links are resolved [11].

14.5 Case Study: Satellite Launch Systems RCM Gone Wrong: Early 1990's Satellite Launch

The launch system discussed here had the ability to launch multiple satellites from a single launch vehicle. However, one of the launches had only one spacecraft on board. To provide a launch platform for multiple satellites, the launch vehicle was wired with a harness that could accommodate multiple spacecraft, with the bottom spacecraft attached to the lower harness connector and continued to the top spacecraft connected to the top part of the wiring harness, where the top spacecraft would be the first to separate from the vehicle [13].

This specific launch had only one spacecraft to be carried to space with the launch vehicle. Given there was only one spacecraft, any of the harness connectors could be utilized if the hardware and flight software agreed on which harness connector was to be connected. Initially in the design, the top wiring harness connector was to be used to connect the spacecraft to the launch vehicle and the flight software would issue the command to separate from the top connector of the wiring harness.

Somewhere in the implementation process, the hardware group decided to utilize the bottom connector of the wiring harness, because it was a better fit for the configuration of the satellite to be launched (i.e., it reached easier). Unfortunately, the Configuration Management change to the wiring harness connectivity was never put under Configuration Management control; therefore, the Change Management notice was never generated to tell the software group that the different wiring harness connector was to be used for the launch. Adding to this problem was the fact that to save money, the Launch Vehicle-to-Satellite interface was not tested, since it was assumed to be correct. The result was that the launch vehicle's second stage failed to separate from the spacecraft (satellite).

The only way to separate the satellite from the launch vehicle was to release the satellite from its apogee motor that would have taken the satellite up to its geosynchronous orbit. Instead, the smaller perigee motor was used to put the satellite into a useless low-earth orbit. Later that same year, a new perigee motor was attached to the satellite (Space Shuttle crew) and it was able to then take itself to its geosynchronous orbit. All in all, this was an approximately $300,000,000.00 mistake, all for the lack of a $0.25 change notice. MDRE engineers ***MUST*** keep track of details, for small mistakes can have major effects.

14.6 Final Thoughts on Change Management

While it may not seem difficult to keep track of information related to requirement changes, it is more difficult to manage the process of change. There are many factors that govern change management. Some system designs have tight regulatory controls (government regulations) while others have tight contractual limitations on changes to software requirements if the system has life-threatening potential (e.g., medical equipment). An effective RCM tool should support changes to existing requirements as well as proposed creation and/or deletion of requirements. The RCM tool should also allow identification of the source of the requirement change or deletion (e.g., customer, design engineer, MDRE engineer). As explained above, the RCM tool should also allow explanation for the change/creation/deletion of any requirement. In short, a good RCM tool should provide a thorough change process for all information related to requirements, and their related information (e.g., attributes and links) [12].

Question to ponder

Think about the types of controls that would be required for changing requirements on medical equipment like pacemakers?

The bottom line is to never take the importance of RCM lightly. Change management, versioning, baselining, traceability, and tracking of requirements can easily make the difference between success and failure of system development and deployment.

14.6.1 Design Problem

MDRE is a life-long journey, and anyone who tells you they have nothing left to learn does not truly understand the discipline or has never been involved in the development of a complex system. With the recent interest in alternative energies and talk of the "Green New Deal," along with the recent issues with ice storms in Texas taking out the power grid, the assignments for this textbook will follow the theme of effectively dealing with both local alternative energy power grids and the national infrastructure required to facilitate large-scale alternative energy production, sustainment, and management.

14.6.2 Design Problem Description: Alternative Energy Power Grid

The purpose of this project is to, over the course of the semester, develop a feasibility study, requirements analysis, requirements elicitation, derivation, and allocation, for a system that will facilitate a "local" alternative energy power grid capable of powering a neighborhood of 500 homes and a large biosphere capable of supporting growing year-round fruits and vegetables for the 500-home neighborhood. The student will have to deal with issues of choosing multiple alternative energies that can provide power throughout all seasonal difficulties, from intense summer heat to freezing winter temperatures and possible ice storms. Control, viability, sustainability, and maintainability will be major factors to understand the entire possibly 30-year life cycle of such a system. How to effectively utilize, store, and manage power distribution, along with when to upconvert "extra" energy to the national power grid, will be issues that drive many of the assignments. Simply relying on the current technologies and energy infrastructures will not move the country and world forward. Hopefully, this will provide insight to the students using this textbook to understand some of the issues that face everyone looking out over the next many decades.

14.6.3 DC Power Grid Requirements Discussion

In the initial days of power generation and distribution, DC power was far less practical than the AC distribution system that soon supplanted it. However, even after Tesla's AC system became the predominant system (and rightly so), Edison's DC power did not become obsolete for some time. Eventually power plants eliminated their delivery of DC power, but the overall concept of a DC power grid is not dead. Even our computer systems (large server systems) have uninterruptible power supplies which contain batteries that are charged off the grid and used to upconvert to AC to power devices when the AC power grid goes off-line. So, the notion of a DC power grid is not dead, but generally used as a backup system. The point is, the notion of a DC power grid has remained and now that renewable energies are viable, we should re-examine their use for widespread, local, power distribution.

One of the advantages of a locally generated DC station is energy savings as well as allowing a facility to run more easily off various DC sources, such as Solar Arrays. And for power-hungry installations, the notion of generating the power locally is growing in popularity. Part of the reason is that waste heat from the power generation generators (solar, geothermal, etc.) can be used to warm nearby buildings. By combining the production of heat and power, local power grids can squeeze much more useful energy out of the renewable energy they use, so this approach will certainly become more widespread as time goes on. The purpose of this design problem is to look ahead and provide an architecture and specification for a local DC power generation station that can serve a local community.

Currently many systems exist to control and handle alternative energies. There are systems to control and manage solar power, wind power, geothermal power, and others, but very few are designed to control, manage, store, and distribute energy from a power grid consisting of multiple alternative energy sources. As we saw the issues recently caused by the intense ice storms in Texas, one alternative energy source is not the answer to the growing energy needs/wants in our society. Each alternative energy source has limitations and issues associated with its use. By harnessing multiple alternative energies, and managing the storage of the energies produced, communities can be worry free from power generation. To facilitate utilizing multiple alternative energy sources, a control center must be included in the design that can manage the energy creation, distribution, storage, and return unused energy to the power grid.

14.6.4 High-Level Technical Description of the DC Power Grid System

The purpose of the multiple alternative energies DC power grid is to:

- Supply DC power to a customer community of 500 homes and a biosphere for growing food.
- Provide a storage device to store energy in case of a system failure.
- Determine the amount of surplus energy that is above and beyond what is required to power the grid and charge the backup system, and then deliver that surplus energy to the national power grid systems.

Throughout the semester you will be asked to analyze, design, and architect different components of the overall DC power grid system. This will include research into alternative designs, like lithium nanowire storage systems for backup power and/or supercapacitors (also called ultracapacitors). Fig. 14.1 illustrates these technologies connected to the DC Power Grid Regulator Control Center.

14.6.5 High-Level Technical Requirements for the DC Grid System

- Design shall utilize at least two alternative energies, with a maximum of 4.
- Total power generation shall be 3.0 megawatts minimum.
- Each of the alternative energy sources shall be capable of generating at least 60% of minimum wattage.
- DC power grid shall generate DC power, with Grid voltage being 48, 24, or 12 volts.
- DC power grid shall be capable of servicing at least 500 single family homes and one 4-acre biosphere with hydroponics capable of growing vegetables year-round. (See Fig. 14.2 for a block diagram of a DC-powered home).
- All lights, appliances, and facilities in each home and the biosphere shall be configured for DC operations, if possible. Things like electric ovens may need DC/AC conversion to accommodate their 220v. requirements.
- Only allowed exceptions are the HVAC system and the Washer/Dryer, although DC is preferred.
- There shall be energy storage capability capable of operating the grid of 500 single family dwellings for 7 days if all alternative forms of energy are not producing.
  - Compliance Criteria: Energy storage devices shall be capable of producing 750,000 watts for the entire 7-day period. Nominal energy storage should be batteries or supercapacitors.
- The DC power grid system shall include controlling h/w and s/w to manage the multi-alternative energy strategies and production.
- All energy produced above the required home and energy storage usage shall be upconverted to 220vac for release to the national power grid.
- The DC power grid shall be capable of operating in a temperature range of −40°F. to 125°F. without degradation of power generation.
- The DC power grid must meet OSHA regulations.

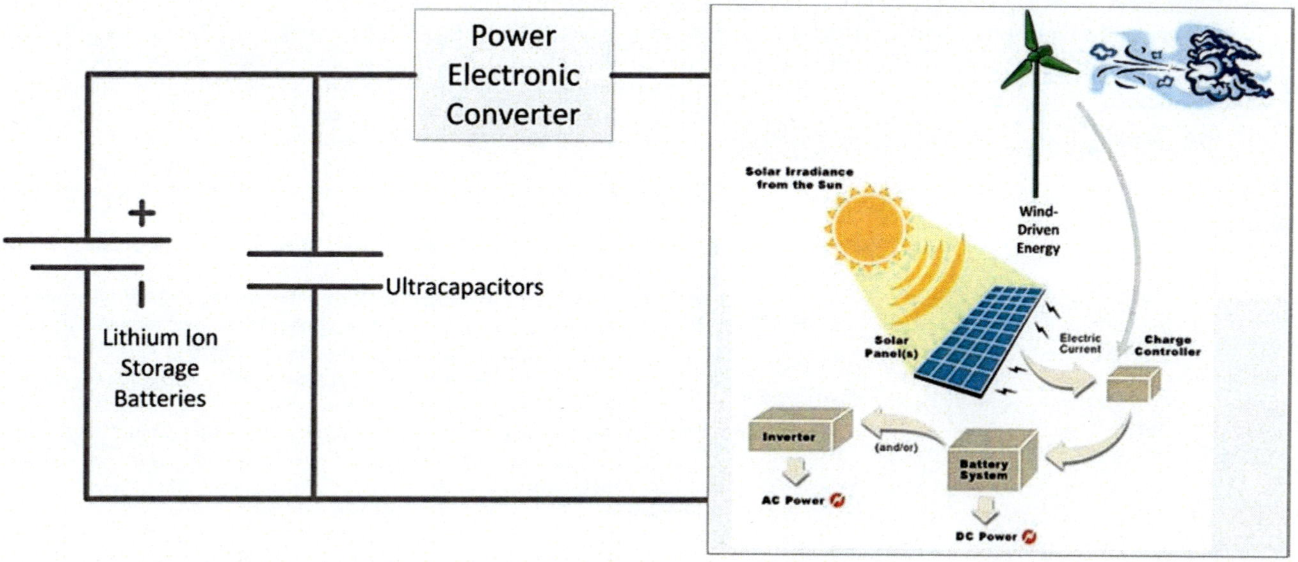

Fig. 14.1 Alternative power management system overview

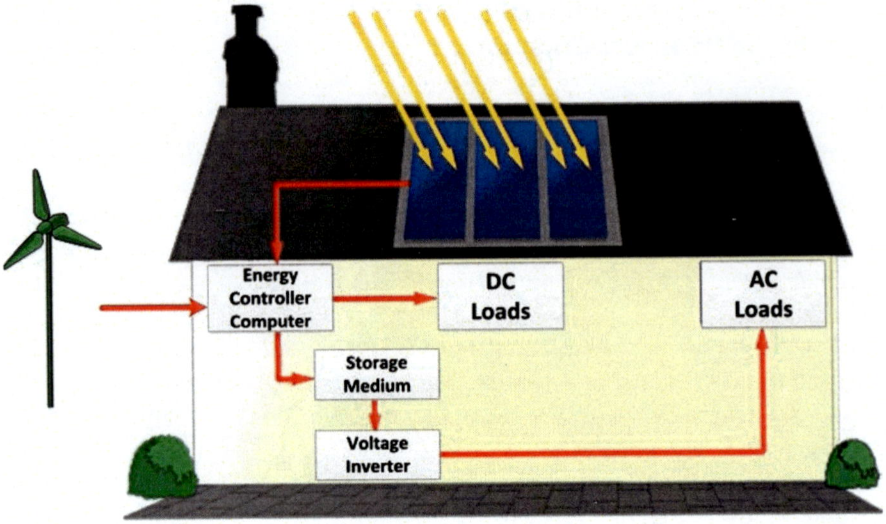

Fig. 14.2 Example of a whole house power grid

14.6.6 Assignments

- Investigate Requirement Change Management tools. Decide which tool would be appropriate for your project. Write a 5–10-page report describing how you will manage requirements for your multi-alternative energy system development.
- Investigate the EPA requirements that are applicable for your system design. Review changes to the EPA requirements that have happened over the last 25 years pertaining to alternative energies. Answer the following questions:
  - List three requirements relative to your system design that would have been affected by changing EPA constraints and regulations from 25 years ago till today.
  - Describe how sensitive your design is to government regulations and why.
  - List any local state and county regulations (based on your state and county) that would affect the design, development, and deployment of your multi-alternative energy system.

References

1. Hull, E., Jackson, K., & Dick, J. (2005). *Requirements engineering* (Vol. 9–13, 2nd ed., pp. 131–151). Springer. ISBN 978-1-85233-879-4.
2. *IEEE Guide for Information Technology – System Definition – Concept of Operations (ConOps) Document. IEEE STD 1362–1998. 1998-12-01. Pp. 1–24.* doi:https://doi.org/10.1109/IEEESTD.1998.89424. ISBN 978-0-7381-1407-1.
3. Gotel, O. C. Z.; Finkelstein, C. W. *(1994). An analysis of the requirements traceability problem. Proceedings of IEEE International Conference on Requirements Engineering. Pp. 94–101.* CiteSeerX 10.1.1.201.7137. doi:https://doi.org/10.1109/icre.1994.292398. ISBN 978-0-8186-5480-0. S2CID 5870868.
4. Gotel, O., Cleland-Huang, J., Hayes, J. H., Zisman, A., Egyed, A., Grünbacher, P., Dekhtyar, A., Antoniol, G., & Maletic, J. (2012). In J. Cleland-Huang, O. Gotel, & A. Zisman (Eds.), *Software and systems traceability* (pp. 3–22). Springer. https://doi.org/10.1007/978-1-4471-2239-5_1. ISBN 9781447122388.
5. Pinheiro, F. A. C., & Goguen, J. A. (1996). An object-oriented tool for tracing requirements. *IEEE Software, 13*(2), 52–64.
6. Mäder, P., Jones, P. L., Zhang, Y., & Cleland-Huang, J. (2013). Strategic traceability for safety-critical projects. *IEEE Software, 30*(3), 58–66. https://doi.org/10.1109/MS.2013.60. ISSN 0740-7459. S2CID 16905456.
7. Li, Y., Li, J., Yang, Y., & Li, M. (2008). *Requirement-centric traceability for change impact analysis:A case study* (pp. 100–111). Springer. ISBN 978-3-540-79587-2.
8. Wiegers, K. (2013). "Requirements traceability: Links in the requirements chain, Part 1". Jama. Retrieved 2016-12-14.
9. Wiegers, K., & Beatty, J. (2013). *Software requirements*. Microsoft Press.
10. Bouillon, E., Mäder, P., & Philippow, I. (2013). In J. Doerr & A. L. Opdahl (Eds.), *Requirements engineering: Foundation for software quality. Lecture notes in computer science* (pp. 158–173). Springer. https://doi.org/10.1007/978-3-642-37422-7_12. CiteSeerX 10.1.1.659.3972. ISBN 9783642374210.
11. Mäder, P., & Egyed, A. (2015). Do developers benefit from requirements traceability when evolving and maintaining a software system? *Empirical Software Engineering, 20*(2), 413–441. https://doi.org/10.1007/s10664-014-9314-z. ISSN 1382-3256. S2CID 2514618.
12. Rempel, P., & Mäder, P. (2016). Preventing defects: The impact of requirements traceability completeness on software quality. *IEEE Transactions on Software Engineering, 99*, 777–797. https://doi.org/10.1109/TSE.2016.2622264. ISSN 0098-5589. S2CID 1959772.
13. Meyers, R. (2009), Encyclopedia of Complexity and Systems Science. Springer Publishing, New York, NY, ISBN 978-0-387-75888-6.

15 Requirements Agility

Abbreviations

AC	Alternating current
DC	Direct current
EAC	Estimate at Completion
ETC	Estimate to Complete
F	Fahrenheit
HVAC	Heating, ventilation, and air conditioning
IT	Information technology
MDRE	Multidisciplinary requirements engineering
OSHA	Occupational Safety and Health Administration
RCM	Requirements change management
RTM	Requirements Traceability Matrix
V	Volts
Vac	Volts AC
Vdc	Volts DC
VTC	Video teleconferencing

Success in the modern, agile development era requires MDRE personnel who truly understand what agile development means, how agile teams operate, collaborate, cooperate, and function in various situations. In addition, they must understand the agile development cycle vs. the traditional development cycle [1]. Fig. 15.1 illustrates the classical program/project development cycle.

This program/project development cycle has been used for many decades. Requirements are decomposed, allocated and/or derived, program plans are made, budgets and schedules established, and you march forward. For this development cycle, change is bad, as discussed in Chap. 14, for it causes rework of the entire development cycle, driving up cost and schedule [2]. And the farther into the development cycle you are when you discover problems, or changes happen (new or changing requirements), the costlier they are to "fix" both from a cost and schedule perspective. Fig. 15.2 illustrates the increasing costs associated with changes, depending on when in the development cycle the changes or problems are encountered [3].

As mentioned, when problems arise, or requirements changes happen, it drives a re-plan of the program schedule, a re-estimation of costs to complete the program/project, either as an Estimate at Completion[1] (EAC), an Estimate to Complete[2] (ETC) or both. A new program plan must be put in place, new budgets and schedule, and the development cycle restarts, as illustrated in Fig. 15.3.

Changes occurring (particularly requirements changes) 1/3 or more through the development cycle create major rework for the entire program, as the requirement changes must be flowed throughout all parts of the system and those affected systems, subsystems, configuration items, components, etc. must be reworked, re-planned, rescheduled, and a new baseline budget and schedule created for execution. This can extend by 50% or more the overall timeline and budget for the program/project [4].

15.1 The Agile Development Cycle

In contrast to classical development cycles, the agile development cycle is broken up into small Sprints that are typically 2–6 weeks in length, although there is much debate over the proper length of a Sprint [5]. Fig. 15.4 illustrates a typical agile program development cycle. In Fig. 15.4 we see the typical Sprint rhythm of initiating, planning, and execution of agile Sprints during the agile development process.

Again, the length of the Sprints is dependent on the program/project, the complexity of the software, total duration,

[1]Estimate at Completion is the current expectation of total cost at the end of a project. The EAC represents the final project cost given the costs incurred to date and the expected costs to complete the project.

[2]Estimate To Complete is an **estimation of funds required to complete the remaining work in a project.**

J. A. Crowder, C. W. Hoff, *Requirements Engineering: Laying a Firm Foundation*,
https://doi.org/10.1007/978-3-030-91077-8_15

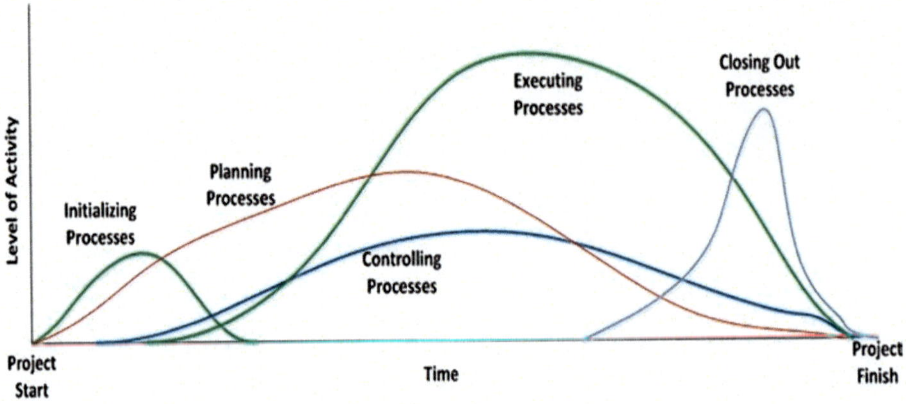

Fig. 15.1 The classical project development cycle

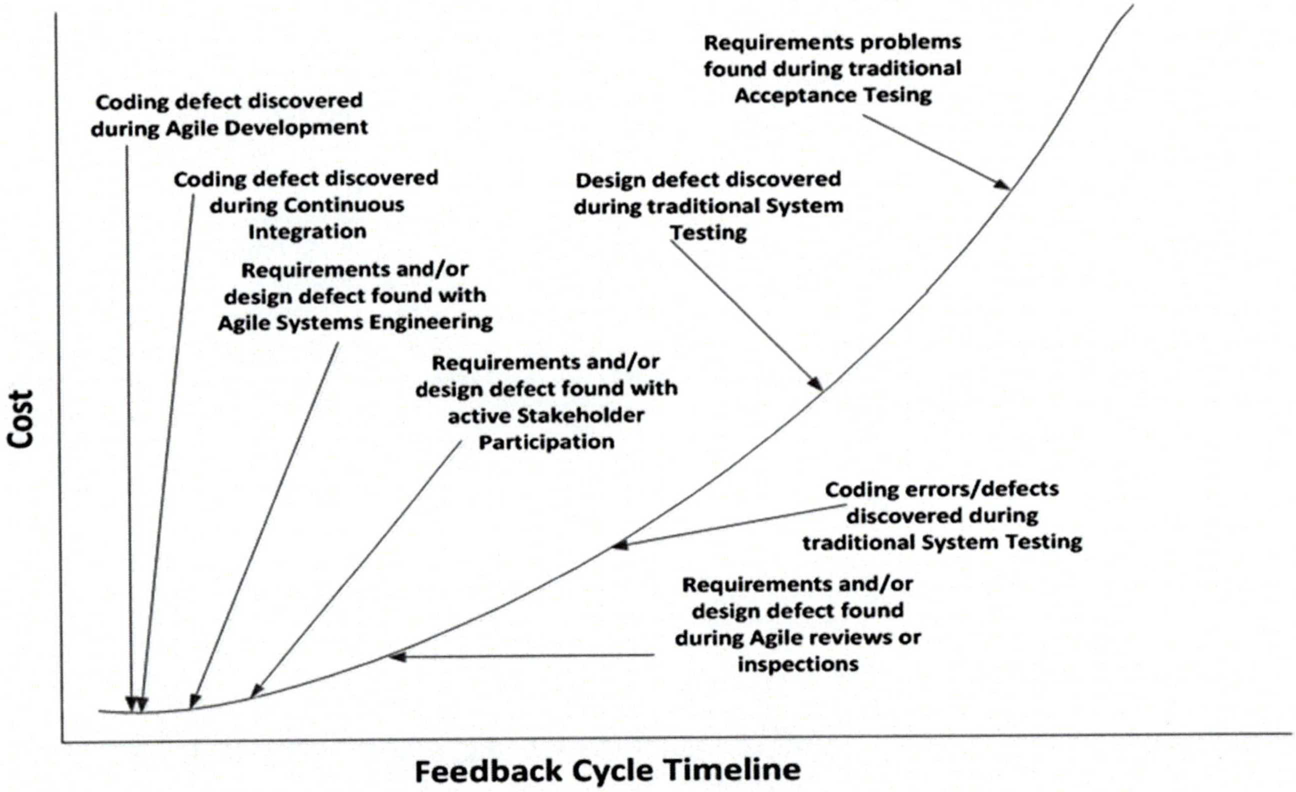

Fig. 15.2 Costs associated with changes along the development timeline

requirements to be satisfied during the current sprint(s), and team membership. The duration of the Sprints is not relevant to this discussion. One of the big advantages to agile development is due to the adaptive nature of the Sprints. If a requirements problem is discovered it can be rolled into the next and subsequent Sprint planning sessions and rolled into the schedules and does not require a major re-plan of the entire project, the way it does in classical development. Fig. 15.5 demonstrates this process. Normally, because there is working software after each Sprint, requirement problems are discovered earlier in the development process than in conventional or classical development methodologies. Customer and management feedback after each Sprint provide the opportunities to re-vector the development efforts before major cost and schedule have been expended.

We will not try to argue here over the length of Sprints, except to mention that the length of Sprints is tied to the overall length and complexity of the program/project development effort, based on the requirements that drive the sprint development effort. Typically Sprints build in some sort of demo or evaluation to demonstrate the goals of the Sprint have been achieved. If the Sprint is too short, the work/evaluation ratios will be small; therefore, the accomplishment/cost will be low. The other problem with short Sprints is that it will be difficult to achieve goals for complicated capabilities. Typically, 30–45 days is a reasonable length for a Sprint,

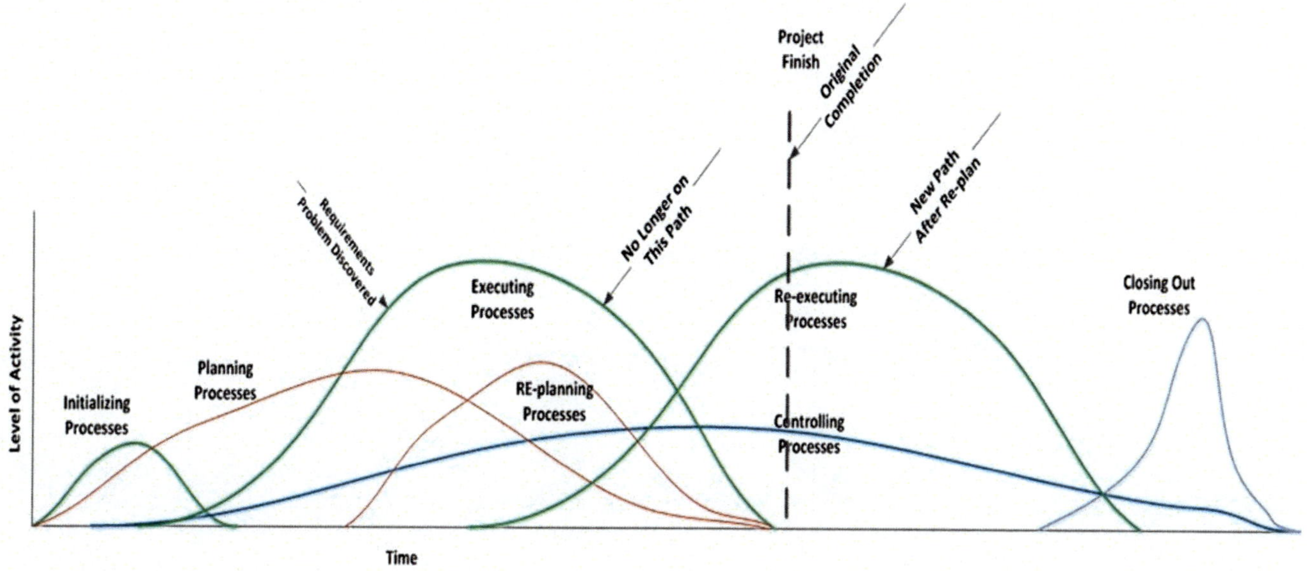

Fig. 15.3 Re-planning of the classical development cycle

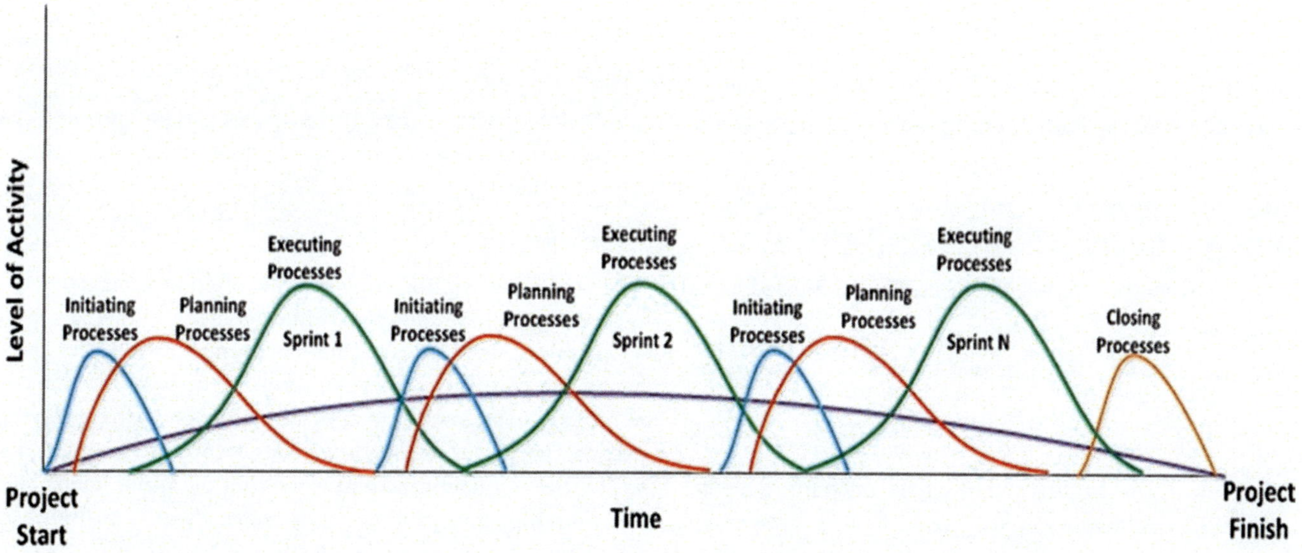

Fig. 15.4 Typical agile program development cycle

again, depending on the requirement complexities and overall goals for the software. The goal is to find the right rhythm for the project. Sprints should be long enough to produce working software that can be properly evaluated against requirements to determine the program/project is progressing along the correct paths [6]. But if the Sprints are too long, too many capabilities will be included in the Sprint, making end-of-Sprint demos and requirements evaluations lengthy, and any issues discovered during the Sprint make it more difficult to recover. Also, if Sprints are too long, the retrospectives evaluation at the end of the Sprint is more difficult as it is more difficult to remember what happened at the beginning of the Sprint. Also, the clients/customers will be flooded with features and requirements to evaluate and provide feedback; in other words, if the Sprints are too long, they are not "agile" enough. Also, there should not be time between Sprints. The demonstration/evaluation, retrospective, backlog evaluations, etc. should be done and then the planning for the next Sprint should commence immediately. Again, Sprint length is tied into the quality and abilities of the teams, driven by the MDRE analysis of development against requirements, the technology and architecture of the systems, the complexity of the software to be developed. Sprint length should be kept as constant as possible, since changing the Sprint length disrupts the rhythm of the project and the teams must constantly re-evaluate what can get

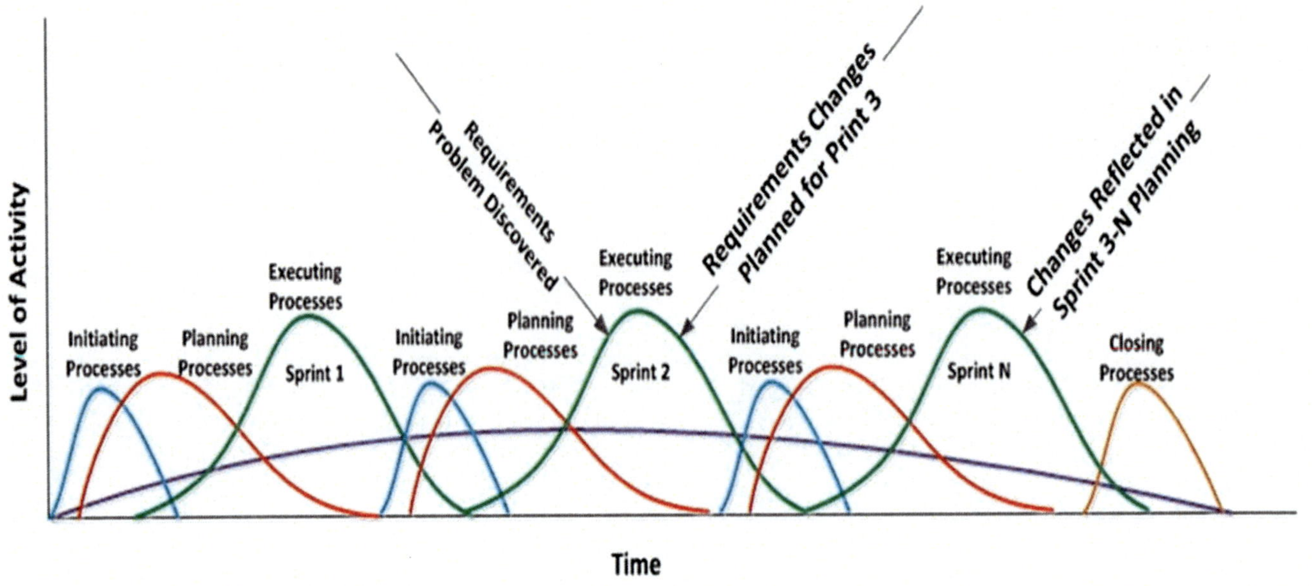

Fig. 15.5 How requirements changes are handled in the agile process

accomplished during any given Sprint. This must involve the MDRE team to assess which requirements capabilities are being developed and tested and how those feed into subsequent requirement capabilities during subsequent sprints [7].

> **Question to ponder**
>
> What type of development process do you envision for your multi-alternative energy system? Waterfall, spiral, or agile?

This new Agile MDRE paradigm is uncomfortable to many MDRE engineers, who may feel like they have given that central control and authority is given to the team, not to the MDRE personnel. However, the Agile MDRE engineer is as busy as or busier in this environment than in a classical development effort [8]. The role of the MDRE engineer across the agile teams is a critical role to the overall success of the development program/project. Many of the responsibilities of the Agile MDRE engineer are not all that different from those in classical management, except that in the Agile Development world, the Agile MDRE must focus on the managing sprint requirements to enable the sprint team's ability to deliver working software, rather than focus on traditional program/project measurement metrics [9]. The Agile MDRE engineer must lead, inspire, and provide empowerment in terms of seamless requirements management to the agile development team. The Agile MDRE engineer must be involved in and facilitate supporting functions that allow the agile development team to be effective.

> **Question to ponder**
>
> What requirements management tool are you using? Think about how it will work with an agile development methodology.

For the Agile MDRE engineer, this often involves adjusting one's personal requirement management style and may require the Agile MDRE engineer to even work to affect and change the company culture in terms of how requirements are managed and tracked to allow the agile teams to be effective and successful. It means working to acquire and utilize the abilities to understand and work with the sprint teams on what requirements produce agile value-based capabilities [10]. Understanding and being able to assess developers' skills, tied to requirements capabilities, in designing, coding, and testing is essential for the MDRE team. The Agile MDRE engineer must ensure that they, as well as the development team, understand the agile style, cultural dynamics, and team dynamics [11]. It is essential that the team understand and be comfortable with the rhythm of the agile development style, Sprint after Sprint, each with its own cycle of planning, coding, testing, lessons learned, and then immediately moving on the next Sprint. Fig. 15.6 illustrates this recurring agile Sprint development rhythm.

Figure 15.6 illustrates the recursive nature of the agile development process. The program MDRE process must mirror this development process to keep requirements always up-to-date and relevant to each sprint. Once the initial project vision and initial scope has been established (with stakeholder buy-in), the teams are selected, the initial number and

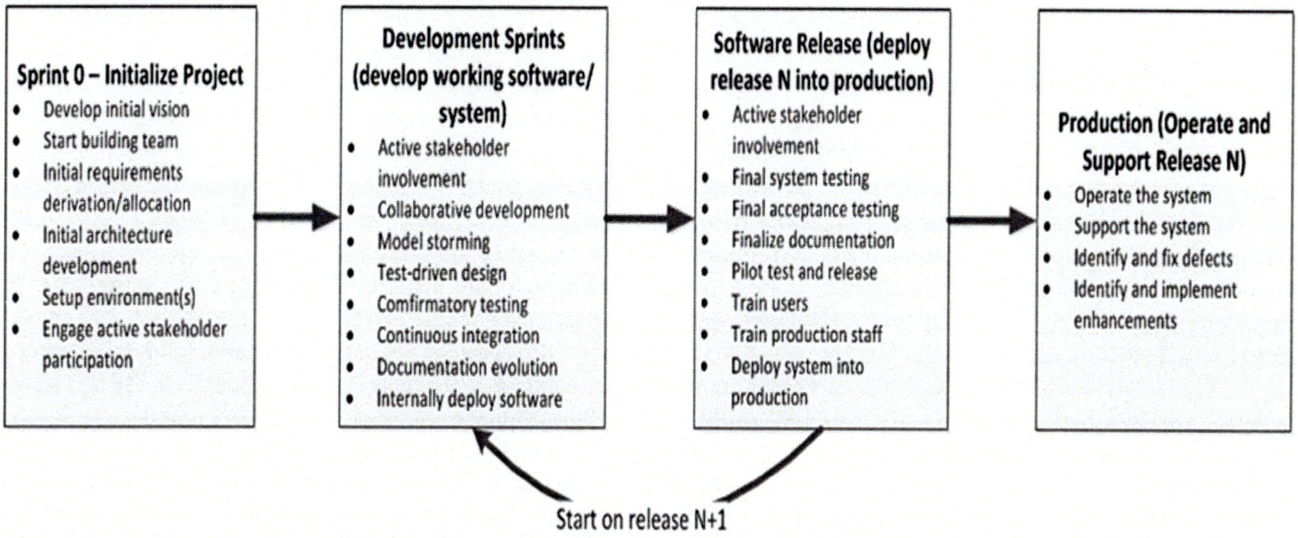

Fig. 15.6 The agile development process rhythm

length of Sprints are determined, IT environments established, and the initial Sprint is initiated. From Sprint 0, the agile development program/project continues through the center two blocks shown in Fig. 15.6 until the program/project is completed and it is time for final product release. Blocks 2 and 3 shown in Fig. 15.6 are the heart and soul of agile development. Block 2 represents the Sprint software development, which includes working with the stakeholders to establish the priorities and capabilities (requirements) needed for the current Sprint. Brain/Model storming is used to determine the test-driven software designs required for the current Sprint capabilities, keeping in mind that the current Sprint software must integrate with the previous volume of software [12]. This is accomplished with a continuous integration process that ensures working software at the end of the Sprint. Block 3 includes the final Sprint system/acceptance testing, releasing software through Sprint N into production and demonstrating the software to ensure functionalities are acceptable and requirements can be validated. Stakeholder involvement at each step ensures the final software will be accepted at the end of the project, ensuring all requirements have been met by the end of the final sprint. After the final Sprint, the final software is released and final documentation is delivered, along with any training that is required [13].

Question to ponder

What kind of software do you envision for your system? How might you divide it up into sprints?

This agile development process rhythm is important to the overall success of the program/project. Without the MDRE team's understanding, facilitating, and protecting this rhythm, it will be difficult to achieve overall success from an agile development program/project. Keeping the teams functioning smoothly is the best thing the Agile MDRE engineer can do to promote a successful project. Case Study #2 illustrates this.

15.2 Case Study: The Disruptive MDRE Engineer

Project Length:	**24 months**
Number of Sprints:	**9 Sprints**
Number of Teams:	**5 Teams**
Average Sprint Duration:	**8 weeks**

Description: This project was struggling early on as personnel from each of the agile development teams were hesitant to take on the role of MDRE engineer for the Sprints, not wanting to be tagged as responsible if the team did not meet their capability development goals for any given Sprint. To solve this, the MDRE team brought in a Senior MDRE engineer from one of the company's "showcase" programs/projects. The teams assumed they would be coming in to train them on how to be effective as MDRE engineers, and to teach management how to allow the MDRE engineer to handle each development team across the Sprints, and to facilitate the teams choosing the requirements to be included in each Sprint. Instead, the Senior MDRE engineer took on the role of MDRE control for all five teams. This was inherently a bad idea, as having one person act as the MDRE engineer for five agile development teams violates many of the roles that the MDRE engineer needs to take on during each of the development Sprints. The first problem with one person acting as MDRE lead for all the teams across all the Sprints is that the person taking on the role of MDRE engineer should change at each Sprint, depending on the complexity points and requirements required to be satisfied for that Sprint and the expertise of the team members, based on the complexity points required for that Sprint. The role of the MDRE engineer includes, but is not restricted to:

1. Assists the product owner with the backlog, both in its creation and with its maintenance since this drives which requirement capabilities are developed when.
2. Works with the agile development team to define completion of the Sprint. What are the criteria for "done?" (i.e., how are the requirements sold off?).
3. Manage requirements to remove stumbling blocks or impediments to progress, whether they are internal to the agile development team, or outside the team (e.g., support needed from another agile development team).
4. Helps the agile development team follow the MDRE process. As such, the MDRE engineer should have an excellent understanding of the agile development requirements management structure and must be willing to teach or mentor junior developers in the process.
5. In the end, the MDRE engineer should spend the Sprint making the rest of the development team as productive as possible.

Question to ponder

What skills do you believe a development team writing the software for your system must have?

Having described the role(s) of the MDRE engineer, having one person act as the MDRE engineer across multiple agile development teams and across all Sprints within the development process, is inherently inefficient, violates the precepts of agile development teams, and produces major problems that will ultimately hurt efficiencies. The issues that developed with the overall project are:

1. Each individual agile development team did nt have access to the MDRE engineer throughout the development cycle, since the MDRE engineer's time and attention were spread over five development teams. This delayed problem resolution, road-block removal, etc., and led to a sloppy requirements management process across all five sprint teams.
2. The MDRE engineer was not a domain expert on any part of the system being developed, and therefore was unavailable as part of the development team throughout all the Sprint development cycles.
3. Much time was wasted each morning as a single MDRE engineer had to cycle through five Scrum meetings, trying to keep track of which team expressed which requirement issues, and then having to prioritize whose requirement issues got worked first. This resulted in much wasted time across the Sprint development teams and fostered a sense of resentment for the MDRE engineer and the entire MDRE process.
4. Since the MDRE engineer was not a domain expert in any part of the system being developed, he could not act as mentor for junior engineers.
5. Having the same MDRE engineer across five agile development teams kills the notion of self-organization and moves the teams Locus of Control outside to the outside MDRE team and not within the team. This causes a general lack of ownership by the development teams [14].
6. Since there was only one MDRE engineer, they tended to fix problems by moving requirements from one development team to another, destroying the cohesiveness of the sprint development efforts. The teams never knew which requirements were going to be developed during which sprint, and frequently the requirements were changed out in the middle of the Sprint.

The result was that by the third Sprint, efficiency had dropped to below 70% and the entire program was behind at least one full Sprint. Attempts to make the developers work 15–20 h of overtime was unsuccessful, since the environment was not conducive to self-organization [15]. Even those that agreed to work the overtime were not as productive as they should have been, since it took too long to overcome any obstacles the teams encountered, as the MDRE engineer was beholding to all five agile development teams.

In the end, program management had to step in, remove the MDRE master engineer, and regroup all the development teams, putting the project behind three full Sprints. The program/project recovered, but not before increasing cost and schedule by over 40%. The result was that management declared Agile Development "no better that when we did it the old way. Might as well go back to a development method we all understand." Agile development was abandoned, and many people left the organization.

Lessons Learned: "If you go out of your way to violate the precepts of agile development, don't expect great results."

15.3 Establishing Agile Requirements Management Goals

For effective agile requirements management, it is crucial early on to establish requirement goals and objectives that establish the atmosphere for each sprint development team.

Understanding how much independence each developer is allowed, how much interdependence each team member and each team should expect, and creating an environment that supports the agile development style will provide your teams with the best chance for success. A careful analysis of requirement independence and interdependence should be done to establish the sprint timeline of which requirements are to be addressed in which sprint and by which sprint team. If it is possible, the requirements being addressed across the sprint teams for a given sprint should be as independent as possible so as not to hold up development efforts across sprint teams when one sprint is having difficulty meeting a requirement capability. In addition, requirements with interdependencies should be worked in successive sprints by the same sprint team, if possible, so that either requirements change or not completing a given sprint software does not affect multiple sprint teams. It may not be possible in all cases, but the MDRE team should endeavor to work with the sprint teams to accommodate this agile requirement process to the maximum extent possible.

Below is a list of agile MDRE characteristics and constraints that must be defined for the teams to establish a requirements development "rhythm" throughout the agile development cycle for the program/project [16]:

1. **Define and Create Independence**: Independence is something many developers crave. For agile development to be successful, there must be a large degree of independence and need to feel an atmosphere of empowerment, where the developers are free to create and code the capabilities laid out during the planning phase of each sprint. This requires a level of trust in the MDRE process. Trust that the developers and the MDRE engineers all have stakeholders in mind. Trust that the developer is working toward the end-product [17] and that the requirements can be met by the planned development efforts. Empowerment and trust at every level provides structure and clear expectations [18]. At the individual level it allows for creativity, and if the requirements management allocates independent requirements across the sprint teams, each team can have the independence they crave.
2. **Define and Create Interdependence**: While independence is a desired and necessary atmosphere for agile teams, the agile MDRE engineer must also establish the boundaries where individual developers, and development teams, must be interdependent on each other, given that the goal is to create an integrated, whole system, not just independent parts. Interdependence is being able to rely on team members [19]. The end goal will require a level of commitment from each person with a common mission in mind. The trust that all individuals on the team have all stakeholders in mind. This gets the whole team to the common goal and reduces each member motivated solely for their own end goal.
3. **Establish Overall, Individual, and Team Requirement Goals and Objectives**: The project/program overall requirement timeline goals, team goals, and individual goals and objectives should be established up-front as well as at the beginning of each sprint. This helps each team and individual team member to work successlyu at all levels of the program/project and makes requirements management of the agile process possible.This can help to identify strengths of individuals so that the team can use its assets to their highest production. It also sets up clear expectations of the overall and team requirement goals. There may be some individual development that is between the leader and the developer that stays between them. This would also build individual trust between members of the team and between the leader and the developers.
4. **Establish Self-Organization Concepts**: Keeping the requirement capability development independence between sprint teams allows each team a level of self-organization. Self-organizing teams is one of the holy grails of agile development teams. However, self-organization is sometimes a myth, mostly because teams are not trained into how to self-organize across sets of requirements. People do not just inherently self-organize well and self-organizing around requirements strategies is even less inherent to people. If not trained, the stronger personalities will always run the teams and subsequent requirements processes, whether they are the best candidates or not [20]. Self-organization can be nearly impossible when there are very structured people coupled with not-so-structured people. Forcing a requirements management process that works well for classical development would be destructive in an agile development environment. There may be some work that the MDRE team can do to promote self-organization. Part of that is opening communication, building dyads, calling behavior what it is, and being transparent so that others will follow. It may be helpful for team members to get to know strengths of other members and how each member can be helpful to everyone.
5. **Establish Requirements Feedback and Collaboration Timelines and Objectives:** Given the loose structure and nature of agile development, feedback early and often is crucial to allowing the teams to adapt to changing requirements or other changes (e.g., hardware). Also, customer collaboration and feedback at each level in the development allows the teams to adjust and vector their development efforts, especially requirements, to match customer expectations at all points in the development cycle. Feedback timelines can increase trust and clarify all expectations. It is nice to know when you need to change a direction (i.e., change, add, or delete one or more

requirements), when you need to change it, instead of later when you had already put so much work into the project. The more feedback is modeled and practiced, the more natural it becomes and becomes more automatic [21]. This builds on the independence and interdependence of the team and individual stakeholders and facilitates effective requirements management during the agile development.

6. **Establish Team's Ability to Challenge and Question Requirements Covered in Sprints**: If the teams are going to be allowed individual and team empowerment, then they must be allowed to challenge and question sprint capabilities (requirements) and content across the development cycle. Forcing solutions on the teams fosters resentment and a lack of commitment to the program/project. If you've built the right team, you should listen to them. It seems more productive to work on something that makes sense to you, instead of handed down by others. The ability to challenge and question will lead to better understanding and more commitment to the end goal [22].
7. **Establish an Environment of Mentoring, Learning, and Creativity**: Invariably, MDRE teams are composed of a combination of experience levels. This provides an excellent atmosphere of mentoring and learning if the agile MDRE team allows this. This must be built into the sprint schedules, understanding that an atmosphere of mentoring, learning, and creativity will increase efficiencies as the team progresses; not just on this project, but on future projects as well, as the team members learn from each other. Keep in mind that experienced developers and MDRE engineer(s) can learn from junior developer too, as the more junior developer may have learned techniques and skills that were not previously available to more senior developers and/or MDRE engineers. The learning environment promotes growth. An environment that fosters learning decreases negative feelings of oneself, and thus other people and especially the overall MDRE process. An environment that fosters learning isn't run by guilt, or feelings of not being good enough, or doing something wrong. A learning environment allows people to grow, and the mentor helps the individuals self-determine the direction they want to develop. The learning environment will foster older members learning from younger members as well. People will want to learn more and more and reduce competitiveness that can destroy a team. The competitiveness can come out as a good product not team dynamics. Transparency can help individuals feel more comfortable with learning. This can show that is ok to have areas of development and that everyone has room to grow [23].
8. **Keep Mission and Requirements Vision Always Out in Front of Teams**: Many believe that an established architecture and requirements baseline is not required for agile development. This is wrong; a solid architecture and a solid requirement starting baseline is even more important during agile development, so each team and team member understands the end goals for the system. However, for the architecture, software, and requirements to stay in sync, the MDRE team must also be agile enough to change as the requirements and subsequent architecture and design are adapted over time [18]. Agility is not free from structure but the ability to move about within the structure.

15.4 Case Study: Scaling MDRE Processes within Large Agile Development Projects

Many believe that the MDRE agile development philosophy works for small to medium programs/projects but will ultimately fail for large development efforts. However, if agile MDRE methodologies are applied correctly, and if there is a commitment throughout the project to agile, even very large programs/projects can benefit from the agile MDRE paradigm. While many people envision agile development as a small group of developers huddled in the same room handling everything within their own team; not required to communicate or collaborate with other. And while many may think this is nirvana for agile development, it is not reality, even for small programs/projects, especially if care is not taken to manage requirements and requirements change management effectively. But is it possible to realize the benefits of agile development for a large-scale development program (e.g., satellite system, commercial space flight software)?

The real challenge for MDRE for large-scale agile development programs/projects is keeping the requirements big picture out in front of all the teams. Since requirements, architecture, use cases, database designs, and dependencies, by definition, evolve in an agile development effort, it is incumbent upon the Agile MDRE team to ensure that the architects, systems engineers, and MDRE engineers within each team communicate and collaborate often to ensure that all teams are on the same page. Admittedly this resembles herding cats, but the goal is to make sure the teams aren't too far afield of each other, and are generally headed in the same direction, as specified by the overall changing requirements baselines, systems and software architecture. According to Joe McKendrick, "…agile has a fractal, scalable nature that allows for growth.[3]"

[3] http://www.zdnet.com/yes-agile-works-in-larger-enterprise-projects-too-7,000,020,875/

In one large-scale agile development program/project (launch vehicle), most of the teams were housed in an aircraft hangar, with a very large white board at one end of the hangar that had the entire system and software architecture drawn on the white board along with the major system-level requirements. The program/project consisted of 10 on-site and 2 off-site Scrum teams, each with 20 team members, consisting of a system and software architect, one systems engineer, one MDRE engineer, two test engineers, and 14 developers. Every morning the system and software architects held a 15–20-min Scrum to review the overall architecture to ensure that everyone was on the same page, including any requirements changes that had occurred. Then each Scrum team held their own 15-min Scrum to plan the day, find out what roadblocks existed, and get ready to roll for the day. The reason the team was all housed in a hangar is that the actual launch vehicle was being built at the other end of the hangar. Not only was the system and software architecture and high-level requirements always kept in front of the teams, but the actual vehicle was also at the end of the hangar and could be viewed by any engineer at any time, even for the teams that were off-site (as described there were two of the teams that were off-site) but continually connected by several closed-circuit cameras and video teleconferencing (VTCs) so they could be involved all throughout every day and throughout the entire development process.

This seems like an extreme example of scaled agile development, but the result was startling, radically reducing the time to delivery, and vastly reducing the amount of rework that is typical with large-scale space vehicle projects.

15.5 Requirements Agility in an Agile World

Requirements management for Agile development projects forms the bases for the project, just as it does in any development effort. Whether waterfall, spiral, or agile development, the MDRE team should own the requirements and requirements management. One big difference is that in Agile development and with Scrums, the product owner represents the needs of the customers, users, and stakeholders. These groups are responsible for determining how the requirements change, based on the demonstration of the working software at the end of each sprint, and therefore which requirements need to be implemented in what order, but the MDRE team needs to own the change and management process [24].

During the Agile development process, requirements are identified during Scrums and are repeatedly re-prioritized and therefore must be managed by the MDRE team and MDRE requirements management process. The bottom line is that agile does not mean void of requirements management, but it does mean that the MDRE team must always recognize that changes are going to occur and most likely very often during the Agile development sprint process. It is important that the MDRE team does not get behind in managing changing requirements and be able to produce a solid requirements baseline anytime it is needed or asked for [25].

15.5.1 Design Problem

MDRE is a life-long journey, and anyone who tells you they have nothing left to learn does not truly understand the discipline or has never been involved in the development of a complex system. With the recent interest in alternative energies and talk of the "Green New Deal," along with the recent issues with ice storms in Texas taking out the power grid, the assignments for this textbook will follow the theme of effectively dealing with both local alternative energy power grids and the national infrastructure required to facilitate large-scale alternative energy production, sustainment, and management.

15.5.2 Design Problem Description: Alternative Energy Power Grid

The purpose of this project is to, over the course of the semester, develop a feasibility study, requirements analysis, requirements elicitation, derivation, and allocation, for a system that will facilitate a "local" alternative energy power grid capable of powering a neighborhood of 500 homes and a large biosphere capable of supporting growing year-round fruits and vegetables for the 500-home neighborhood. The student will have to deal with issues of choosing multiple alternative energies that can provide power throughout all seasonal difficulties, from intense summer heat to freezing winter temperatures and possible ice storms. Control, viability, sustainability, and maintainability will be major factors to understand the entire possibly 30-year life cycle of such a system. How to effectively utilize, store, and manage power distribution, along with when to upconvert "extra" energy to the national power grid, will be issues that drive many of the assignments. Simply relying on the current technologies and energy infrastructures will not move the country and world forward. Hopefully, this will provide insight to the students using this textbook to understand some of the issues that face everyone looking out over the next many decades.

15.5.3 DC Power Grid Requirements Discussion

In the initial days of power generation and distribution, DC power was far less practical than the AC distribution system that soon supplanted it. However, even after Tesla's AC system became the predominant system (and rightly so), Edison's DC power did not become obsolete for some time. Eventually power plants eliminated their delivery of DC power, but the overall concept of a DC power grid is not dead. Even our computer systems (large server systems) have uninterruptible power supplies which contain batteries that are charged off the grid and used to upconvert to AC to power devices when the AC power grid goes off-line. So, the notion of a DC power grid is not dead, but generally used as a backup system. The point is, the notion of a DC power grid has remained and now that renewable energies are viable, we should re-examine their use for widespread, local, power distribution.

One of the advantages of a locally generated DC station is energy savings as well as allowing a facility to run more easily off various DC sources, such as Solar Arrays. And for power-hungry installations, the notion of generating the power locally is growing in popularity. Part of the reason is that waste heat from the power generation generators (solar, geothermal, etc.) can be used to warm nearby buildings. By combining the production of heat and power, local power grids can squeeze much more useful energy out of the renewable energy they use, so this approach will certainly become more widespread as time goes on. The purpose of this design problem is to look ahead and provide an architecture and specification for a local DC power generation station that can serve a local community.

Currently many systems exist to control and handle alternative energies. There are systems to control and manage solar power, wind power, geothermal power, and others, but very few are designed to control, manage, store, and distribute energy from a power grid consisting of multiple alternative energy sources. As we saw the issues recently caused by the intense ice storms in Texas, one alternative energy source is not the answer to the growing energy needs/wants in our society. Each alternative energy source has limitations and issues associated with its use. By harnessing multiple alternative energies, and managing the storage of the energies produces, communities can be worry free from power generation. To facilitate utilizing multiple alternative energy sources, a control center must be included in the design that can manage the energy creation, distribution, storage, and return unused energy to the power grid.

15.5.4 High-Level Technical Description of the DC Power Grid System

The purpose of the multiple alternative energies DC power grid is to:

- Supply DC power to a customer community of 500 homes and a biosphere for growing food.
- Provide a storage device to store energy in case of a system failure.
- Determine the amount of surplus energy that is above and beyond what is required to power the grid and charge the backup system, and then deliver that surplus energy to the national power grid systems.

Throughout the semester you will be asked to analyze, design, and architect different components of the overall DC power grid system. This will include research into alternative designs, like lithium nanowire storage systems for backup power and/or supercapacitors (also called ultracapacitors). Fig. 15.7 illustrates these technologies connected to the DC Power Grid Regulator Control Center.

15.5.5 High-Level Technical Requirements for the DC Grid System

- Design shall utilize at least two alternative energies, with a maximum of 4.
- Total power generation shall be 3.0 megawatts minimum.
- Each of the alternative energy sources shall be capable of generating at least 60% of minimum wattage.
- DC power grid shall generate DC power, with grid voltage being 48, 24, or 12 volts.
- DC power grid shall be capable of servicing at least 500 single family homes and one 4-acre biosphere with hydroponics capable of growing vegetables year-round. (See Fig. 15.8 for a block diagram of a DC-powered home).
- All lights, appliances, and facilities in each home and the biosphere shall be configured for DC operations, if possible. Things like electric ovens may need DC/AC conversion to accommodate their 220v. requirements.
- Only allowed exceptions are the HVAC system and the Washer/Dryer, although DC is preferred.
- There shall be energy storage capability capable of operating the grid of 500 single family dwellings for 7 days if all alternative forms of energy are not producing.
  - Compliance Criteria: Energy storage devices shall be capable of producing 750,000 watts for the entire

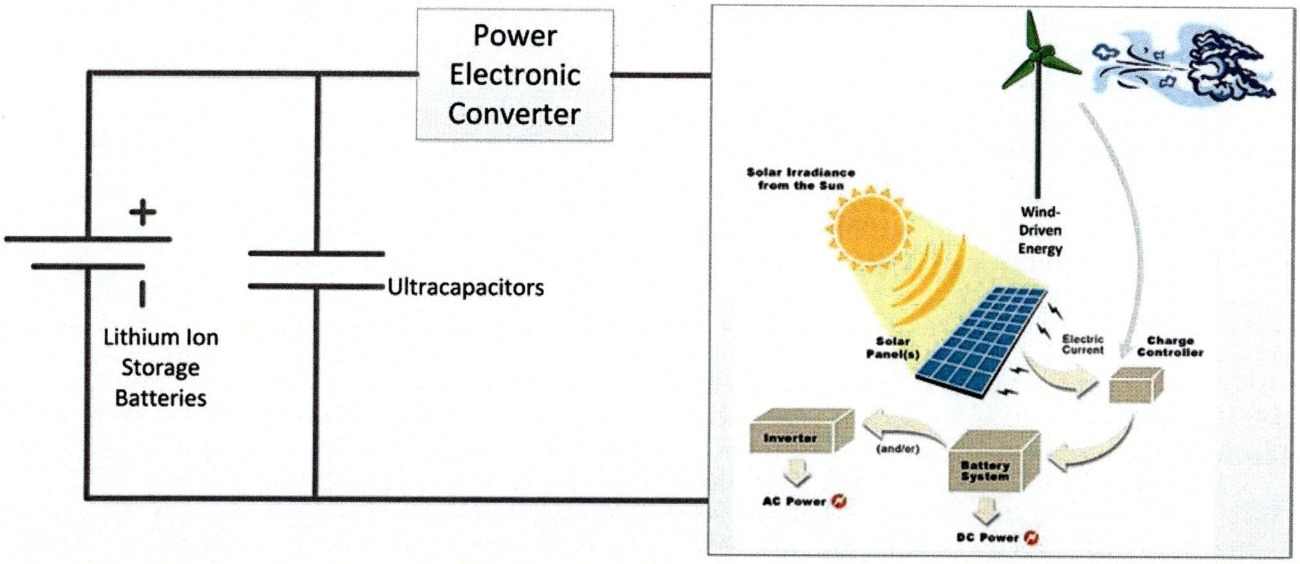

Fig. 15.7 Alternative power management system overview

Fig. 15.8 Example of a whole house power grid

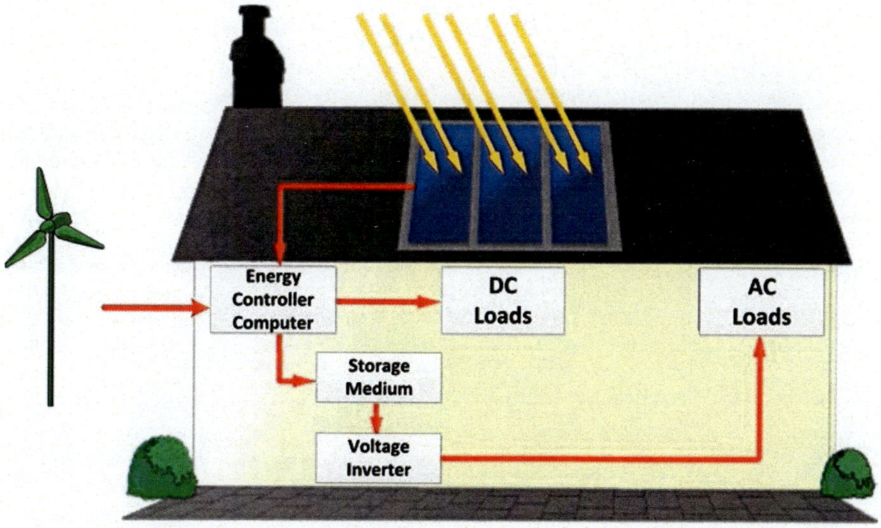

7-day period. Nominal energy storage should be batteries or supercapacitors.

- The DC power grid system shall include controlling h/w and s/w to manage the multi-alternative energy strategies and production.
- All energy produced above the required home and energy storage usage shall be upconverted to 220vac for release to the national power grid.
- The DC power grid shall be capable of operating in a temperature range of −40°F to 125°F without degradation of power generation.
- The DC power grid must meet OSHA regulations.

15.5.6 Assignments

- Study the two case studies in this chapter and answer the following questions:
  - In the first case study:
    - What was the main issue with using one MDRE engineer across multiple sprint teams?
    - Would you have had a separate MDRE engineer for each sprint team or have an MDRE team oversee all sprint teams?
  - In the second case study:
    - Describe how you would scale requirements management to handle multiple (say more than 5) sprint

teams to ensure requirements changes are captured and handled properly.

- Study Figs. 15.1 and 15.4. Describe the similarities and differences between the methodologies and how each would affect requirements management.
- Write a 5–10-page report on how you would manage requirements on your multi-alternative energy system if you chose:
  - Spiral development strategy.
  - Agile development strategy.

References

1. Abran, A., Ormandjieva, O., & Abu Talib, M. (2001). *Information-theory-based functional complexity measures and function size with COSMIC-FFP*. Université du Québec à Montréal.
2. Allen, B. (1995). Diversity and organizational communication. *Journal of Applied Communication Research, 23*, 143–155.
3. Bandura, A. (1997). *Self-efficacy: The exercise of control*. WH Freeman.
4. Beck, K. (1999). *Extreme programming explained—Embrace change*. Addison-Wesley.
5. Brooks, F. (1975). *The mythical man-month*. Addison-Wessley. ISBN 0-201-00650-2.
6. Crowder, J., & Friess, S. (2013). *Systems engineering agile design methodologies*. Springer Publishing. ISBN-10: 1461466628.
7. Francisco, M., Holcombe, M., and Gheorghe, M. (2003). A formal experiment comparing extreme programming with traditional software construction. In Proceedings of the Fourth Mexican International Conference on Computer Science.
8. Glib, T. (2013). Evolutionary project management. http://ebook-browsee.net/document-evoprojectmanagement-pdf-d357611912
9. Goodpasture, J. (2004). *Quantitative methods in project management* (pp. 173–178). J. Ross Publishing. ISBN 1-932159-15-0.
10. Highsmith, J. (2002). *Agile software developments ecosystems*. Addison-Wesley.
11. Jeffries, R., Anderson, A., & Hendrickson, C. (2001). *Extreme programming installed*. Addison-Wesley.
12. Kanter, R. M. (1992). *The challenges of organizational change*. Simon and Schuster.
13. Peters, J., & Pedrycx, W. (2000). *Software measures in software engineering: An engineering approach*. J. Wiley and Sons.
14. Roope, K. (2003). Efficient Authoring of Software Documentation Using RaPiD7. icse, pp.255, 25th International Conference on Software Engineering (ICSE'03).
15. Schwaber, K. (1995). The scrum development process. Processings of the OOPSLA'95 Workshop on Business Object Design and Implementation.
16. Sumara, J. and Goodpasture, J. (1997). Earned value—The next generation—A practical application for commercial projects. Retrieved 2006-10-26.
17. Tom, G. (1998). *Principles of software engineering management* (pp. P133–P158). Addison-Wesley.
18. Wienberg, G. (2003). Iterative and incremental development: A brief history. *Computer, 36*(6), 47–56.
19. Williams, L., Succi, G., Stefonovic, J., Marchesi, M.. (2003). A metric suite for evaluating the effectiveness of an agile methodology. In Extreme Programming Perspectives, Addison-Wesley, Boston, MA.
20. Booch, G. (1995). *Object solutions: Managing the object-oriented project*. Pearson Education. ISBN 0-8053-0594-7.
21. Rumbaugh, J., Blaha, M, Premerlani, W., Eddy, F., and Lorensen, W. (1990). Object-oriented modeling and design. Prentice Hall, ISBN 0-13-629842-9.
22. Stanhope, D. S., Samuel, B., & Surface, E. A. (2013). Core self-evaluations and training effectiveness: Prediction through motivational intervening mechanisms. *Journal of Applied Psychology, 98*(5), 820–831.
23. Spreitzer, G. M. (1995). Psychological empowerment in the workplace: Dimensions, measurement and validation. *Academy of Management Journal, 38*(5), 1442–1465.
24. Lappo, P., & Andrew, H. (2004). Assessing agility. In extreme programming and agile processes in software engineering lecture notes in computer science. *Springer, 3092*, 331–338.
25. Kurian, T. (2006). Agility metrics: A quantitative, fuzzy-based approach for measuring agility of a software process. ISAM-Proceedings for the International Conference on Agile Manufacturing'06 (ICAM-2006), Norfolk, VA.

16 Requirements Anti-Patterns

Abbreviations

AC	Alternating current
DC	Direct current
F	Fahrenheit
MDRE	Multidisciplinary requirements engineering
OSHA	Occupational Safety and Health Administration
RCM	Requirements Change Management
V	Volts
Vac	Volts AC
Vdc	Volts DC
VTC	Video teleconferencing

MDRE engineers have access to a host of requirements analysis, elicitation, and derivation techniques for a given project. The issue of deciding on an appropriate requirements creation and management process can be daunting. The question is: "How much is too much for a given project?" The tendency is to either apply no process (too little) or apply too much process than the project requires; managers and customers ***LOVE*** documentation and process. The main thing the MDRE engineers need to keep in mind is that while any requirements analysis, elicitation, management and tracking techniques and tools can be applied to almost any program using any development methodology (e.g., waterfall, agile, eXtreme), the key to successful MDRE is choosing the appropriate methodologies and tools, eliminating those that are doomed to failure, and finding those constructs and methods that minimize risk to your program. Any program is subject to change as problems and issues evolve during development, test, deployment, etc., and the MDRE engineer should see requirements management in this light [1]:

> *The concept of failure is central to the requirements management process, and it is by thinking in terms of obviating failure that successful requirements management is achieved. Although often an implicit and tacit part of the methodology of design, failure considerations and proactive failure analysis are essential for achieving success. And it is precisely when such considerations and analyses are incorrect or incomplete that requirements errors are introduced, and actual failures occur.*

The questions that must be answered by the MDRE engineer are:

- How much is enough requirements documentation?
- What tracking and management techniques are applicable to this specific development program?
- How does the overall MDRE and systems engineering organization ensure that the requirements keep up with design changes as the system design evolves through discovery of the emergent behavior of the system (particularly important in an agile development) as the system is tested and evaluated?

Answers to these questions are central to the MDRE engineers in learning how to choose requirements management pattern and techniques to reduce design and implementation risks across the system development, deployment, and sustainment life cycle. The balance of process vs. efficiency is often abandoned for traditional approaches.

16.1 High-Level Requirements Management Anti-Patterns

There are scores of wrong approaches that have been used over and over throughout the last many decades, none of which have proved very effective [2]:

Ad Hoc Requirements Management: Here, the MDRE team reacts real-time to project needs and decides at each increment in the development how many requirements and design to go forward with. This approach is, unfortunately, popular, even though it usually results in a haphazard design that has missing or incorrect requirements and is very subjective in nature, rarely capturing the needs of the customer and/or meeting the mission/business requirements [3]. Many

J. A. Crowder, C. W. Hoff, *Requirements Engineering: Laying a Firm Foundation*,
https://doi.org/10.1007/978-3-030-91077-8_16

use this approach and call themselves "agile." This is ***NOT*** agile; it is just wrong.

Question to ponder

Why do think some might consider Ad Hoc a way to do Agile requirements management?

Weighted Requirements Management: Here, the time and money allocated to the MDRE process is based on the amount of software to be developed. Everything, including systems design, requirements development and management, test, verification, etc., is estimated as a % of the software development effort. An example that we have seen used is:

15% of schedule and budget allocated to requirements development/derivation/allocation and management.

15% of schedule and budget allocated to design.

50% of schedule and budget to development (H/W and S/W).

20% of schedule and budget to system test and verification.

Invariably, the software estimates are wrong, as they are based on VERY optimistic views of the software, developers' efficiency, the number of issues (or bugs) that will be found in the code, etc. Very few programs bid this way come in on-cost, on-schedule, and on-performance. This methodology helps make planning easier, even though the notion that for every system requirements and architecture development/design you can utilize the same weightings is very short sited; basing everything on the software is equally problematic with today's modern languages, coding techniques, and code generators. What is typically shorted as things go wrong is requirements management and the system ends up with major rework through missing and/or incorrect requirements, since the requirements review process is cut short for the sake of cost and schedule concerns. The thought is always, "We understand the requirements, just get the software written." This is never the case, and it never ends well for the program [4].

Question to ponder

What happens in a weighted style when the estimates are completely wrong?

No requirements management: This may seem like the Ad Hoc style but is different and even worse. Development begins immediately (both H/W and S/W), based on the experience of the staff, since they "know what to develop" and the requirements are developed later, based on the "as-built" system. And yes—the system always meets requirements since the requirements are written after the fact. However, the system is never developed to the correct requirements and rarely does the customer accept what was developed. This methodology for requirements is particularly risky, since there is no guarantee that the system development will magically evolve into an integrated, coherent system that meets the customer objectives and high-level requirements [5].

Developing a "complete" requirement documentation set: While you may think this is the overall goal of MDRE, development organizations that take this approach and try to nail down every nuance of the system requirements to the last detail required major upfront time and energy. If this approach also included time and money for comprehensive management and tracking of requirements things would be great. However, often these turn out to be static documents that do not change and evolve as the system design evolves and changes, which is almost always the case. The result is that the developed system and the requirements that should have driven the development are not in sync and very much out of touch with the system implementation. This creates a maintenance and sustainment nightmare since the system documentation does not reflect the system implementation [6].

Use of standard processes/templates: The use of standard processes and requirements management templates is not inherently bad as they can provide guidance and conformity across the program. However, they are often utilized poorly [7]. A template is not a substitute for requirements elicitation, analysis, and decomposition/allocation; they are merely useful to help with the mechanics of documenting the requirements once a requirements management strategy and tool has been chosen.

Using industry standards in non-standard ways: Often, industry standards are chosen but implemented and used in non-standard ways (e.g., making up your own terms, views, or meanings for industry standard terms). This makes communicating the requirements and their ties to design, test, etc. difficult if not impossible and does not ensure consistency across the program. If the MDRE team has a different meaning for a standard term, it will affect the overall design and development, resulting in the system not conforming to the customer's objectives and expectations. Communicating non-standard artifacts and terms to customers and to the rest of the development organization means each person can bring their own interpretation to each element where you have introduced a non-industry standard entity [8].

Question to ponder

What terms used in the design of your multi-alternative energy system might be subject to different interpretations?

The bottom line is, the purpose of requirements management is to communicate and manage the requirements that drive the design, development, test, deployment, and sustainment and maintenance of the system. Communicating requirements and ensuring they are correct throughout the system life cycle is a difficult but necessary task. Here, the MDRE engineer must ask some basic questions [9]:

What requirements creation/management/dissemination methods should be used for a given program?

What visualization medium does the customer and development team need to communicate and collaborate effectively?

A final note: ***PowerPoint is NOT a requirements documentation medium***. Too many programs get wrapped around the axle in arguing over requirements and requirements management tools. Too many MDRE engineers fail to realize the goal of requirements documentation and management is to communicate and allocate the needs (voiced through requirements) for buy-in/consensus from the customer and development team to support the development process. That doesn't mean tools like Microsoft Word®, Excel®, and PowerPoint® aren't useful for presenting information, but they should not be the requirements management tool(s) of choice.

16.1.1 Design Problem

MDRE is a life-long journey, and anyone who tells you they have nothing left to learn does not truly understand the discipline or has never been involved in the development of a complex system. With the recent interest in alternative energies and talk of the "Green New Deal," along with the recent issues with ice storms in Texas taking out the power grid, the assignments for this textbook will follow the theme of effectively dealing with both local alternative energy power grids and the national infrastructure required to facilitate large-scale alternative energy production, sustainment, and management.

16.1.1.1 Design Problem Description: Alternative Energy Power Grid

The purpose of this project is to, over the course of the semester, develop a feasibility study, requirements analysis, requirements elicitation, derivation, and allocation, for a system that will facilitate a "local" alternative energy power grid capable of powering a neighborhood of 500 homes and a large biosphere capable of supporting growing year-round fruits and vegetables for the 500-home neighborhood. The student will have to deal with issues of choosing multiple alternative energies that can provide power throughout all seasonal difficulties, from intense summer heat to freezing winter temperatures and possible ice storms. Control, viability, sustainability, and maintainability will be major factors to understand the entire possibly 30-year life cycle of such a system. How to effectively utilize, store, and manage power distribution, along with when to upconvert "extra" energy to the national power grid, will be issues that drive many of the assignments. Simply relying on the current technologies and energy infrastructures will not move the country and world forward. Hopefully, this will provide insight to the students using this textbook to understand some of the issues that face everyone looking out over the next many decades.

16.1.1.2 DC Power Grid Requirements Discussion

In the initial days of power generation and distribution, DC power was far less practical than the AC distribution system that soon supplanted it. However, even after Tesla's AC system became the predominant system (and rightly so), Edison's DC power did not become obsolete for some time. Eventually power plants eliminated their delivery of DC power, but the overall concept of a DC power grid is not dead. Even our computer systems (large server systems) have uninterruptible power supplies which contain batteries that are charged off the grid and used to upconvert to AC to power devices when the AC power grid goes off-line. So, the notion of a DC power grid is not dead, but generally used as a backup system. The point is, the notion of a DC power grid has remained and now that renewable energies are viable, we should re-examine their use for widespread, local, power distribution.

One of the advantages of a locally generated DC station is energy savings as well as allowing a facility to run more easily off various DC sources, such as Solar Arrays. And for power-hungry installations, the notion of generating the power locally is growing in popularity. Part of the reason is that waste heat from the power generation generators (solar, geothermal, etc.) can be used to warm nearby buildings. By combining the production of heat and power, local power grids can squeeze much more useful energy out of the renewable energy they use, so this approach will certainly become more widespread as time goes on. The purpose of this design problem is to look ahead and provide an architecture and specification for a local DC power generation station that can serve a local community.

Currently many systems exist to control and handle alternative energies. There are systems to control and manage solar power, wind power, geothermal power, and others, but very few are designed to control, manage, store, and distribute energy from a power grid consisting of multiple alternative energy sources. As we saw the issues recently caused by the intense ice storms in Texas, one alternative energy source is not the answer to the growing energy needs/wants in our society. Each alternative energy source has limitations and issues associated with its use. By harnessing multiple alternative energies, and managing the storage of the energies produces, communities can be worry free from power generation. To facilitate utilizing multiple alternative energy sources, a control center must be included in the design that can manage the energy creation, distribution, storage, and return unused energy to the power grid.

16.1.1.3 High-Level Technical Description of the DC Power Grid System

The purpose of the multiple alternative energies DC power grid is to:

- Supply DC power to a customer community of 500 homes and a biosphere for growing food.
- Provide a storage device to store energy in case of a system failure.
- Determine the amount of surplus energy that is above and beyond what is required to power the grid and charge the backup system, and then deliver that surplus energy to the national power grid systems.

Throughout the semester you will be asked to analyze, design, and architect different components of the overall DC power grid system. This will include research into alternative designs, like lithium nanowire storage systems for backup power and/or supercapacitors (also called ultracapacitors). Fig. 16.1 illustrates these technologies connected to the DC Power Grid Regulator Control Center.

16.1.1.4 High-Level Technical Requirements for the DC Grid System

- Design shall utilize at least two alternative energies, with a maximum of 4.
- Total power generation shall be 3.0 megawatts minimum.
- Each of the alternative energy sources shall be capable of generating at least 60% of minimum wattage.
- DC power grid shall generate DC power, with grid voltage being 48, 24, or 12 volts.
- DC power grid shall be capable of servicing at least 500 single family homes and one 4-acre biosphere with hydroponics capable of growing vegetables year-round. (See Fig. 16.2 for a block diagram of a DC-powered home).
- All lights, appliances, and facilities in each home and the biosphere shall be configured for DC operations, if possible. Things like electric ovens may need DC/AC conversion to accommodate their 220v. requirements.
- Only allowed exceptions are the HVAC system and the Washer/Dryer, although DC is preferred.
- There shall be energy storage capability capable of operating the grid of 500 single family dwellings for 7 days if all alternative forms of energy are not producing.
  - Compliance Criteria: Energy storage devices shall be capable of producing 750,000 watts for the entire 7-day period. Nominal energy storage should be batteries or supercapacitors.
- The DC power grid system shall include controlling h/w and s/w to manage the multi-alternative energy strategies and production.

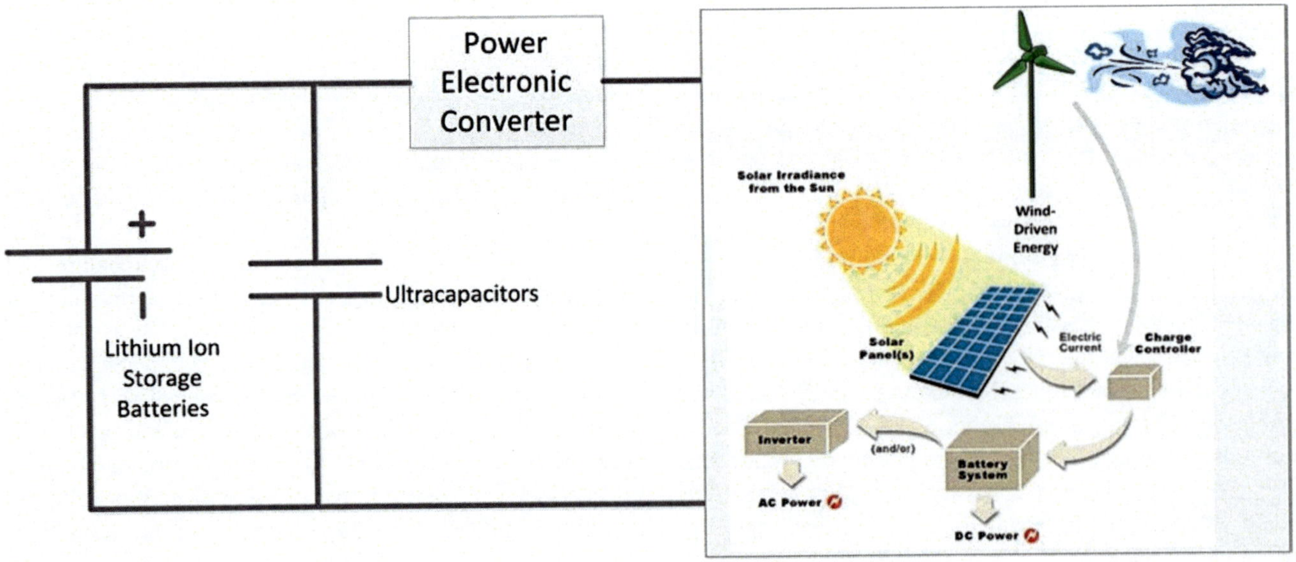

Fig. 16.1 Alternative power management system overview

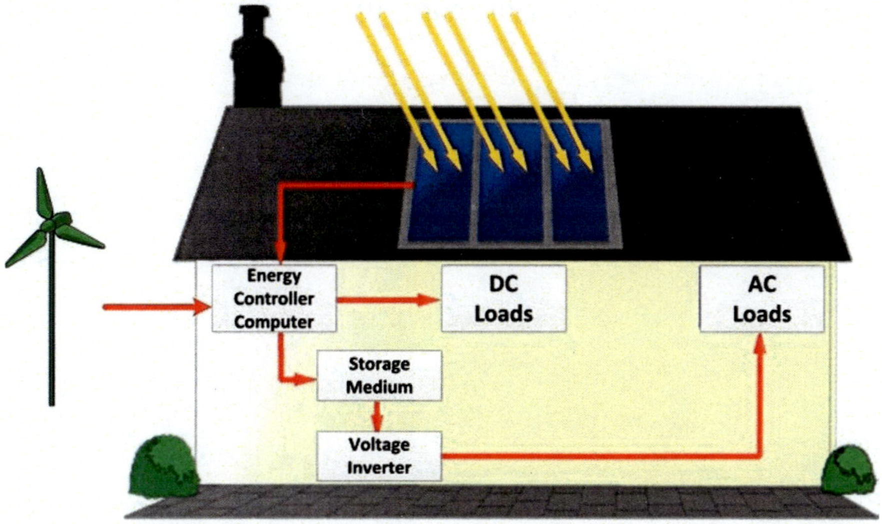

Fig. 16.2 Example of a whole house power grid

- All energy produced above the required home and energy storage usage shall be upconverted to 220vac for release to the national power grid.
- The DC power grid shall be capable of operating in a temperature range of −40°F to 125°F. without degradation of power generation.
- The DC power grid must meet OSHA regulations.

16.1.2 Assignments

- Study the anti-patterns discussed in Sect. 16.1 and answer the following questions:
  - Why do you think the Ad Hoc method is so popular and, considering MDRE, what requirement management concepts does it violate?
  - What methodology for estimating the cost/schedule of requirements development/management would you suggest other than the weighted method?
  - If you eliminate requirements management and simply tell those implementing your multi-alternative energy system design to build what they think is right, describe where you believe the biggest disconnect between needed requirements and system implementation might occur.
  - Discuss whether you would use standard templates for your requirements management and why or why not.

References

1. Petroski, H. (1992). *To engineer is human: The role of failure in successful design*. Vintage Books. ISBN-10: 0679734163.
2. Dwyer, M., Avrunin, G., and Corbett, J. (1999). Patterns in property specifications for finite-state verification. In Proceedings of the International Conference on Software Engineering, Los Angeles, CA.
3. Edvardsson, J. (1999). A survey on automatic test data generation. Proceedings of the Second Conference on Computer Science and Engineering in Linkoping. CiteSeerX: 10.1.1.20.963.
4. Egyed, A., Grunbacher, P., and Medvidovic, N. (2001). Refinement and evolution issues in bridging requirements and architecture-the CBSP approach. In International Software Requirements to Architecture Workshop, Toronto, CAN.
5. Egyed, A., Grunbacher, P., Heindl, M., and Biffl, S. (2009). Value-based requirements traceability: Lessons learned. Design Requirements Engineering: A Ten-Year Perspective. Springer-Verlag, Berlin Heidelberg, GER, ISBN 978-3-540-92966-6.
6. Erdobmus, H., & Torchiano, M. (2005). On the effectiveness of test-first approach to programming. *Proceedings of the IEEE Transactions on Software Engineering, 31*(1), NRC 47445.
7. Feathers, M. (2004). *Working effectively with legacy code*. Prentice Hall.
8. Ferguson, R., & Korel, G. (1996). The chaining approach for software test data generation. *ACM Transactions of Software Engineering and Methodology, 5*(1), 63–86.
9. Fiadeiro, J., Lopes, A., & Bocchi, L. (2006). A formal approach to service component architecture. *Web Services and Formal Methods, 4148*, 193–213.

Model-Based Systems Engineering

17

Abbreviations

AADL	Architecture Analysis Design Language
AC	Alternating current
ADM	Architecture Development Method
CONOPS	Concept of Operations
COTS	Commercial off the shelf
DAU	Defense Acquisition University
DC	Direct current
DoD	Department of Defense
DoDAF	Department of Defense Architecture Framework
EIA	Engineering Industries Alliance
F	Fahrenheit
HMI	Human machine interface
HVAC	Heating, ventilation, and air conditioning
IDEAS	International Defense Enterprise Architecture Specification
IEEE	Institute of Electrical and Electronic Engineers
INCOSE	International Council on Systems Engineering
ISO	International Organization for Standardization
M&SBSE	Model and simulation-based systems engineering
MBD	Model-Based Development
MBSAE	Model-Based System Architecture Enterprise
MBSE	Model-based systems engineering
MDRE	Multidisciplinary requirements engineering
MODAF	Ministry of Defense Architecture Framework
MODEM	MODAF Ontological Data Exchange Mechanism
NAF	NATO Architectural Framework
NASA	National Aeronautics and Space Administration
NATO	North Atlantic Treaty Organization
NDIA	National Defense Industrial Association
OMG	Object Management Group
OSHA	Occupational Safety and Health Administration
RCM	Requirements change management
SoO	Statement of Objectives
SysML	Systems Modeling Language
TAFIM	Technical Infrastructure Framework for Information Management
TOGAF	The Open Group Architecture Framework
UML	Unified Modeling Language
V	Volts
Vac	Volts AC
Vdc	Volts DC
VTC	Video teleconferencing

The concept of model-based systems engineering (MBSE) began to become popular in 2007 when the International Council on Systems Engineering (INCOSE) introduced the initiative to the international community [1]. According to INCOSE, at a high level, MBSE is a formalized application of modeling to support systems requirement generation and management, system design, analysis, verification, and validation activities that begin in the system conceptualization phase and continue throughout the life cycle of the developed system [2]. Under the MBSE methodology, focus is placed on creating and utilizing system domain models as the primary means of system design, requirements, development, test, deployment, and sustainment and maintenance, rather than a document-based system information exchange [3]. In the 2014 timeframe, the focus of MBSE began to cover computer simulations for the MBSE model execution to combine system model specifications and system simulation software. During this period the term MBSE began to

J. A. Crowder, C. W. Hoff, *Requirements Engineering: Laying a Firm Foundation*,
https://doi.org/10.1007/978-3-030-91077-8_17

expand to include simulations, and many now refer to MBSE as M&SBSE or model and simulation-based systems engineering. According to the National Defense Industrial Association (NDIA):

"Model-Based Engineering (MBE) is an approach to engineering that uses models as an integral part of the technical baseline that includes the requirements, analysis, design, implementation, and verification of a capability, system, and/or product throughout the acquisition life cycle.[1]"

Until the introduction of MBSE, most systems engineering, and therefore requirements engineering, was a document-based process. Document-based requirements engineering relies on documents like a Concept of Operations (CONOPS) to define the system and how it supports the overall customer objectives or needs. Typically, a functional analysis would be performed to decompose the system into separate capabilities or functions and allocate them to the applicable components of the system [4]. A recurring issue over many decades with document-based systems engineering is that it can suffer from:

- A lack of precision, since it is hard to instill design precision into document descriptions.
- Inconsistencies between documents, since different groups typically produce different documents.
- Difficulties in maintaining, updating, and using the information in many volumes of either paper or on-line documents.

This does not mean that document-based system engineering does not work, for this method has been responsible for untold thousands of successful programs for hundreds, possibly thousands of years. However, the MBSE approach puts computerized models at the center of the systems design process, and the digital modeling environments available over the last 10 years has led to an increasing adoption of MBSE for systems engineering, and MDRE processes must be adapted to this new MBSE paradigm for systems development [5].

17.1 The Basics of Model-Based Systems Engineering

For the MBSE approach to system development to be effective, formal application of models to support system requirements (MDRE), analysis, design, test, and sustainment activities must be created and managed throughout all phases of a system's life cycle. The use of models has been used for decades, including functional models to support requirements, simulation models to analyze system behavior(s), and analytical models to predict and analyze reliability, safety, availability, power consumption, and a host of other factors. Even so, the results and all other aspects of the system requirements, design, development, etc. rely on document-based artifacts to capture and manage the system development, test, deployment, interface control, and sustainment. The information is contained across multiple documents created and managed by different engineers/groups, each with a combination of text, spreadsheets, drawings, etc. Any change to the system must be coordinated across multiple models, documents, views, etc. and the changes reviewed and approved by multiple systems engineers working across multiple discipline groups. This causes difficulties in maintaining and reusing the information, since no one document is sufficient to relay all the information on a given system design [6].

In contrast to multiple documents to contain a design, in MBSE, the design is contained within a set of descriptive and analytical models that are utilized throughout the systems design, development, deployment, and sustainment life cycle, including models at the System-of-Systems level down to the component level [7]. In MBSE, the digital system models are used by all engineering disciplines, especially specialty engineering, for all aspects of the system life cycle. In MBSE digital modeling there is a single source of truth for system design/development where discipline-specific views can be created and generated using the same model elements. Typically, there are three types of models utilized for the MBSE approach:

- **Descriptive Models**: These are used for the system architecture for requirements and overall system design. An MBSE descriptive model describes the system and its interaction and relationship to the environment for which the system is intended. It is used to understand what the system is, what it does, and how it does it. Once example would be a geometric or geospatial model that represents spatial relationships [8].
- **Analytical Models**: These are used to analyze system performance, based on executable code that matures throughout the development process.
- **Physical Models**: These are used for analyzing quality characteristics like reliability, maintainability, safety, etc. An example is the use of a flowchart model to unambiguously define the components of a system (H/W and/or S/W) and their related requirements and functions.

The models described above are utilized to provide model views into the four quadrants of the MBSE process (see Fig. 17.1):

- Conceptual Description: Logical system view.
- Conceptual Design: Logical solution view.
- Actual Description: Physical system view.

[1]Final Report, Model-Based Engineering Subcommittee, NDIA, Feb. 2011.

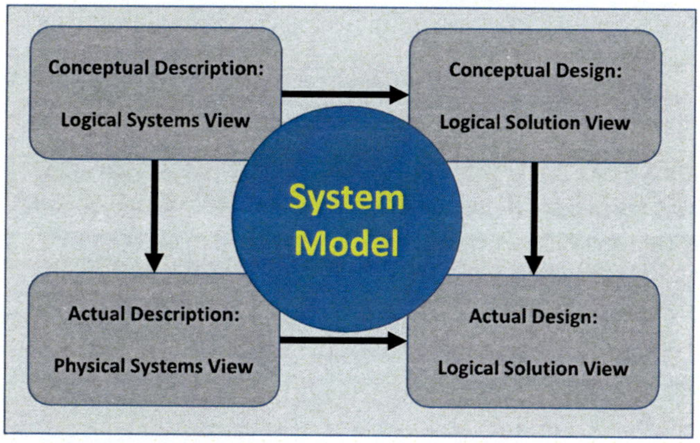

Fig. 17.1 MBSE model quadrants

- Actual Design: Physical solution view.

The MBSE concept shown in Fig. 17.1 ties to the modeling domains that are the heart of MBSE [9]. Overall, MBSE does not dictate specific processes or procedures, but rather allows the systems engineering organization for the program to choose processes and/or procedures that cover four basic engineering domains [10]. These domains, listed below, are not defined within a set of documents, but rather are defined, stored, and managed from within a set of models that are formalized within a modeling language we will discuss in Sect. 17.2. The models represent the details of how the system must be designed, implemented, tested, deployed, and sustained for the program to be successful [11]. These are:

- Requirements.
- Behaviors.
- Architecture/design.
- Verification and validation.

Creating models for the system that provide an integrated modeling environment allows open communication and collaboration among engineering disciplines within the program organization since they are all getting information from one source of truth. By allowing customers and stakeholders access to the models, and therefore design, architecture, requirements, and other program information, it fosters open communication between the MBSE engineers, the MDRE team, and the customers. Any discrepancies between interpretations can be dealt with since all the information is contained within the set of MBSE models. What many in the systems engineering community mistake is the notion that if I create a specific view within a "modeling tool," that it constitutes a model. This is not the case. In fact, it happens so often, there is even a term for it, namely, a "zombie model" [12]. But a set of views can provide a representation of some part of the system. One of the big differences between a set of views and an MBSE model is that they are connected, consistent, and traceable across the entire system. Different views are merely one or more facets of the overall system design and just a part of an overall system model [13]. A true MBSE model should be able to produce views for all parts of the system, including pictorial views, matrix views, analytical views, and simulation views. Especially important in MBSE are 3-D system models instead of 2-D drawings. The three-dimensional aspect of systems is important to understand the physical nature of the design but also uses flowchart models to describe the control side of the system (embedded software) along with the physical layouts to test their interactions. And while mechanical systems can be driven by equations and formulas, the MBSE equivalent can be done with a full multi-physics simulation of the completed 3-D model. MBSE is in its infancy in adoption by development firms but will become more prevalent over the next decade. Again, it is important to remember that an MBSE system model allows various views to be created and displayed for a system, but a group of views is not a model. A separate group of system views that was not generated from a central set of models generally is neither interconnected nor are dependencies identified between elements of the views [14]. Since the MBSE philosophy and practice does not specify a modeling language of choice, Section 17.2 will discuss several modeling languages that could be used, depending on the domain and customer system for which the system is being developed. Many engineers will assume that the System Modeling Language (SysML) is the default choice for MBSE, but many others are useful and can be utilized to form coherent, connected system models. All of them can be used incorrectly and not be useful in an MBSE environment as well [15]. The point to remember is that all of these are just modeling languages and are based on human language, perspectives, and understandings. The formality of each of these modeling languages adds discipline and clarity to the system descriptions. All of them map to parts of language components (parts of speech), such as nouns, verbs, adjectives, and adverbs. Examples include, but are not limited to [16]:

- **Nouns**: requirements, functional blocks, actors, and components.
- **Verbs**: capabilities, operational activities, use cases.
- **Adjectives**: functional attributes.
- **Adverbs**: interfaces, needs, relationships, information exchanges.

Section 17.2 will highlight several common modeling languages that are available. There are similarities and differences between each of the modeling languages. No reference will be made to better or worse, but it is up to the MBSE organization within a program organization to decide which modeling language is most appropriate for a given project, based on its requirements, complexity, size, and customer/stakeholder wishes and needs.

17.2 MBSE Modeling Languages

System modeling, and thus MBSE, is a technique for abstracting a system to understand the reality of the system. Models are pervasive and utilized in every discipline from artificial intelligence, to neuroscience, to automobile manufacturing. Models have become the foundation of systems development and operation [17]. Based on Rumbaugh [18], models should perform the following functions:

- Capture requirements and domain knowledge so the MBSE team and customer/stakeholder(s) understand and agree on their interpretation.
- Allow the MBSE team to think about and envision the system design.
- Produce usable products during all phases of a system's life cycle.
- Organize, find, filter, manipulate, and allow editing of information about the system being developed.
- Allow the MBSE team to explore multiple solutions to a given set of objectives, being able to trade off operational, economical, and environmental factors in deriving solutions.
- Allow complex systems to be broken down and understood.

To deal with the complexity of MBSE designs for modern systems, formalized methods and tools have been created to allow documentation, communication, and comparison of systems and software architectures [19]. These methods and tools all focus on the creation of different drawings, each with their own set of standards, each designed to aid in the design, implementation, and manufacturing phases of overall system development, operations, and maintenance. As has been discussed throughout this textbook, the system must be defined through a complete multidisciplinary life cycle perspective to capture all the requirements, needs, and goals of the customer. The overall objective is to develop a complete life cycle-balanced system. As system complexities have increased over the decades, the Systems and Software Engineering tools have evolved to keep up with changing system designs (e.g., service-based systems, agile development methods) [20].

Process standards have been developed which aid the multidisciplinary systems engineer in establishing baselines for system design. One example is the IEEE 29148-2011–Systems and Software Engineering–Lifecycle Processes–Requirements Engineering standards. Others include the IEEE 12207–Systems and Software Engineering–Software Lifecycle Processes, which is an international standard for establishing a common framework for software life cycle processes. The EIA 632 Processes for system engineering was established by the Engineering Industries Alliance (EIA) in conjunction with the International Council on Systems Engineering (INCOSE) to focus on standards for conceptualization and implementation of complex systems engineering designs [21].

Large amounts of design data are created during the life cycle of any complex system. Different engineering methodologies are utilized during each life cycle phase, each with their own modeling tools and defined "views" of the overall system, all from a common MBSE set of interconnected models. No one tool or MBSE view captures the entire systems, software, operations, and maintenance views of the system. It is incumbent on the MBSE team to manage all the design and implementation documentation to provide complete traceability, horizontally and vertically through any stage in the life cycle process. This includes not just initial design views, but systems integration, verification and validation plans, operations and maintenance specifications, and overall human-machine interface plans [22]. Management of all these different model views and specifications is vital for the engineering of the overall system. Such processes and methodologies are not linear, requiring multiple iterations between each set of specifications to ensure a viable, usable, working system. The MBSE multidisciplinary approach is necessary to coordinate and facilitate the entire system traceability across the entire product life cycle.

Each set of design and implementation models requires computer-based tools to produce the required design model views, each based on a different set of standards, whether U.S. Department of Defense (DoD), U.S. Commercial standards, or international standards. There are many systems and software design tools available in the market. Each is based on a given methodology and focus (e.g., systems vs. software engineering) [23]. The following discussion will describe the main methodologies for each of the computer-aided productivity tools are based upon. One of the important points to understand is that no one tool or methodology will give a complete model view or complete description of a complex systems and software engineering design. It should be expected that development information for any complex system will be distributed over multiple interconnected models, covering multiple engineering disciplines, hence the need for the multidisciplinary approach. The first methodology to be discussed will be the U.S. Department of Defense Architectural Framework (DoDAF).

Question to think about

Does having formal models help or hurt a system development?

17.3 System Architecture Modeling Frameworks

For the MBSE designer to create the architectures required for system development (systems, software, information, data, etc.), there must be an established framework (or structure) that can be used to communicate the architecture to the rest of the program and to the customer. An architecture modeling framework defines how to create a given architecture, providing the principles, practices, and standards for creating, using, and communicating an architectural description of an engineered system. It allows the architect to structure their thinking and architecture vision by allowing the MBSE analyst to divide up the architecture into separate model layers or views and provides standards methodologies for creating and documenting the architecture in terms of diagrams, matrices, and word documents. A high-level view of an architectural layer view is shown in Fig. 17.2 [24].

17.3.1 The Zachman Modeling Framework

The Zachman Framework represents a classification methodology for system architecture descriptions [25]. The Zachman Framework matrix, shown in Table 17.1, is constructed to classify the architecture documentation by the audience the documentation is intended for, and the content focus of the artifact. The Zachman Framework does not specify which of the architecture modeling frameworks (e.g., DoDAF, TOGAT, MODAF, SysML) to use for model artifact creation, it just lays out a detailed analysis of which artifacts are applicable to which perspective and which models (audience) [26]. For instance, row 2, the Conceptual Models can be thought of as the Owner or Customer's perspective since these artifacts relate to the conceptual design and capabilities for the design of the system. The customer and development team should come to concurrence at this level, to ensure that the system being developed is in line with the customer's Concept of Operation (CONOPS), technical and performance needs, as well as usability by the end users. Understand, that by customer, we mean who ever has commissioned, or is paying for the system. This might be an outside customer like NASA, the Department of Defense, NATO, etc., or it might be an internal customer where upper management is paying for the system (e.g., Information Technology computer and network upgrades) [27]. Row 1 defines the scope of each perspective, where the information contained in each perspective will drive the conceptual views and will drive the requirements specified for the system. The artifacts specified for each perspective in row 3, called the Designers Perspective, lay out the overall architectural views and designs that take the concepts in row 2, translates them into a design, allowing the perspectives in row 4 to define what is technically and physically feasible for the project. Row 5 disassembles the designs into physical parts (hardware and networks), and software components (services), that are required to instantiate the overall design; that is, how to physically put the system together. Row 6 represents the final instantiation of row 2. Row 6 is the demonstration to the customer that the system meets their needs, as specified in the requirements and CONOPS [28].

17.3.2 The Department of Defense Architecture Framework (DoDAF)

The Department of Defense Architecture Framework (DoDAF) provides an Architecture Framework standard for defense contractors to organize and document system architectures for proposals, acquisitions, and system design/development. It allows comparisons of different defense contractors' designs without having to translate between different architectural frames or reference. The DoDAF 2.2 Framework specifies artifacts and documentation that pro-

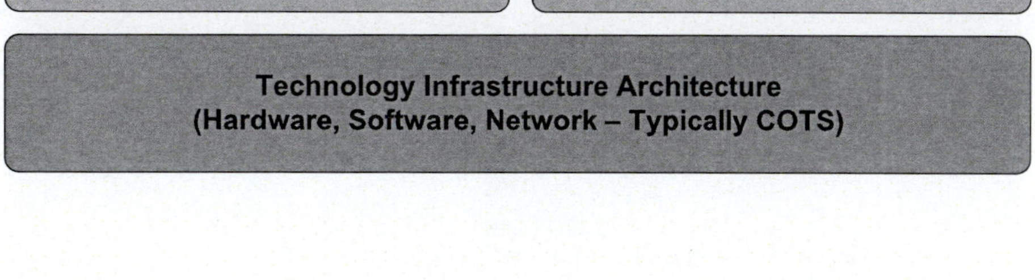

Fig. 17.2 High-level systems architecture MBSAE layer view

Table 17.1 High-level Zachman Framework matrix

	Data perspective	Functional perspective	Network perspective	Organizational perspective	Temporal perspective	Motivational perspective
Contextual model	Mission/business data/ information	Mission/business processes	Mission/business physical locations	Mission/business organizational structures	Important business/ Mission events	Mission/business strategies
Conceptual model	Conceptual information/data model	Mission/business process model	Mission/business logistics model	Mission/business workflow/ orchestration model	Mission/business master schedule	Mission/business functional model
Logical model	Logical data model	System architecture model	Distributed network/ systems model	Human-systems Interface model	Information/data processing architecture	Mission/business rules
Technology model	Physical information/ data model	Technology model (plans)	Technology infrastructure model	Presentation/visualization architecture	Process control model/ architecture	Process rule strategy
Representational model	Data item descriptions	Program execution model	Network topology	Security architecture	Timing diagrams	Processing rule specification
Functional model	Data instantiation	Functional system architecture	Functional network architecture	Functional organization	Functional schedule	Functional working/business/ mission strategies

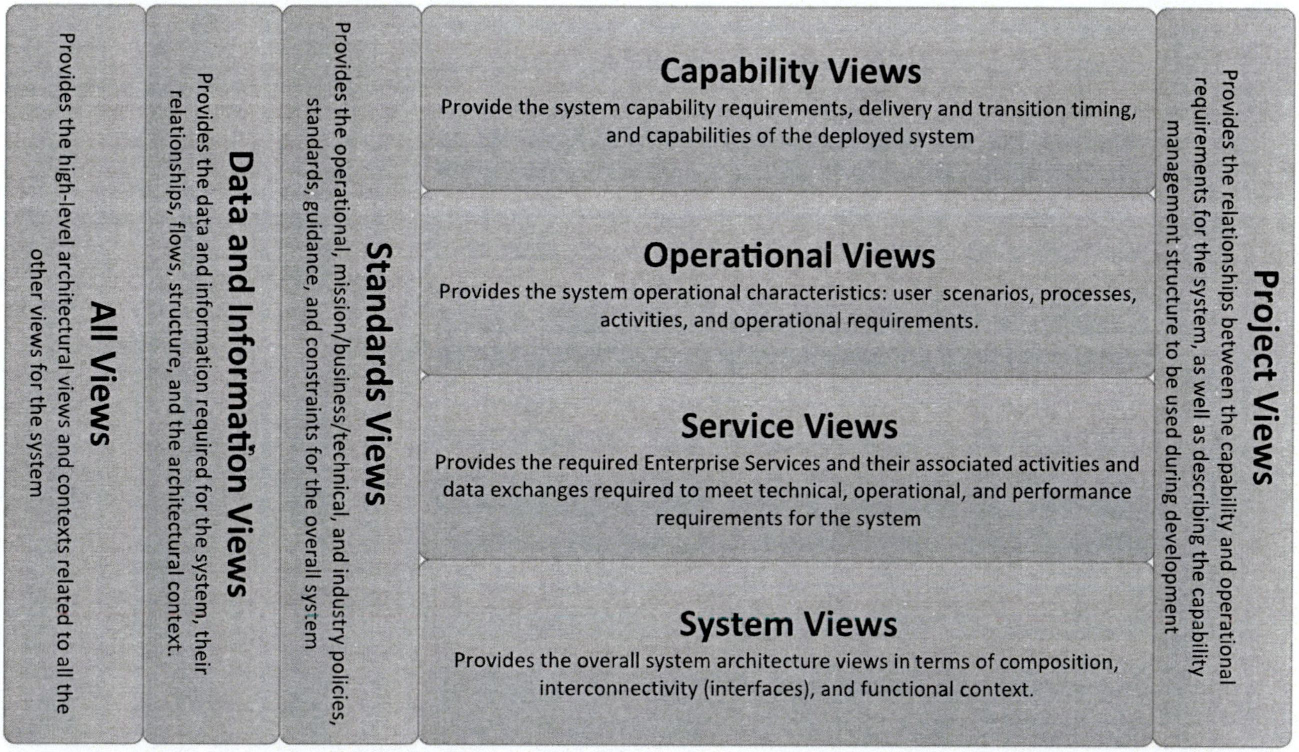

Fig. 17.3 DoDAF 2.2 architecture framework model views

vide views from various perspectives for the system (e.g., Data Views, Capability Views, Service Views). These are described at a high level in Fig. 17.3 [29].

The DoDAF 2.2 Framework contains the following architectural views and documentation shown in Tables 17.2–17.6 below.

The DoD Architecture Framework (DoDAF) is the US Department of Defense's customized framework for developing integrated system architectures. DoDAF describes the rules and conventions that defense and defense contractor systems architects use when developing architectures for DoD systems. DoDAF was created so that the resulting work is compatible, at least at a basic level. DoDAF was created to support many facets of the DoD and was intended to be useful even when limited parts of the architecture modeling framework are implemented. The current instantiation of DoDAF, DoDAF 2.2 is based on a six-step architecture design process. The DoDAF 2.2 guidelines differs from previous versions of DoDAF in that 2.2 emphasizes relationships between data rather than products (data-based vs. view-based). This emphasis on a data-centric architecture helps to ensure consistency between the systems, software, and information architectures, as well as providing proper data/information flows in/out of both external and internal interfaces [30]. Fig. 17.4 illustrates the DoDAF 2.2 systems architecture model development process. Following the model development process outlined in Fig. 17.4 helps to define the model views that are appropriate for the system to be designed. Only those model views that are required to represent the applicable data/information in the system should be baselined for the architectural design. The architecture model views represent the architecture for a given time-period, realizing that the needs of the system, the requirements, and the operational concepts may change over time and the architecture models must evolve as these characteristics change. Agile systems engineering methodologies and techniques should be built into the overall process accommodate changes to the system [31].

It is important, when utilizing the DoDAF structure, to determine which views and artifacts are relevant to the system being designed. In one recent case, a program chose to create ALL the DoDAF views and artifacts, which left the development team confused as to which were relevant and confused on how to prioritize the respective views. It also challenged the architecture team in the maintenance of the views and artifacts as the design matured.

17.3.3 The Open Group Architectural Framework (TOGAF)

While DoDAF models concentrate on providing architecture various model views, as shown in Tables 17.2–17.6, it does not specify the methodology for producing these model

Table 17.2 DoDAF 2.2 all, data, standard, and capability model views

Views	Description
All views	
AV-1	Describes the project goals, objectives, activities and events, plans, constraints, measures of effectiveness and expected outcomes (products)
AV-2	Repository that includes all terms with definition used throughout the program
Data/information views	
DIV-1	Conceptual data model(s)
DIV-2	Logical data model(s)
DIV-3	Physical data model(s)
Standard views	
Stdv-1	List of standards that apply across the program
Stdv-2	List of possible emerging standards and their potential impacts on the program
Capability views	
CV-1	Provides the strategic context for capability development for the program. Should provide the context for the capabilities shown in the architecture views
CV-2	Models a hierarchy of capabilities (the capabilities taxonomy)
CV-3	Shows time-phasing of capabilities development across the program, including the activities, conditions, desired effects, and necessary resources
CV-4	Describes the dependencies between the planned capabilities. Also provides the definitions for logical groupings of capabilities
CV-5	Planned capability development. Shows the planned capability phasing in terms of personnel and locations and their concept of capability development
CV-6	Provides a mapping between the capabilities and the operational activities throughout the program
CV-7	Provides a mapping between the capabilities and proposed service development across the development efforts

views. In contrast, TOGAF focuses on the methodologies available to provide architecture models and views, without specifying which model views, or constructs, should be used to design system architectures. TOGAF evolved out of the Department of Defense Technical Infrastructure Framework for Information Management (TAFIM). Version 3.0 of TAFIM came out in 1996, was base-lined, turned over to TOGAF, and retired. It is utilized by the TOGAF group to develop their TOGAF Architecture Development Method (ADM). The purpose of the ADM is to provide a guide to systems architects to lead them through the system architecture life cycle. The TOGAF ADM provides a mapping between their methodologies and the DoDAF architecture

Table 17.3 DoDAF 2.2 operational model views

Operational views	
OV-1	Operational concepts. Graphical and textual description of the overall program operational concepts
OV-2	Operational resource flows. Description of the resources exchanged between operational activities
OV-3	Operational resource attributes. Provides a description of the resource attributes relevant to the resources exchanged in the OV-2
OV-4	Describes the relationships between different organizations being utilized across the program (e.g., subcontractors)
OV-5a	Operational activity decomposition. Hierarchical view of the proposed operational activities for the system(s)
OV-5b	Operational activity model. Describes the proposed system operational capabilities and their relationships to inputs, outputs, and system activities
OV-6a	Operational rules model. Describes the mission/business rules that constrain the overall system activities
OV-6b	State transition model. Describes the mission/business processes and their responses to events and/or activities experienced in system operations
OV-6c	Event trace model. Describes the traces between operational activities and sequences of events across the proposed system operators

views. Fig. 17.5 illustrates the TOGAF model views and Fig. 17.6 illustrates this model view mapping [32] from TOGAF to DoDAF.

TOGAF is a high-level approach to model creation and design. It is typically modeled at four levels: business, application, data, and technology. It relies heavily on modularization, standardization, and already existing, proven technologies and products. It is a set of tools which can be used for developing a broad range of different architectures. It should:

- Describe a method for defining an information system model in terms of a set of building blocks.
- Show how the building blocks fit together.
- Contain a set of modeling tools.
- Provide a common model vocabulary [33].
- Include a list of recommended model standards.
- Include a list of compliant products that can be used to implement the model building blocks.

17.3.4 The Ministry of Defense Architecture Framework (MODAF)

MODAF is an international architecture description standard that has been utilized within the international community to

Table 17.4 DoDAF 2.2 operational service views

Service views	
Svcv-1	Service context. Identifies the proposed system services and their interconnectivity/interdependencies
Svcv-2	Service resource flow. Describes how resources flow between proposed system services
Svcv-3	System service matrix. Describes the relationships between the system, subsystems, and components, and the proposed system services for each architectural view
Svcv-4	Service functionality description. Describes the functionality of each proposed system service and the associated data flows
Svcv-5	Operational activity to services matrix. Maps the proposed services to proposed system operational activities
Svcv-6	Services resource flow matrix. Describes the resources and resource flow between proposed services and the attributes of those resource exchanges
Svcv-7	Service measures matrix. Describes the measures of effectiveness for proposed services and the timing of the measures
Svcv-8	Service evolution. Describes the incremental development of proposed services. Includes a description of the evolution of services towards a more efficient overall system operations in the future
Svcv-9	Services technology & skills forecast. Describes the emerging technologies possible for use on the program, including hardware, software, personnel skills and the associated time frames for each. This is specific to how it affects and/or improves services throughout the system.
Svcv-10a	Service rules model. Identifies constraints that are imposed on the proposed system service functionality, cased by the proposed system design and/or implementation
Svcv-10b	Service state transitions. Describes the service state transitions caused by service responses to system events
Svcv-10c	Service event-traces. Describes service-specific sequence of events driven by the operational views

support the delivery of military capabilities across the world and has been adopted by NATO to form the core of their NATO Architectural Framework (NAF). Currently the United States Department of Defense, the Canadian Department of National Defense, the Australian Department of Defense, and the Swedish Armed forces are working together to utilize MODAF and DoDAF to form the International Defense Enterprise Architecture Specification (IDEAS) which will converge TOGAF, MODAF, and DoDAF into a single, unified architecture framework. Fig. 17.7 illustrates the basic MODAF model views [34].

17.3.5 The International Defense Enterprise Architecture Specification (IDEAS)

IDEAS is a multinational organization aimed at forming a unified System Architecture Framework and data exchange format for military and civilian system architecture MBSE modeling and design use. Creation of a unified architecture model framework will allow seamless sharing of data, architectures, and designs between

Table 17.5 DoDAF 2.2 operational system views

System views	
SV-1	System interfaces. Describes the systems, subsystems, and system items and their proposed interactions
SV-2	System resource flows. Describes the resource flows between system elements
SV-3	System-system matrix. Describes the relationships between systems, elements, subsystems, etc., given in the architectural views. This may include planned vs. existing interfaces, both internal and external
SV-4	System functionalities. Describes the functionality of the systems, subsystems, etc., and the data flows between the proposed system activities
SV-5a	Operational activity to functionality matrix. Describes the mapping between the proposed operational capabilities and operational activities
SV-5b	Operational activity to system tracing. Describes the mapping of systems, subsystems, etc., to operational activities
SV-6	System resource flows matrix. Describes resource exchanges between the systems and the attributes of those exchanges
SV-7	System measures. Describes the metrics that will be utilized for the system elements and their time frame(s)
SV-8	System evolution. Describes the program's plans for migrating from the current design to a more efficient system, or towards evolving the current system to a future, better implementation
SV-9	System technology and skills assessment. Describes emerging technologies, such as new hardware and software (usually COTS) that are expected to be available in the future (with time frame estimates) and how that will affect the overall system activities and performance
SV-10a	System rules model(s). Describes the constraints that are imposed on the system and its overall design. These are in terms of system functionality
SV-10b	System state transition. Describes the service effects and transitions in response to expected events
SV-10c	System event trace. Describes system-specific critical sequences-of-events that are documented in the operational views of the system

Table 17.6 DoDAF 2.2 operational project views

Project views	
PV-1	Describes the relationships between the program organizations of outlines the program organizational structures that are required to manage the program and maps these to program capabilities
PV-2	Program timelines. Describes key milestones, their timing throughout the program, and their interdependencies
PV-3	Program capability mapping. Describes the mapping between program capabilities to illustrate how program elements and/or specific projects within the program are utilized to achieve specific capabilities required by the system

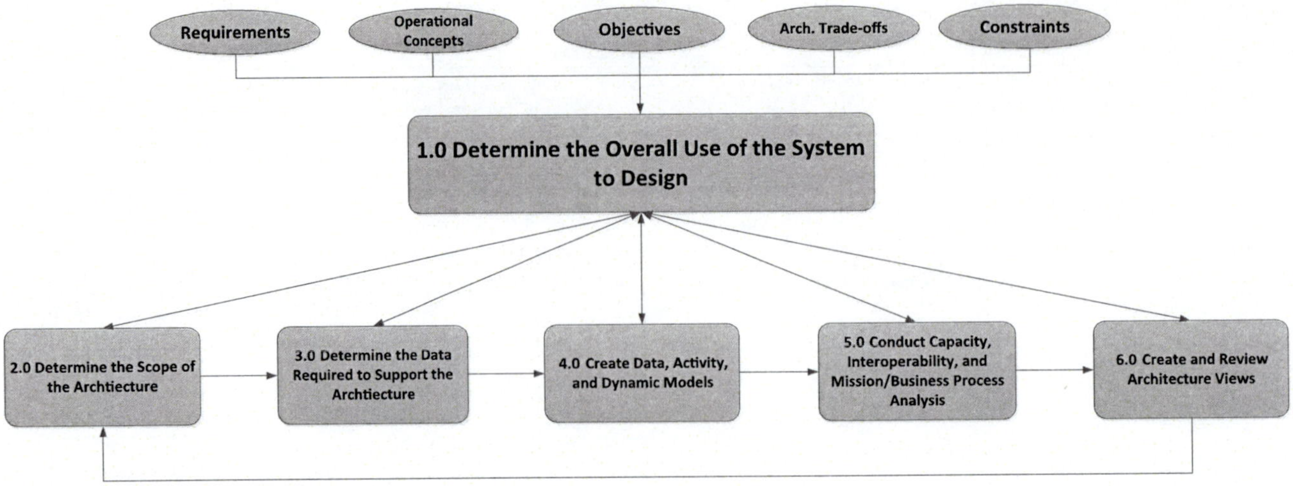

Fig. 17.4 Simplified view of the DoDAF MBSE architecture model development process

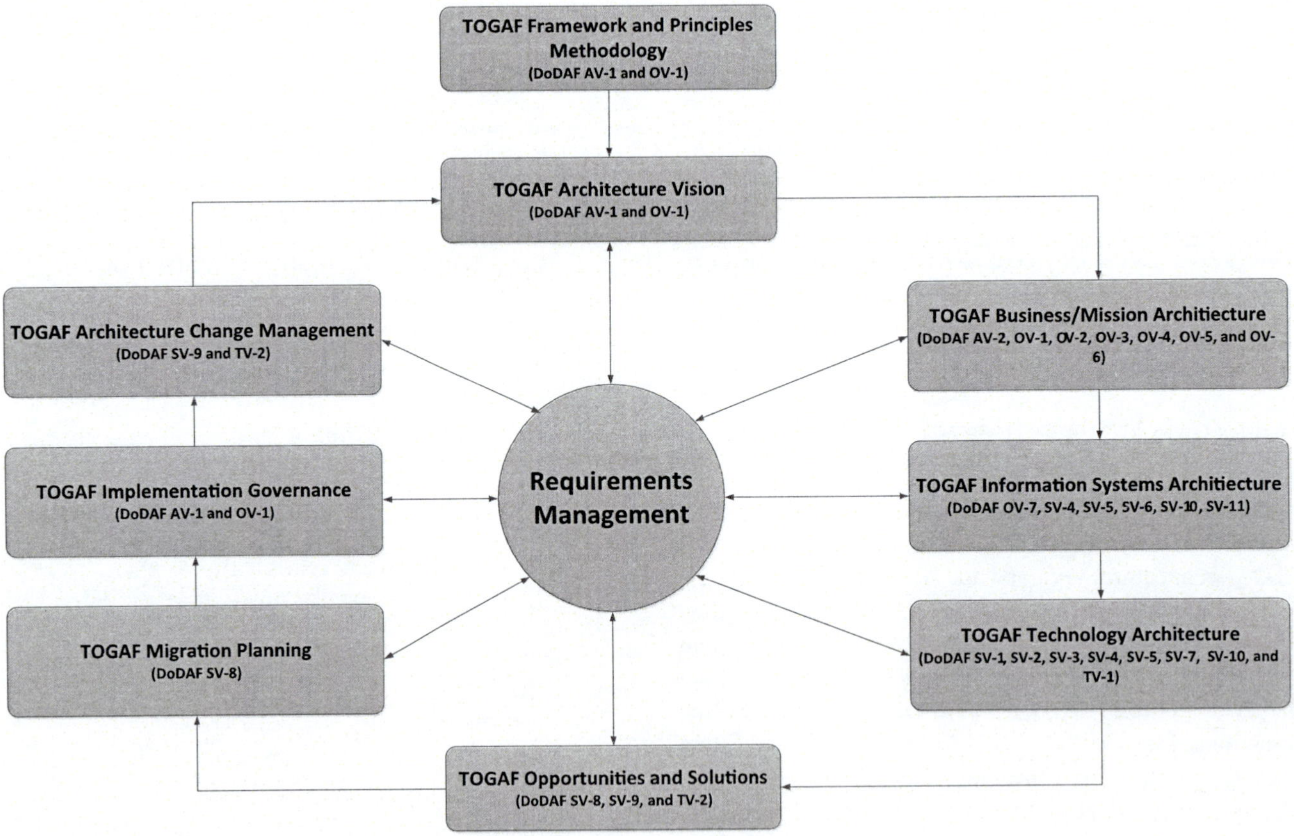

Fig. 17.5 MBSE TOGAF model views

partner nations. The initial scope of the IDEAS group is to provide translational capabilities between architecture frameworks (e.g., DoDAF and TOGAF), and the initial charter of the group is for exchange of systems architectural data to support multi-nation coalition planning and operations. Long-term plans are to look at merging the DoDAF 2.2 meta-model and the MODAF Ontological Data Exchange Mechanism (MODEM) into a unified architectural modeling tool (like how UML unified software architecture) [35].

17.3.6 The Unified Modeling Language (UML)

The **Unified Modeling Language** (UML) is a general-purpose modeling language in the field of software engineer-

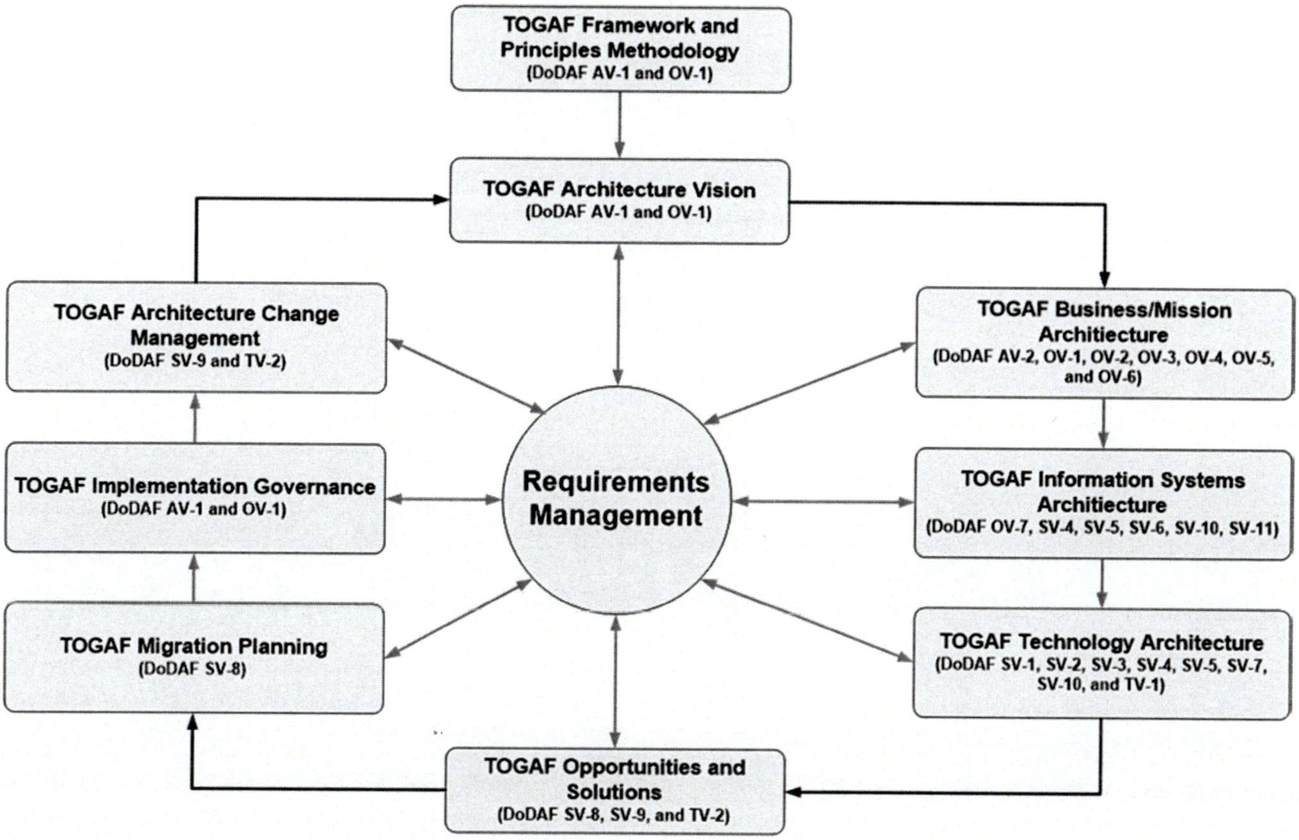

Fig. 17.6 TOGAF to DoDAF model view mapping

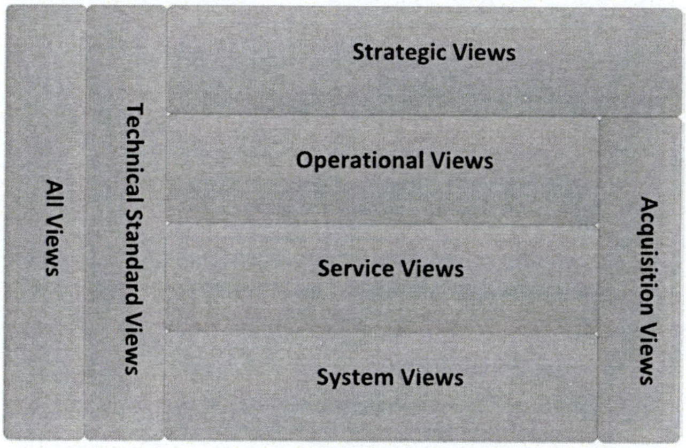

Fig. 17.7 MBSE MODAF model framework views

ing, which is designed to provide a standard way to visualize the design of a system. It was created and developed by Grady Booch, Ivar Jacobson, and James Rumbaugh at Rational Software during 1994–95 with further development led by them through 1996 [36]. In 1997 it was adopted as a standard by the Object Management Group (OMG), and has been managed by this organization ever since. In 2000 the Unified Modeling Language was also accepted by the International Organization for Standardization (ISO) as an approved ISO standard. Since then, it has been periodically revised to cover the latest revision of UML [37].

17.3.7 The System Modeling Language (SysML)

SysML is based on UML and involves modeling blocks instead of modeling classes, thus providing a vocabulary that's more suitable for systems engineering. A block encompasses software, hardware, data, processes, personnel, and facilities. SysML reuses a subset of UML2 (UML4SysML) and defines its own extensions. Therefore, SysML includes nine diagrams instead of the thirteen diagrams from UML2, making it a smaller language that is easier to learn and apply. SysML can be easily understood by the software community, due to its relation with UML2, while it is accessible to other communities, SysML makes it possible to generate specifications in a single language for heterogeneous teams, dealing with the realization of the system hardware and software modeling blocks (see Fig. 17.8). Knowledge is thereby captured through models stored in a single repository, enhancing communication throughout all teams. In the long term, blocks can be reused

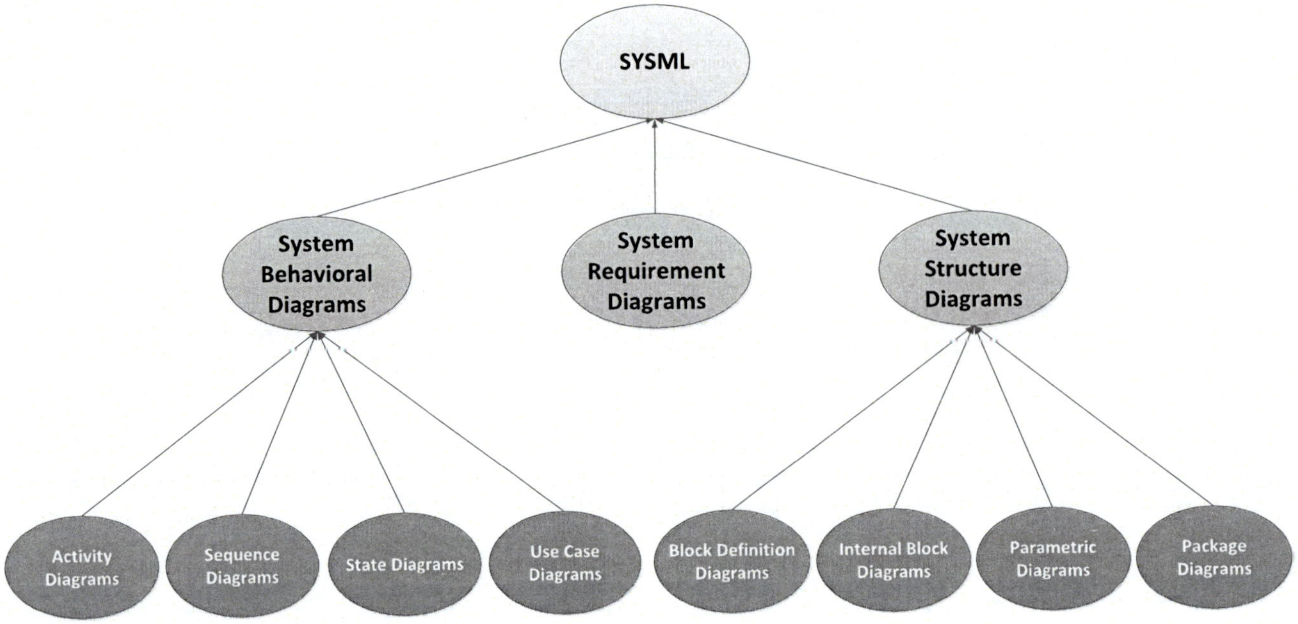

Fig. 17.8 MBSE SysML model process

as their specifications and models enable suitability assessment for subsequent projects [38].

> **Questions to think about**
>
> *Do the models need to be maintained in relation to the approved changes on the program?*
>
> *What happens when the models are allowed to decay (become outdated)?*
>
> *What happens to the system life cycle cost (LCC) model when the decay has occurred in the models?*

So now we've laid the foundation for MBSE, and the modeling languages utilized for modeling systems and producing common views for different disciplines within a program organization. But, the question remains, how does all this affect the MDRE organization and the creation and management of requirements? Next, we will explore MDRE considering the MBSE environment.

17.4 MDRE in a Model-Based Systems Engineering Environment

As we have discussed throughout the book, the starting point for any system development, whether you are using MBSE or not, is requirements. In the MBSE MDRE world, requirements should align with the four quadrants shown in Fig. 17.1 in describing the logical and physical aspects and characteristics of the system to be developed. As with any system to be developed, the initial set of requirements can come from many sources, customer Statement of Objectives (SoO), analytical processes, a formal change request, of possibly verbal requests during reviews [39]. Once the requirements are captured, they must be put into the MBSE model(s). Before ingesting requirements into the MBSE model(s), they should first go through a process of classification, deduplication, and then ensuring the phrasing is consistent with all other requirements in the MBSE models and system descriptions [40].

In the MBSE world the MDRE team must classify the requirements into three main MBSE requirements groups:

- **Mission/Business Requirements**: These are high-level requirements that are presented expressed in terms of overall system objectives from the customer organization and may come as a set of requirements or in form of a Statement of Objectives (SoO).
- **User Requirements**: These requirements generally come from stakeholder(s) who either use the system or represent those that will use/operate the system. These requirements are used to describe how the system is accessed, used, and the kinds of interactions users or other systems will have with the system to be developed.
- **System Requirements**: These requirements describe the required system capabilities, behaviors, information flows, etc. This includes requirements that indicate quality attributes (e.g., reliability, availability) which indicate how well or how long or under what conditions the system must operate withing or performance criteria it must meet.

The system requirements must be classified into the categories discussed in Chap. 11, based on functionality, subsystem, and component decomposition, and then reviewed, as discussed throughout the book, to ensure the language is consistent, concise, and unambiguous [41]. To develop a well-organized set of requirements that will drive the development of the MBSE system models, the MBSE methodology suggests requirements be organized in requirements packages, hierarchically traced, and tracked. One of the most important issues in building the requirements packages and hierarchical structure is creating structures that allow requirements modeling traceability to sources, processes, and to each element in the system architecture models. A key element of the MBSE digital model thread is that it can greatly simplify traceability by tracing across models. A requirement diagram is one view that can enable the MDRE to enumerate about each requirement, where it came from, how it drives the system models, and ultimately the system design, development, and test. The hierarchical structure requirements diagram (or view) should reveal:

- The requirements baseline associated with the requirements.
- Which requirements (if any) were derived from a given requirement.
- The use cases associated with each requirement.
- The test cases associated with each requirement.
- The part(s) of the system this requirement satisfies.

When creating the associated requirements traceability matrices to other artifacts like use cases and test cases, care should be taken not to create traceability matrices that are so complex that they inhibit model navigation and the creation of useful views of the model(s) [42]. Modeling packages like SysML provide the tools required to create requirements diagrams, traceability matrices, and other useful views of the MBSE-derived system models.

One of the big differences between classical MDRE and MBSE-driven MDRE is in the continued development and maintenance of the system models throughout the development life cycle and its effects on requirements management. Fig. 17.9 illustrates a classical iterative development timeline. In Fig. 17.9, initial system models and simulations are created to help in the design of the system during the analysis phase of design. Once these simulations and models have been tested, they are normally shelved and not used during development, verification, deployment, and transition to sustainment. MDRE engineers maintain the requirements baseline documentation and eventually hand the requirements set off to the sustainment group once the system is transitioned to the customer [43].

Figure 17.10 illustrates the changes to the timeline for MBSE in the same iterative development timeline. The differences may be subtle but are important. In Fig. 17.10 sys-

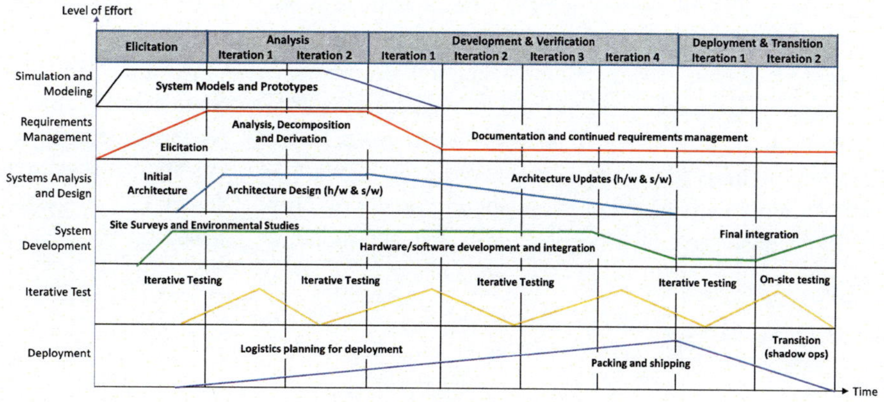

Fig. 17.9 Classical iterative development effort timeline

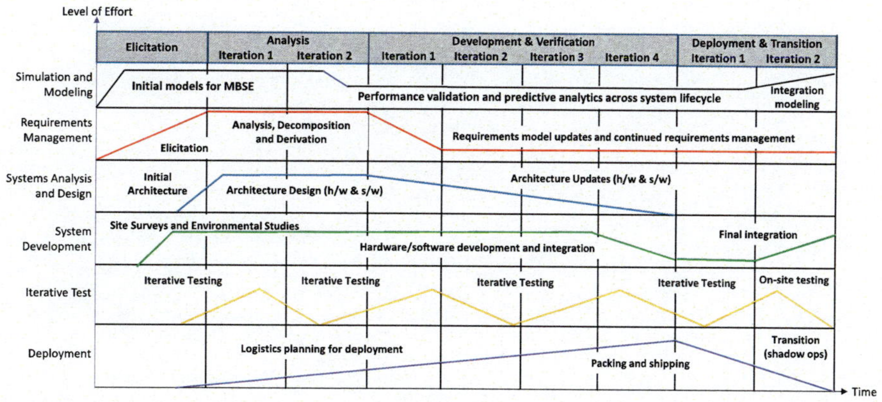

Fig. 17.10 MBSE iterative development effort timeline

tem models and simulations again are created during the analysis phase of development, but in the MBSE timeline, they are continually updated, and performance is validated against the models and the models updated to reflect the actual design and performance of the system [44]. This includes predictive analytics that are used during integration and test, driving new integration models to be created and used during deployment and final transition testing of the system. These models and simulations are then used by the sustainment team throughout the operational life cycle of the system to predict how the system will operate under changing environmental conditions the system may be put under that were not part of the original design. These can be valuable tools that support reduced maintenance and sustainment cost over years of system operations [45].

Similar differences would be present for waterfall or agile development timelines as well. The continued MDRE efforts pertain to updating and managing the requirements descriptive models, not maintaining purely requirements documentation. Fig. 17.11 illustrates the MBSE MDE requirements prototyping and modeling activities that are a continual part of the MBSE system development life cycle.

Information on the system domain (environment) is used to develop MBSE system models, which drive the creation of prototypes and mock-ups, based on early requirements elicitation and analysis efforts. Peer reviews of the system models and prototype scenarios are used to elicit feedback from the customer/stakeholder(s) to assess the validity of the requirements and subsequent models. One area that is important to capture in the early prototype efforts is consideration of human-centered design. Human-centered design is an ISO-standard approach to design and problem-solving commonly used in design and management frameworks that develops solutions to problems by involving the human perspective in all steps of the problem-solving process. Part of MBSE is considering all aspects of the system and modeling all aspects of the system, including human-systems interfaces. The use of Human-Centered Design and Human-Centered Thinking can help create MBSE system models of human interaction with the system that provide system designs better suited and more usable by those who must operate/interact with the system. Based on the feedback from the customer/stakeholder(s), adjustments are made to the requirements and a current requirement specification descriptive model is maintained [46]. As the system design matures through the development and test life cycle, system performance is vetted against the system/ requirement models to verify the system meets the overall objectives and is satisfying the requirement models for the system. This continues throughout, with adjustment made to the requirements and associated system models throughout the life cycle, including operations and sustainment. This continual comparison of the models to system performance is an important aspect of MBSE and the MDRE team becomes an integral part of the overall success of the system development and not just requirement document managers. When we put all these activities together, we see the entire MBSE process and how the MDRE team is involved throughout system development, test, deployment, and sustainment [47]. Fig. 17.12 illustrates the overall MBSE high-level context diagram. This is true regardless of the development style (waterfall, iterative, agile, etc.). Again, the difference between classical system development and MBSE system development involves the MDRE process of requirements refinement throughout the MBSE model-driven development. Fig. 17.13 illustrates this process.

As the MDRE process evolves though analysis of the Statement of Objectives, through the elicitation and analysis to derive/decompose/allocate system requirements, and the subsequent system design process, each stage supports a different set of system models and simulations. Feedback, based on model and simulation analyses, is used to refine

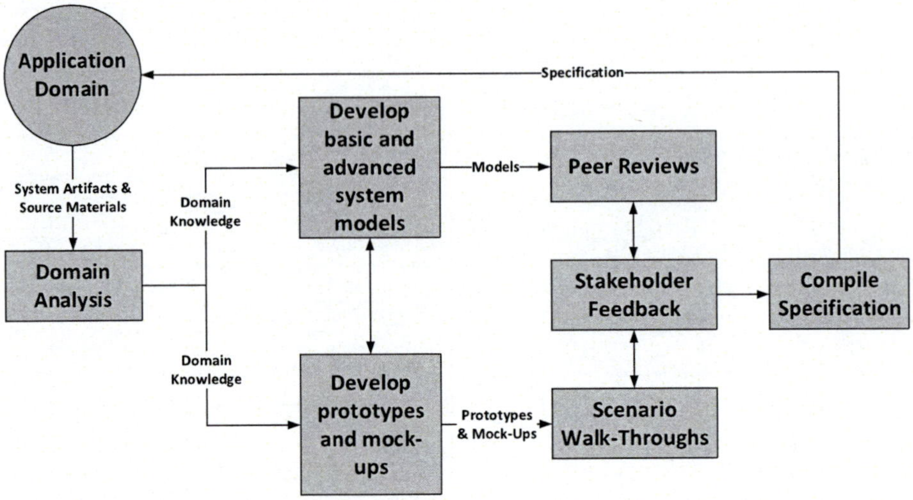

Fig. 17.11 MBSE MDRE requirements prototyping and modeling

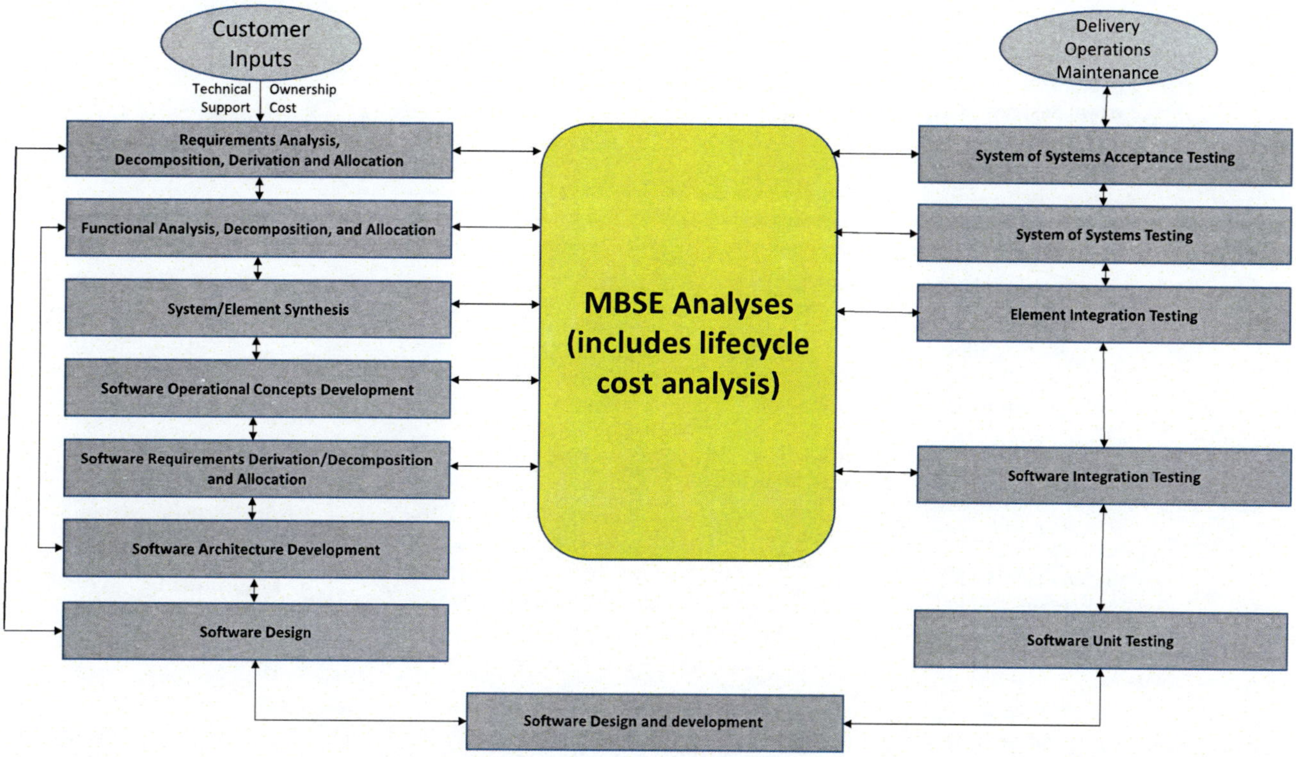

Fig. 17.12 MBSE high-level context diagram

both the models and the requirements throughout the system life cycle [48]. The MBSE MDRE requirements process, and life cycle, involves multiple disciplines to facilitate effective requirements engineering within the MBSE process. Fig. 17.14 illustrates this in terms of a Venn diagram of disciplines and functions that culminates in an effective MDRE model-driven approach for MBSE.

The use of modern MBSE development tools allows the systems architecture and software architecture to be linked to requirements. This then allows the architecture to be linked to class diagrams required for the MBSE process. By comparing the resultant software design (based solely on the code) with the system performance models, an analysis can be made to determine if the architecture of the code written has resemblance to the original software architecture, i.e., "does the code do what it was architected to do?" [49]. Fig. 17.15 illustrates this process.

The overall integration of MBSE tools, models, simulations, and views combined with a collaborative framework, like the process shown in Fig. 17.15, can create an effective MDRE process to fully support the MBSE process.

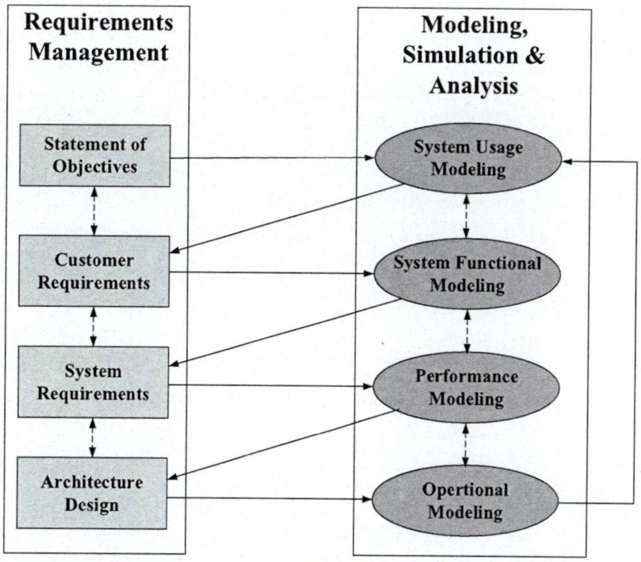

Fig. 17.13 MBSE MDRE feedback-driven requirements analysis process

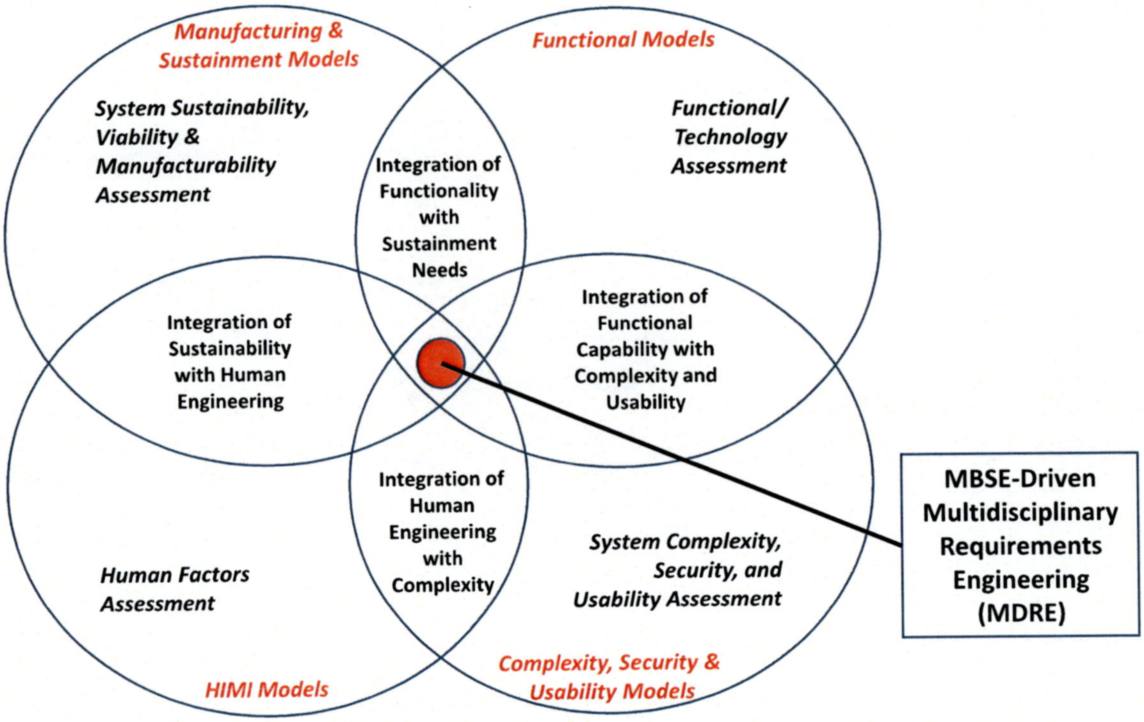

Fig. 17.14 MBSE MDRE cross-disciplinary approach

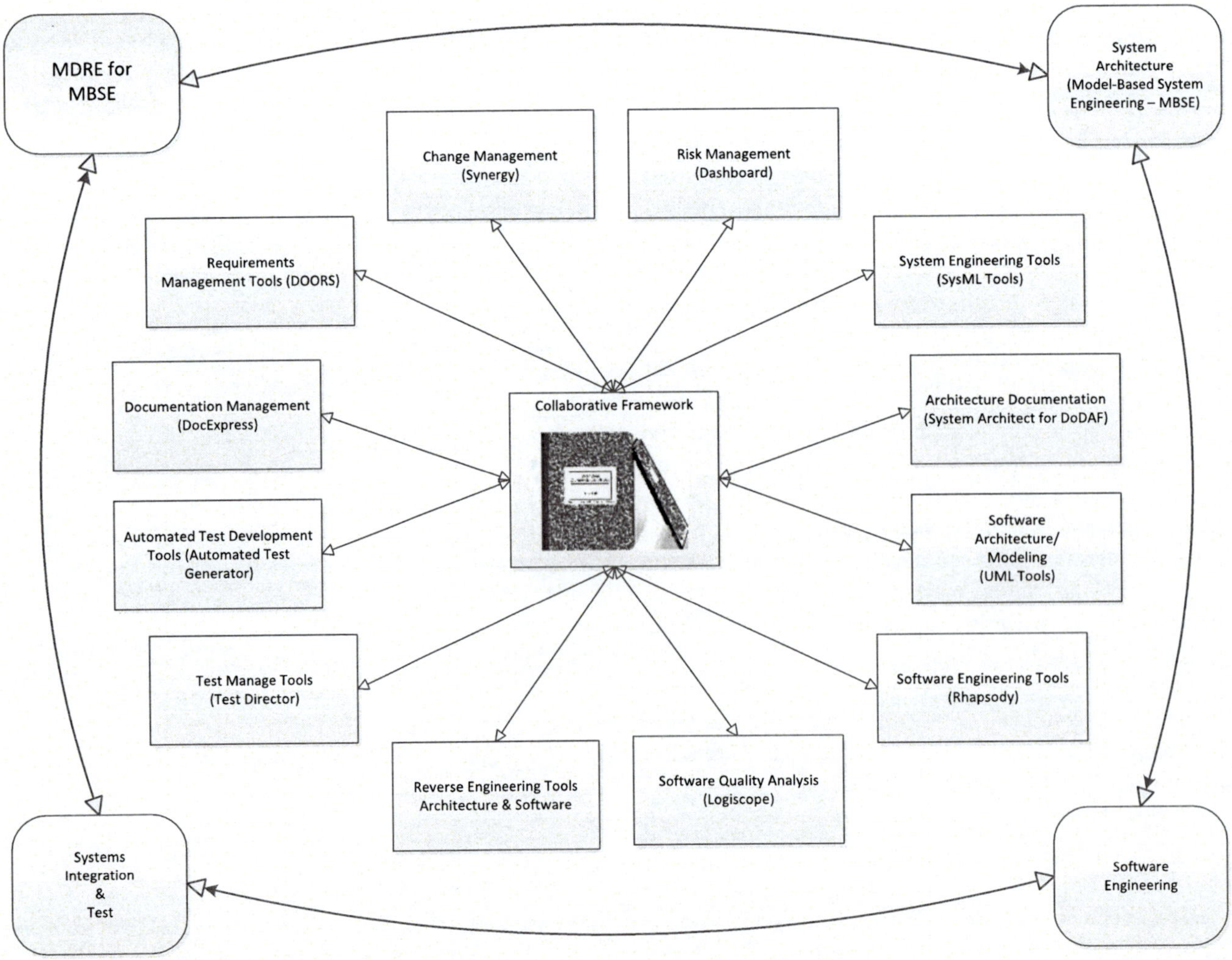

Fig. 17.15 MBSE MDRE collaborative framework

17.4.1 Design Problem

MDRE is a life-long journey, and anyone who tells you they have nothing left to learn does not truly understand the discipline or has never been involved in the development of a complex system. With the recent interest in alternative energies and talk of the "Green New Deal," along with the recent issues with ice storms in Texas taking out the power grid, the assignments for this textbook will follow the theme of effectively dealing with both local alternative energy power grids and the national infrastructure required to facilitate large-scale alternative energy production, sustainment, and management.

17.4.2 Design Problem Description: Alternative Energy Power Grid

The purpose of this project is to, over the course of the semester, develop a feasibility study, requirements analysis, requirements elicitation, derivation, and allocation, for a system that will facilitate a "local" alternative energy power grid capable of powering a neighborhood of 500 homes and a large biosphere capable of supporting growing year-round fruits and vegetables for the 500-home neighborhood. The student will have to deal with issues of choosing multiple alternative energies that can provide power throughout all seasonal difficulties, from intense summer heat to freezing winter temperatures and possible ice storms. Control, viability, sustainability, and maintainability will be major factors to understand the entire possibly 30-year life cycle of such a system. How to effectively utilize, store, and manage power distribution, along with when to upconvert "extra" energy to the national power grid, will be issues that drive many of the assignments. Simply relying on the current technologies and energy infrastructures will not move the country and world forward. Hopefully, this will provide insight to the students using this textbook to understand some of the issues that face everyone looking out over the next many decades.

17.4.3 DC Power Grid Requirements Discussion

In the initial days of power generation and distribution, DC power was far less practical than the AC distribution system that soon supplanted it. However, even after Tesla's AC system became the predominant system (and rightly so), Edison's DC power did not become obsolete for some time. Eventually power plants eliminated their delivery of DC power, but the overall concept of a DC power grid is not dead. Even our computer systems (large server systems) have uninterruptible power supplies which contain batteries that are charged off the grid and used to upconvert to AC to power devices when the AC power grid goes off-line. So, the notion of a DC power grid is not dead, but generally used as a backup system. The point is, the notion of a DC power grid has remained and now that renewable energies are viable, we should re-examine their use for widespread, local, power distribution.

One of the advantages of a locally generated DC station is energy savings as well as allowing a facility to run more easily off various DC sources, such as Solar Arrays. And for power-hungry installations, the notion of generating the power locally is growing in popularity. Part of the reason is that waste heat from the power generation generators (solar, geothermal, etc.) can be used to warm nearby buildings. By combining the production of heat and power, local power grids can squeeze much more useful energy out of the renewable energy they use, so this approach will certainly become more widespread as time goes on. The purpose of this design problem is to look ahead and provide an architecture and specification for a local DC power generation station that can serve a local community.

Currently many systems exist to control and handle alternative energies. There are systems to control and manage solar power, wind power, geothermal power, and others, but very few are designed to control, manage, store, and distribute energy from a power grid consisting of multiple alternative energy sources. As we saw the issues recently caused by the intense ice storms in Texas, one alternative energy source is not the answer to the growing energy needs/wants in our society. Each alternative energy source has limitations and issues associated with its use. By harnessing multiple alternative energies, and managing the storage of the energies produces, communities can be worry free from power generation. To facilitate utilizing multiple alternative energy sources, a control center must be included in the design that can manage the energy creation, distribution, storage, and return unused energy to the power grid.

17.4.4 High-Level Technical Description of the DC Power Grid System

The purpose of the multiple alternative energies DC power grid is to:

- Supply DC power to a customer community of 500 homes and a biosphere for growing food.
- Provide a storage device to store energy in case of a system failure.
- Determine the amount of surplus energy that is above and beyond what is required to power the grid and charge the backup system, and then deliver that surplus energy to the national power grid systems.

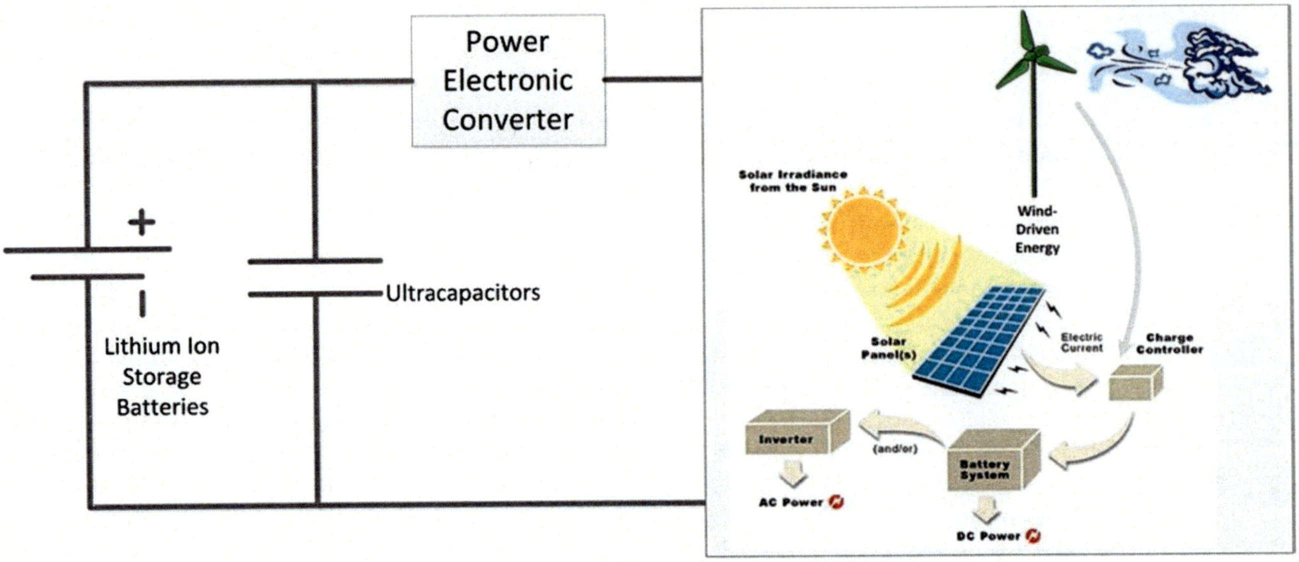

Fig. 17.16 Alternative power management system overview

Throughout the semester you will be asked to analyze, design, and architect different components of the overall DC power grid system. This will include research into alternative designs, like lithium nanowire storage systems for backup power and/or supercapacitors (also called ultracapacitors). Fig. 17.16 illustrates these technologies connected to the DC Power Grid Regulator Control Center.

17.4.5 High-Level Technical Requirements for the DC Grid System

- Design shall utilize at least two alternative energies, with a maximum of 4.
- Total power generation shall be 3.0 megawatts minimum.
- Each of the alternative energy sources shall be capable of generating at least 60% of minimum wattage.
- DC power grid shall generate DC power, with grid voltage being 48, 24, or 12 volts.
- DC power grid shall be capable of servicing at least 500 single family homes and one 4-acre biosphere with hydroponics capable of growing vegetables year-round. (See Fig. 17.17 for a block diagram of a DC-powered home).
- All lights, appliances, and facilities in each home and the biosphere shall be configured for DC operations, if possible. Things like electric ovens may need DC/AC conversion to accommodate their 220v. requirements.
- Only allowed exceptions are the HVAC system and the Washer/Dryer, although DC is preferred.
- There shall be energy storage capability capable of operating the grid of 500 single family dwellings for 7 days if all alternative forms of energy are not producing.
  - Compliance Criteria: Energy storage devices shall be capable of producing 750,000 watts for the entire 7-day period. Nominal energy storage should be batteries or supercapacitors.
- The DC power grid system shall include controlling h/w and s/w to manage the multi-alternative energy strategies and production.
- All energy produced above the required home and energy storage usage shall be upconverted to 220vac for release to the national power grid.
- The DC power grid shall be capable of operating in a temperature range of −40°F to 125°F without degradation of power generation.
- The DC power grid must meet OSHA regulations.

17.4.6 Assignments

- Write a 20–30-page report explaining the following:
  - Which system models would you create for your multi-alternative energy system and why (there must be at least 2)?
    - Write a test case for each system model you described.
    - Create a use case for each system model you described.
  - Which performance models would you create for your multi-alternative energy system and why (there must be at least 2)?
    - Write a test case for each performance model you described.
    - Create a use case for each performance model you described.

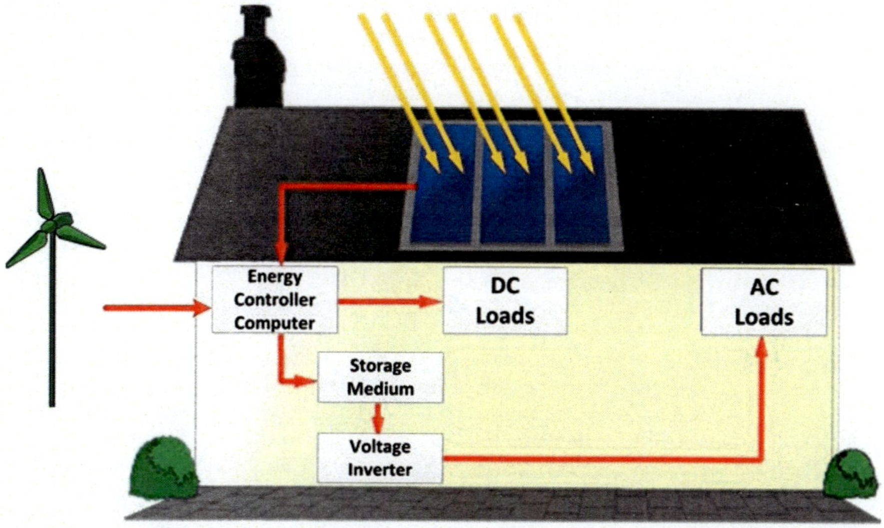

Fig. 17.17 Example of a whole house power grid

- Create a mock-up for the multi-alternative system control screen.
- Research model-based systems engineering. Write a 5–10-page report on your perception of whether MBSE helps or hinders the overall MDRE process and why?

References

1. Jamshidi, M. (2008). System of systems engineering–new challenges for the 21st century. *IEEE Aerospace and Electronic Systems Magazine, 23*(5), 4–19.
2. Lane, J., & Boehm, B. (2008). System of systems lead system integrators; where do they spend their time and what makes them more or less efficient? *Systems Engineering, 11*, 81–91.
3. Haskins, C. (2008). Using patterns to transition systems engineering from a technological to social context. *Systems Engineering, 11*, 147–155.
4. de Bruijn, H., & Herder, P. (2009). System and actor perspectives on sociotechnical systems. *IEEE Transactions on Systems, Man, Cybernetics A, Systems, Humans, 39*(5), 981–992.
5. Tien, J. (2008). On integration and adaptation in complex service systems. *Journal of Systems Science and Systems Engineering, 17*, 385–415.
6. Rhodes, D. (2008). Addressing systems engineering challenges through collaborative research. In *SEARI—Systems engineering advancement research initiative*. MIT Press. 110 IEEE TRANSACTIONS ON SYSTEMS, MAN, AND CYBERNETICS—PART C: APPLICATIONS AND REVIEWS, VOL. 42, NO. 1, JANUARY 2012.
7. Sheard, S., & Mostashari, A. (2009). Principles of complex systems for systems engineering. *Systems Engineering, 12*, 295–311.
8. Wymore, A. (1993). *Model-based systems engineering*. CRC.
9. INCOSE. (2007). In C. Haskins, K. Forsberg, & M. Krueger (Eds.), *Systems engineering handbook—A guide for system life cycle processes and activities*. Int. Council Syst. Eng.
10. Hybertson, D., & Sheard, S. (2008). Integrating and unifying old and new systems engineering elements. *INSIGHT-INCOSE Journal, 11*, 13–16.
11. Honour, E. (2008). Systems engineering and complexity. *INSIGHT-INCOSE Journal, 11*, 20–21.
12. Rouse, W. (2003). Engineering complex systems: Implications for research in systems engineering. *IEEE Transactions on Systems, Man, Cybernetics C, Applications and Reviews, 33*(2), 154–156.
13. Arnold, S. (2007). Where is standardization guiding us? *INSIGHT-INCOSE Journal, 10*, 41–43.
14. Valerdi, R., & Davidz, H. (2009). Empirical research in systems engineering: Challenges and opportunities of a new frontier. *Systems Engineering, 12*, 169–181.
15. Friedenthal, S., Moore, A., & Steiner, R. (2008). *A practical guide to SysML, the systems modeling language*. Morgan Kaufmann/ OMG, Elsevier.
16. Cloutier, R., & Verma, D. (2007). Applying the concept of patterns to systems architecture. *Systems Engineering, 10*, 138–154.
17. Tang, A., Han, J., and Chen, P. (2004). "A comparative analysis of architecture frameworks," Centre Component Software Enterprise Syst., School Inf. Technol. Swinburne Univ. Technol., Melbourne, Vic., Australia, Technical Rep. SUTIT TR2004.01.
18. Rumbaugh, J., Jacobson, I., & Booch, G. (1999). *The unified modeling language reference manual*. Addison-Wesley Longman.
19. Wymore, A. W. (2004). *"Contributions to the mathematical foundations of systems science and systems engineering," systems movement: Autobiographical retrospectives*. The University of Arizona.
20. Browning, T. (2009). The many views of a process: Toward a process architecture framework for product development processes. *Systems Engineering, 12*, 69–90.
21. Richards, M., Shah, N., Hastings, D., and Rhodes, D. (2007). "Managing complexity with the department of defense architecture framework: Development of a dynamic system architecture model," Eng. Syst. Div., Mass. Inst. Technol., Cambridge, MA, Working Paper ESD-WP-2007-09.
22. Kruchten, P. (1995). The 4+1 view model of architecture. *IEEE Software, 12*(6), 42–50.
23. Ludewig, J. (2003). Models in software engineering—An introduction. *Software System Modelling, 1*, 5–14.
24. Sussman, J. (2000). *Introduction to transportation systems*. Artech House.
25. Buede, D. (2009). *The engineering design of systems—models and methods* (2nd ed.). Wiley.
26. Stanford Encyclopedia of Philosophy. Models in Science.[Online]. (2006). Available: http:/plato.stanford.edu/entries/models science.
27. Hughes, R. (1997). Models and representations. *Philosophy in Science, 64*, 325–336.
28. Di Carlo, T., Khoshnevis, B., & Udwadia, F. (2009). Whole brain thinking in systems architecting. *Systems Engineering, 12*, 265–273.

29. Soliman, S. (2005). Systems and creative thinking. Center for Advancement of Postgraduate Studies and Research in Engineering Sciences, Faculty of Engineering, Cairo University, Cairo, Egypt [Online]. Available: http:/www.pathways.cu.edu.eg/subpages/Creativity-Engine.htm
30. Senge, P. (1990). *The fifth discipline—The art & practice of the learning organization*. Doubleday.
31. Ross, D. (1977). Structured analysis (SA): A language for communicating ideas. *IEEE Transactions on Software Engineering, SE-3*(1), 16–34.
32. Kim, C.-H., Weston, R., Hodgson, A., & Lee, K.-H. (2003). The complementary use of IDEF and UML modelling approaches. *Computers in Industry, 50*, 35–56.
33. Oliver, D., Andary, J., & Frisch, H. (2009). Model based systems engineering. In A. Sage & W. Rouse (Eds.), *Handbook of systems engineering and management* (2nd ed., pp. 1361–1399). Wiley.
34. Booch, G., Rumbaugh, J., & Jacobson, I. (1999). *The unified modeling language user guide*. Addison-Wesley Longman.
35. Dori, D. (2002). *Object-process methodology: A holistic systems paradigm*. Springer.
36. Grobshtein, Y. and Dori, D., "Creating SysML views from an OPM model," in Proc. 2nd Int. Conf. Model Based Syst. Eng., Herzelya and Haifa, Israel, Mar. 2009, pp. 36–45.
37. Wilkiens, T. (2007). *Systems engineering with SysML/UML—Modeling, analysis, design*. Morgan Kaufmann/OMG.
38. Grady, J. (2009). Universal architecture description framework. *Systems Engineering, 12*, 91–116.
39. Mellor, S., Clark, A., & Futagami, T. (2003). Model driven development. *IEEE Software, 20*(5), 14–18.
40. Andary, J., & Oliver, D. (2007). Models in systems engineering and software engineering. *INSIGHT-INCOSE Journal, 10*, 26–27.
41. Bahill, A., & Botta, R. (2008). Fundamental principles of good system design. *Engineering Management Journal, 20*(4), 9–17.
42. Little, J. (1970). Models and managers: The concept of a decision calculus. *Management Science, 16*(8), B466–B485.
43. Valerdi, R. (2008). *The constructive systems engineering cost model (COSYSMO): Quantifying the costs of systems engineering effort in complex systems*. VDM Verlag.
44. Baker, L., Clemente, P., Cohen, B., Permenter, L., Purves, B., and Salmon, P. "Foundational concepts for model driven syst. design," INCOSE Model Driven Syst. Design Interest Group, Int. Council Syst. Eng., San Diego, CA, White Paper.
45. Wymore, A. (2004). Systems movement: Autobiographical retrospectives, contributions to the mathematical foundations of systems science and systems engineering. *International Journal of General Systems, 33*(6), 593–610.
46. J. Estefan, "MBSE methodology survey," INSIGHT-INCOSE Journal, vol. 12, pp. 16–18, Apr. 2009.
47. INCOSE, "Survey of model-based systems engineering (MBSE) methodologies," Model-Based Systems Engineering Initiative of the International Council on Systems Engineering [published online INCOSE TD 2007 003 01], 2008.
48. J. Garcia, Jr., "Executable and integrative whole-system modeling via the application of OpEMCSS and Holons for model-based systems engineering," INSIGHT-INCOSE Journal, vol. 12, pp. 21–23, Dec. 2009.
49. Peak, R., Paredis, C., McGinnis, L., Friedenthal, S., and Burkhart, R. (2010). "Modelling & simulation interoperability team status update," presented at INCOSE Int. Workshop, Phoenix, AZ.

18 Conclusions and Discussion: Putting it all Together

Abbreviations

AC	Alternating current
ADA	American with Disabilities Act
CI	Configuration item
COTS	Commercial off the shelf
CPPI	Continuous product/process improvement
DC	Direct current
EEOC	Equal Employment Opportunity Commission
F	Fahrenheit
HVAC	Heating, ventilation, and air conditioning
MBSE	Model-based systems engineering
MDRE	Multidisciplinary requirements engineering
OSHA	Occupational Safety and Health Administration
RCM	Requirements Change Management
V	Volts
Vac	Volts AC
Vdc	Volts DC
VTC	Video teleconferencing

In the end what we want, what we need, is MDRE that provides value and is useful to the overall life cycle of system development. Requirement engineering is important because it allows systems to be architected, designed, implemented, tested, deployed, operated, and maintained with separable components at all levels of the system (element, subsystems, services, configuration items (CIs), and components), such that the developed system can be managed, operated, and maintained at low cost over a long period of time [1]. Some programs would not benefit from this approach, but those are few and generally small projects. As an MDRE engineer, hopefully the following phrase is never used:

> There's never enough time to do it right, but there's always enough time to over.

This classic phrase described many, many systems, as engineers and managers fail to understand the importance and need to do the requirements engineering right the first time. What follows is a discussion of what happens when MDRE is done right, and what happens when it is down wrong [2].

18.1 When MDRE Goes Right

Almost all programs will encounter problems in development; however, when the MDRE discipline is properly followed, and the full scope of the objectives and requirements provided by the customer are address across all disciplines of systems engineering, the programs are placed into operations. Most well-engineered programs (correct requirements management) meet the cost and schedule targets, in addition to the technical capabilities desired [3].

Properly executed MDRE may encounter pushback by the customer or company management, where the "do more with less" demands are encountered. The purpose of MDRE is to provide the necessary facts and data for the overall scope of the program, and all changes being requested in the development. The negotiations between the MDRE and the customer/management teams typically flow smoothly when proper facts and data are presented.

Well-executed programs establish and enforce the engineering processes and practices to be used during development. All new personnel added to a MDRE effort should be provided training on the contractual objectives, concepts of operations, requirements, and architectural flow down, and procedures being employed on the program. While most organizations have standardized processes and practices, they allow the processes and practices to be tailored based on the program. All startup staff, new supervisors, and new engineers being added to a development should be trained on the program-specific process and practices. Well-run programs do not allow engineers to

J. A. Crowder, C. W. Hoff, *Requirements Engineering: Laying a Firm Foundation*,
https://doi.org/10.1007/978-3-030-91077-8_18

start on the program and continue operating as if they came from another program [4].

Well-executed programs establish milestones with well-defined requirements and objectives for those milestones, as well as on-ramp and off-ramp plans caused by changes in direction, objectives, requirements, and like procurement changes. The entire purpose of following MDRE with top-down engineering discipline is to protect both the customer and development teams from drifting off course and ending up with a failed program.

Some programs are pilot programs that will lead to the overall system of systems development and or integration. The MDRE engineer should ensure that the scope of the requirement efforts match the objectives of the procurement, and not allow unplanned growth to occur during development (scope creep). This requires the MDRE team ensure that the flow down of requirements are properly being implemented in the software and hardware baselines appropriate to the procurement. Incorporation of "nice to haves" or "we know this will be needed" solutions is not executed in a well-managed MDRE environment.

Programs properly executed lead to satisfied customers, company management, development teams, and a low stress work environment, most importantly working, fielded systems. It is the satisfied customer that provides you with more of their business, and referrals to other customers. It is for this reason the engineer should develop MDRE practices and incorporate them into their daily duties [5].

18.2 When MDRE Goes Wrong

As has been discussed throughout this book, when engineers do not follow MDRE practices, with top-down analysis and engineered solutions, the development organization becomes exposed to higher risks and potential for failure. When program staff is not properly trained on the customer objectives, requirements, program plans, processes, and practices, the MDRE efforts will take on a flavor of "multiple different programs within a program." The development will enter a reactionary engineering culture, where development is fragmented across different standards and solution sets, all leading to extensive risk, cost and schedule growth, and potential program cancellation [6].

The decomposition of the system should be uniform across the specific development, which will depend on the depth of the procurement. A simple prototype may have a few tiers of decomposition, whereas an enterprise of System of Systems may have dozens of tiers. Skipping tiers of decomposition typically leads to failure but is acceptable when managed properly on smaller programs. An improper decomposition will eventually lead to bottoms-up and/or reactionary engineering during integration and test phases, where massive cost and schedule overruns will impact the fielding of the system [7].

The practice of reuse product lines, and therefore requirements reuse, comes with risk in new developments. More often than naught, the customer may have selected an organization to perform a new development to add diversity to his vendor base. While the selected vendors may have existing reuse product lines, the MDRE team should ensure proper requirements and architectural flow down occurs. This allows the development staffs to select well-documented solutions from the reuse libraries [8].

The use of bottoms-up development, where reuse should come with libraries of linked artifacts that enable the selection of reuse solutions for inclusion in a new development. Taking for granted, that the older solution solves the new procurement with minor requirement changes, is a frequent cause of program failures. The use of a product line, with changes to support a new development, typically leads to bottoms-up engineering, which then leads to reactionary engineering and then reactionary management and eventual program cancellation. In numerous cases:

- The reuse product lines are not up to the current security requirements demanded for the new system.
- The reuse product lines' COTS products have become obsolete in the marketplace.
- The older human-machine interfaces are replaced by EEOC and/or disability act requirements (example ADA).
- The system will not be maintained by the developer due to customer procurement constraints, etc. It is for these reasons that many developments have extensive cost overruns and/or failures when using reuse product lines.

Throughout this book, examples of failed developments were used to outline why requirements engineering should employ multidisciplinary solutions to the daily activities of the development staff. A failure on a prototype garners low attention, but a failure of a system development becomes a marketing nightmare for the organization and in some cases the entire company [9].

18.3 Final Thoughts on MDRE and Agile Development

As we move toward more complex system designs and as Agile Software development becomes even more prevalent, MDRE must adopt new philosophies and paradigms to keep up with the evolutionary nature of modern systems design (e.g., MBSE). The MDRE team(s) must embrace this change and develop methodologies and structures which allow rapid and accurate evolution of the Systems Architecture as the program progresses through its life cycle [10]. The evolution to Agile Systems Engineering is a natural progression from the pre-industrial age, through the industrial age, and now into the information age. The information age can also be thought of as the "Knowledge Age," with the understanding that data by itself means nothing. Data must be turned into information (data with context) and then into knowledge to support decision analysis in modern systems. The MDRE engineers must be on the leading edge of this "Knowledge Paradigm."

18.4 MDRE and the End of the "Paper Paradigm"

The paper paradigm pervades every aspect of our lives. Efficiency improvements often lead to the generation of more paper as quality improvements are sought. The purpose of modern techniques like MDRE and MBSE is to capture the process using information technology, eliminating paper capture of information and rendering the process repeatable and reusable. Using paper, and even using thousands of disparate files in "electronic collaboration" paradigms, causes us to be awash in information and dependent of quality checks on the product upon its completion. Even the use of standard office automation products provides little help. Many design documents and information are lost in a sea of electronic folders from individual engineers in an ocean of directories. Often information is lost, as no one knows where the information was stored, under what directory, and what naming convention. This process is little better than paper reports in separate filing cabinet. The automation for MDRE engineering functions and disciplines is similar. For functions, such as requirements management require tools that must be self-documenting. If the MDRE has to "copy" information into a flat-file report, then there is opportunity for error and quality control checking will have to be reinstated. The team need only check the original model-based requirements, knowing the new data propagation will be correct. The end goal is integrating modeling and management tools together to form a requirement knowledgebase that captures the requirement management and metrics analysis process cradle to grave. The following attributes have been noted [11]:

- ***Productivity*** increases up to tenfold.
- ***Repeatability*** enables junior engineers and team members to reliably reuse the design.
- ***Shareability*** has enabled the modeling knowledgebase to be deliverable to the customer.
- ***Verification and Validation*** engineers now review the requirements and model knowledgebase for design methodology, code base, results, etc.; a complete design, development, test, and deployment package.

18.5 Organizational Changes for MDRE

People dislike change. It takes them out of their comfort zone. Human nature demands continuity and consistency. For the MDRE engineer, these factors must be taken into consideration if continual improvement is to be realized. There are telltale signs that indicate when change is a natural part of the process:

- ***Quality*** is measured to prove the customer's issues are being met and to provide the basis for forcing change. Changes required for continuous product/process improvement (CPPI) to become an everyday affair.
- ***Quality is implicit*** and is as much a part of the engineering as correct math; if the team cannot generate correct math, then the organization needs a math department. The same is true for quality. So if the organization has a quality department, ***it indicates engineering processes that are not robust cannot integrate change***.
- ***Quality is related to training and education*** because change is perpetual. Quality is a race to be run, not just an objective to complete. Therefore, it is likely that at least one fifth of an engineer's time will be spent in training and education. Without it, the simulations and models required to find improvement opportunities will have low yields.

18.5.1 The MDRE Organization

New organizational structures are required to facilitate that allow the MDRE teams to own the products and resources. In some respects, this is a return to previous organizational structures where the team are the industrial engineers, determining how the resources can best be utilized for process-product improvement. These resources may be allocated to education, faster hardware and software, additional team members, etc. This may require an organizational focus on long-term profit before commitment can be made, as the cultural change involves the engineer as well as the manager. Some short-term losses will no doubt be incurred as the team learns to take advantage of the new paradigm [12]. Change should be an integral component of the MDRE design methodology and affects the whole product and process life cycle. Because the engineer and his tools are part of this, and because computer technology has had such a widespread impact, a holistic view must be taken. The discussion of new organizational structure is a focus on how a change in systems engineering to the new MDRE paradigm requires a change in work organization. Engineering organizations were designed when engineering methods were manual and unsophisticated. This new paradigm is possible using modern information systems technology and Web services to manage the mass of information. The real question to be answered is: will the manager cede control to the team? The solution will most likely come through economic necessity, and the sheer inevitability of the Information Age.

Increasing quality should be closely linked to MDRE. Imagine a process with all waste eliminated. Now design it. In contrast, engineering that is driven by conformance-to-spec is suffering paradigm paralysis. To eliminate waste there must be quality in the process, as this will affect quality of the product. For the MDRE engineer, quality and robustness are often linked, as the measure of performance can be the same for each. Therefore, advanced design methods have become even more valued as a means of reducing variance. The toughest challenge for the MDRE may be to make statistics an intuitive skill; solutions to improving quality will then follow more comfortably.

18.5.1.1 Design Problem

MDRE is a life-long journey, and anyone who tells you they have nothing left to learn does not truly understand the discipline or has never been involved in the development of a complex system. With the recent interest in alternative energies and talk of the "Green New Deal," along with the recent issues with ice storms in Texas taking out the power grid, the assignments for this textbook will follow the theme of effectively dealing with both local alternative energy power grids and the national infrastructure required to facilitate large-scale alternative energy production, sustainment, and management.

18.5.1.2 Design Problem Description: Alternative Energy Power Grid

The purpose of this project is to, over the course of the semester, develop a feasibility study, requirements analysis, requirements elicitation, derivation, and allocation, for a system that will facilitate a "local" alternative energy power grid capable of powering a neighborhood of 500 homes and a large biosphere capable of supporting growing year-round fruits and vegetables for the 500-home neighborhood. The student will have to deal with issues of choosing multiple alternative energies that can provide power throughout all seasonal difficulties, from intense summer heat to freezing winter temperatures and possible ice storms. Control, viability, sustainability, and maintainability will be major factors to understand the entire possibly 30-year life cycle of such a system. How to effectively utilize, store, and manage power distribution, along with when to upconvert "extra" energy to the national power grid, will be issues that drive many of the assignments. Simply relying on the current technologies and energy infrastructures will not move the country and world forward. Hopefully, this will provide insight to the students using this textbook to understand some of the issues that face everyone looking out over the next many decades.

18.5.1.3 DC Power Grid Requirements Discussion

In the initial days of power generation and distribution, DC power was far less practical than the AC distribution system that soon supplanted it. However, even after Tesla's AC system became the predominant system (and rightly so), Edison's DC power did not become obsolete for some time. Eventually power plants eliminated their delivery of DC power, but the overall concept of a DC power grid is not dead. Even our computer systems (large server systems) have uninterruptible power supplies which contain batteries that are charged off the grid and used to upconvert to AC to power devices when the AC power grid goes off-line. So, the notion of a DC power grid is not dead, but generally used as a backup system. The point is, the notion of a DC power grid has remained and now that renewable energies are viable, we should re-examine their use for widespread, local, power distribution.

One of the advantages of a locally generated DC station is energy savings as well as allowing a facility to run more easily off various DC sources, such as Solar Arrays. And for power-hungry installations, the notion of generating the power locally is growing in popularity. Part of the reason is that waste heat from the power generation generators (solar, geothermal, etc.) can be used to warm nearby buildings. By combining the production of heat and power, local power grids can squeeze much more useful energy out of the renewable energy they use, so this approach will certainly become more widespread as time goes on. The purpose of this design problem is to look ahead and provide an architecture and specification for a local DC power generation station that can serve a local community.

Currently many systems exist to control and handle alternative energies. There are systems to control and manage solar power, wind power, geothermal power, and others, but very few are designed to control, manage, store, and distribute energy from a power grid consisting of multiple alternative energy sources. As we saw the issues recently caused by the intense ice storms in Texas, one alternative energy source is not the answer to the growing energy needs/wants in our society. Each alternative energy source has limitations and issues associated with its use. By harnessing multiple alternative energies, and managing the storage of the energies produces, communities can be worry free from power generation. To facilitate utilizing multiple alternative energy sources, a control center must be included in the design that can manage the energy creation, distribution, storage, and return unused energy to the power grid.

18.5.1.4 High-Level Technical Description of the DC Power Grid System

The purpose of the multiple alternative energies DC power grid is to:

- Supply DC power to a customer community of 500 homes and a biosphere for growing food.
- Provide a storage device to store energy in case of a system failure.
- Determine the amount of surplus energy that is above and beyond what is required to power the grid and charge the backup system, and then deliver that surplus energy to the national power grid systems.

Throughout the semester you will be asked to analyze, design, and architect different components of the overall DC power grid system. This will include research into alternative designs, like lithium nanowire storage systems for backup power and/or supercapacitors (also called ultracapacitors). Fig. 18.1 illustrates these technologies connected to the DC Power Grid Regulator Control Center.

18.5.1.5 High-Level Technical Requirements for the DC Grid System

- Design shall utilize at least two alternative energies, with a maximum of 4.
- Total power generation shall be 3.0 megawatts minimum.
- Each of the alternative energy sources shall be capable of generating at least 60% of minimum wattage.
- DC power grid shall generate DC power, with grid voltage being 48, 24, or 12 volts.
- DC power grid shall be capable of servicing at least 500 single family homes and one 4-acre biosphere with hydroponics capable of growing vegetables year-round. (See Fig. 18.2 for a block diagram of a DC-powered home).
- All lights, appliances, and facilities in each home and the biosphere shall be configured for DC operations, if possible. Things like electric ovens may need DC/AC conversion to accommodate their 220v. requirements.
- Only allowed exceptions are the HVAC system and the Washer/Dryer, although DC is preferred.
- There shall be energy storage capability capable of operating the grid of 500 single family dwellings for 7 days if all alternative forms of energy are not producing.
  - Compliance Criteria: Energy storage devices shall be capable of producing 750,000 watts for the entire 7-day period. Nominal energy storage should be Batteries or Supercapacitors.
- The DC power grid system shall include controlling h/w and s/w to manage the multi-alternative energy strategies and production.
- All energy produced above the required home and energy storage usage shall be upconverted to 220vac for release to the national power grid.
- The DC power grid shall be capable of operating in a temperature range of −40°F to 125°F without degradation of power generation.
- The DC power grid must meet OSHA regulations.

18.5.1.6 Assignments

- Answer the following questions:

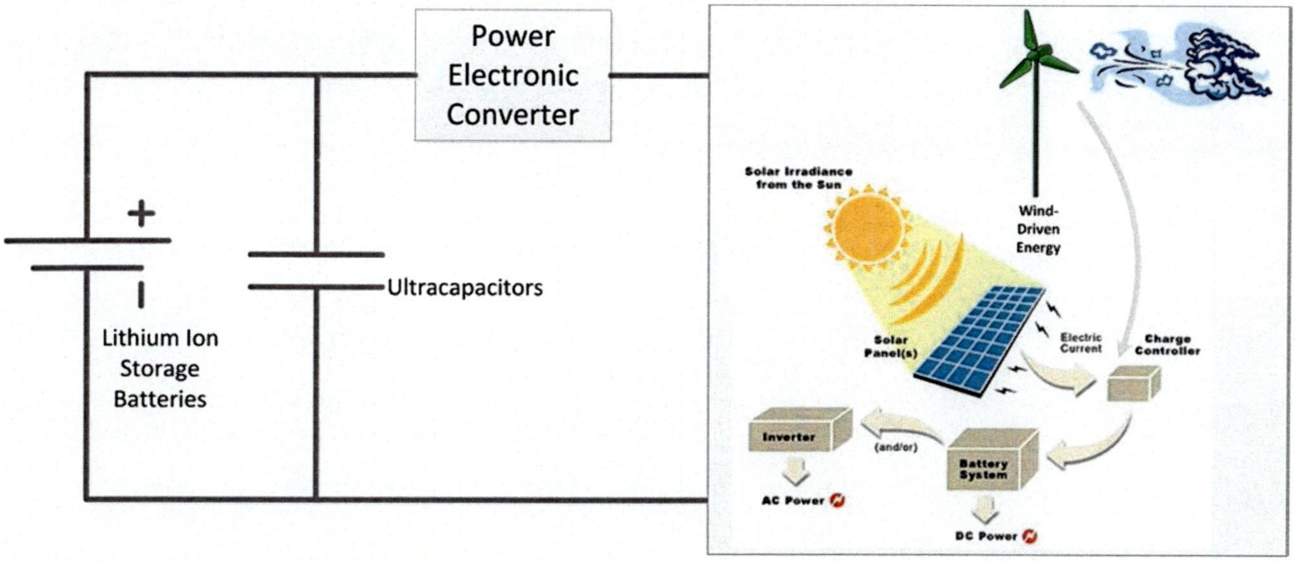

Fig. 18.1 Alternative power management system overview

…xample of a
…use power grid

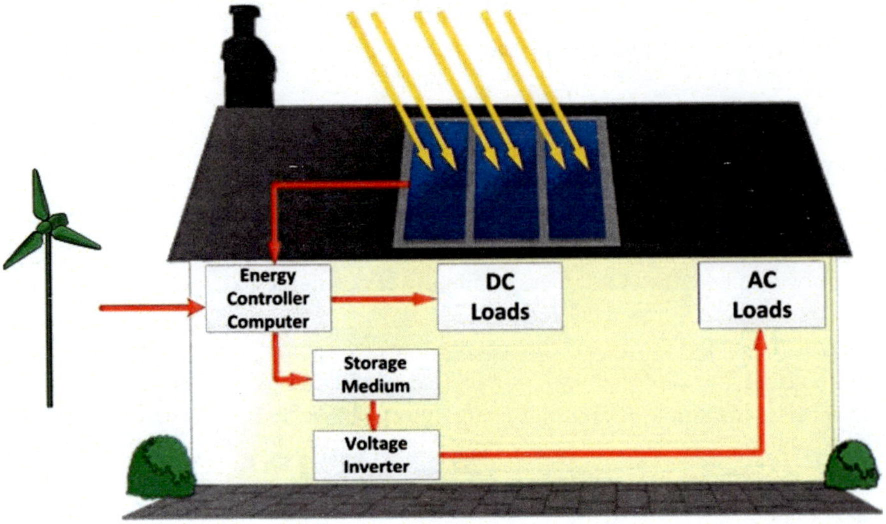

- Do you believe reuse is a realizable concept in large-scale system development? Justify your answer.
- Define "reactionary engineering." When is this appropriate?
- Describe the difference between data, information, and knowledge.
- What are the pitfalls of using information from the Web?
- How would automation help/hurt systems integration and test?

- Create a 25-slide presentation of your multi-alternative energy project design.
  - What MDRE processes did you incorporate?
- Do you believe your multi-alternative energy system is viable? Why or why not?

References

1. Rao, M., Ramakrishnan, S., & Dagli, C. (2008). Modeling and simulation of net centric system of systems using systems modeling language and colored petri-nets: A demonstration using the global earth observation system of systems. *Systems Engineering, 11*, 203–220.
2. M. Butterfield, J. Pearlman, and S. Vickroy, "A system-of-systems engineering GEOSS: Architectural approach," IEEE Systems Journal, vol. 2, no. 3, pp. 321–332, Sep. 2008.
3. Mandutianu, S., Moshir, M., and Donahue, K. (2009). "Conceptual model for space mission systems design," presented at 19th Ann. Int. Symp. INCOSE, Singapore.
4. Simpkins, P., Kleinholz, A., and Maley, J. (2009). "A practical application of MBSE—The automated parking system," presented at 3rd Asia-Pacific Conf. Syst. Eng., Singapore.
5. Soyler, A. and Diakanda, S.. "A model-based systems engineering approach to capturing disaster management systems," in Proc. 4th Ann. IEEE Syst. Conf., San Diego, CA, Apr. 2010, pp. 283–287.
6. Tellioglu, H. "Practicing modelling in manufacturing," in Proc. 2nd Int. Conf. Model Based Syst. Eng., Herzelya and Haifa, Israel, Mar. 2009, pp. 75–82.
7. Andersson, H., Herzog, E., Johansson, G., and Johansson, O.. (2009). Experience from introducing unified modeling language/systems modeling language at Saab aerosystems, Syst. Eng. [Online]: DOI https://doi.org/10.1002/sys.20156.
8. Haan, B. "A model of the competitive advantage of prognostics and health management," in Proc. Ann. Rel. Maintainability Symp., Jan. 2009, pp. 442–447.
9. Cloutier, R. (2009). Introduction to this special edition on model-based systems engineering. *INSIGHT-INCOSE Journal, 12*(4), 7–8.
10. Delp, C., Cooney, L., Dutenhoffer, C., Gostelow, R., Jackson, M., Wilkerson, M., Kahn, T., & Piggott, S. (2009). The challenge of model-based systems engineering for space systems, year 2. *INSIGHT-INCOSE Journal, 12*, 36–39.
11. Friedenthal, S. (2009). SysML: Lessons from early applications and future directions. *INSIGHT-INCOSE Journal, 12*(4), 10–12.
12. A. Bahill and B. Gissing, "Re-evaluating systems engineering concepts using systems thinking," IEEE Transactions Systems Man Cybernetics C Applications Reviews, vol. 28, no. 4, pp. 516–527, Nov. 1998.

Index

J. A. Crowder, C. W. Hoff, *Requirements Engineering: Laying a Firm Foundation*,
Textbooks in Telecommunication Engineering, https://doi.org/10.1007/978-3-030-91077-8

E

F

N

T